Chris Cornelis, Glad Deschrijver, Mike Nachtegael, Steven Schockaert, and Yun Shi (Eds.)

# 35 Years of Fuzzy Set Theory

Studies in Fuzziness and Soft Computing, Volume 261

**Editor-in-Chief**

Prof. Janusz Kacprzyk
Systems Research Institute
Polish Academy of Sciences
ul. Newelska 6
01-447 Warsaw
Poland
E-mail: kacprzyk@ibspan.waw.pl

---

Further volumes of this series can be found on our homepage: springer.com

Vol. 246. Piedad Brox, Iluminada Castillo, Santiago Sánchez Solano
*Fuzzy Logic-Based Algorithms for Video De-Interlacing,* 2010
ISBN 978-3-642-10694-1

Vol. 247. Michael Glykas
*Fuzzy Cognitive Maps,* 2010
ISBN 978-3-642-03219-6

Vol. 248. Bing-Yuan Cao
*Optimal Models and Methods with Fuzzy Quantities,* 2010
ISBN 978-3-642-10710-8

Vol. 249. Bernadette Bouchon-Meunier, Luis Magdalena, Manuel Ojeda-Aciego, José-Luis Verdegay, Ronald R. Yager (Eds.)
*Foundations of Reasoning under Uncertainty,* 2010
ISBN 978-3-642-10726-9

Vol. 250. Xiaoxia Huang
*Portfolio Analysis,* 2010
ISBN 978-3-642-11213-3

Vol. 251. George A. Anastassiou
*Fuzzy Mathematics: Approximation Theory, 2010*
ISBN 978-3-642-11219-5

Vol. 252. Cengiz Kahraman, Mesut Yavuz (Eds.)
*Production Engineering and Management under Fuzziness,* 2010
ISBN 978-3-642-12051-0

Vol. 253. Badredine Arfi
*Linguistic Fuzzy Logic Methods in Social Sciences,* 2010
ISBN 978-3-642-13342-8

Vol. 254. Weldon A. Lodwick, Janusz Kacprzyk (Eds.)
*Fuzzy Optimization,* 2010
ISBN 978-3-642-13934-5

Vol. 255. Zongmin Ma, Li Yan (Eds.)
*Soft Computing in XML Data Management,* 2010
ISBN 978-3-642-14009-9

Vol. 256. Robert Jeansoulin, Odile Papini, Henri Prade, and Steven Schockaert (Eds.)
*Methods for Handling Imperfect Spatial Information,* 2010
ISBN 978-3-642-14754-8

Vol. 257. Salvatore Greco, Ricardo Alberto Marques Pereira, Massimo Squillante, Ronald R. Yager, and Janusz Kacprzyk (Eds.)
*Preferences and Decisions,*2010
ISBN 978-3-642-15975-6

Vol. 258. Jorge Casillas and Francisco José Martínez López
*Marketing Intelligent Systems Using Soft Computing,* 2010
ISBN 978-3-642-15605-2

Vol. 259. Alexander Gegov
*Fuzzy Networks for Complex Systems,* 2010
ISBN 978-3-642-15599-4

Vol. 260. Jordi Recasens
*Indistinguishability Operators,* 2010
ISBN 978-3-642-16221-3

Vol. 261. Chris Cornelis, Glad Deschrijver, Mike Nachtegael, Steven Schockaert, and Yun Shi (Eds.)
*35 Years of Fuzzy Set Theory,* 2010
ISBN 978-3-642-16628-0

Chris Cornelis, Glad Deschrijver,
Mike Nachtegael, Steven Schockaert,
and Yun Shi (Eds.)

# 35 Years of Fuzzy Set Theory

## Celebratory Volume Dedicated to the Retirement of Etienne E. Kerre

Springer

**Editors**

Dr. Chris Cornelis
Ghent University
Department of Applied Mathematics
and Computer Science
Krijgslaan 281
9000 Gehnt, Belgium
E-mail: Chris.Cornelis@ugent.be

Dr. Glad Deschrijver
Ghent University
Department of Applied Mathematics
and Computer Science
Krijgslaan 281
9000 Gehnt, Belgium
E-mail: glad.deschrijver@ugent.be

Dr. Mike Nachtegael
Ghent University
Department of Applied Mathematics
and Computer Science
Krijgslaan 281
9000 Gehnt, Belgium
E-mail: Mike.Nachtegael@ugent.be

Dr. Steven Schockaert
Ghent University
Department of Applied Mathematics
and Computer Science
Krijgslaan 281
9000 Gehnt, Belgium
E-mail: Steven.Schockaert@ugent.be

Dr. Yun Shi
Ghent University
Department of Applied Mathematics
and Computer Science
Krijgslaan 281
9000 Gehnt, Belgium
E-mail: Yun.Shi@ugent.be

ISBN 978-3-642-16628-0 e-ISBN 978-3-642-16629-7

DOI 10.1007/978-3-642-16629-7

Studies in Fuzziness and Soft Computing ISSN 1434-9922

Library of Congress Control Number: 2010937351



*Typeset & Cover Design:* Scientific Publishing Services Pvt. Ltd., Chennai, India.

Printed on acid-free paper

9 8 7 6 5 4 3 2 1

springer.com

# Foreword

The volume "35 Years of Fuzzy Set Theory: Celebratory Volume Dedicated to the Retirement of Etienne E. Kerre", edited by Chris Cornelis, Glad Deschrijver, Mike Nachtegael, Steven Schockaert and Yun Shi, is a fitting tribute to Professor Etienne Kerre as a scientist, educator and a member of the profession. For me, to write a preface to this volume is an opportunity to offer a personal tribute to Professor Kerre – my long time friend. I should like to precede my tribute with an expression of regret. First, because Etienne is retiring from his full professorship at Ghent University and his position as Director of the Fuzziness and Uncertainty Modelling Research Unit of Ghent University; and second, because in the world of science and academia the likes of Etienne are so few and far between.

My acquaintance with Etienne goes back to the seventies when Etienne entered the academic world and shortly after he became interested in fuzzy set theory and its applications. Since those days, Etienne has become a very prominent figure within the fuzzy logic community. His many seminal contributions have brought him both national and international recognition. Particularly worthy of note are his contributions to fuzzy topology, fuzzy algebraic structures, fuzzy relational calculus, fuzzy reasoning and if-then rules, possibility theory, reliability theory and fuzzy image processing. He has authored or co-authored close to ten books and over 390 papers on fuzzy set theory and its applications. It is a remarkable record. Equally impressive are Etienne's contributions as an educator. He has guided the research of over twenty Ph.D.s and many graduate students. In addition, Etienne has played an active role in organizing and participating in international conferences. Etienne has a very warm personality and works very closely with his students. He is a role model in all respects.

I should like to take this opportunity to comment on the profound changes in the world of fuzzy logic since Etienne began his career in the academic world. By 1980 there were 521 papers in the INSPEC database which contained "fuzzy" in the title. Today, there are 68073. In addition, there are 18302 papers with "fuzzy" in the title in the MathSciNet database. Fuzzy logic is employed in a wide variety of products, among them cameras, appliances, hybrid engines and transmissions, TV sets, copiers, elevators, blood pressure meters, etc. In recent years, applications of fuzzy logic within human sciences, notably sociology and political science, began to grow in visibility and importance.

These numbers paint a rosy picture. But beyond the rosy picture one can see some problems which call for attention. A problem which has existed since the birth of fuzzy set theory and fuzzy logic is related to the meaning of "fuzzy". In English, the word "fuzzy" has a pejorative connotation. In the United States, in particular, pejorative connotation of "fuzzy" has seriously impeded acceptance of fuzzy logic. Despite the profusion of products which employ fuzzy logic, what is still visible is a residue of skepticism, particularly in the academic world.

Pejorative connotation of "fuzzy" is rooted in a deep-seated attitude in science – an attitude of respect for precision and lack of respect for imprecision. What this means is that "fuzzy" is pejorative because it is related to imprecision. This is a problem that cuts across most languages and is hard to resolve.

A problem which is more visible in the United States than in other countries is that at many universities courses on fuzzy logic are not taught. As a consequence, most engineering graduates are unfamiliar with fuzzy logic and hence are not in a position to employ it even when fuzzy logic can result in a superior solution to a problem. However, in coming years the problem may be solved, at least in part. More specifically, what is happening is that computational intelligence, which is just another name for soft computing, is rapidly growing in popularity. Computational intelligence (soft computing) is an amalgam of fuzzy logic, neurocomputing and evolutionary computing. Thus, in coming years fuzzy logic will be taught in many universities in courses on computational intelligence.

What we observe today is that fuzzy logic, along with many other fields of science and technology, is developed and applied more actively in Asian countries – China, Japan, South Korea, Taiwan and Singapore – than in Western countries and the United States. This is a manifestation of a global shift in power and influence from the West to the East. That is why Western science needs so badly the likes of Etienne Kerre.

In my view, in years to come fuzzy logic – its name notwithstanding – is certain to grow in visibility, importance and acceptance. There is a basic reason. In essence, fuzzy logic is the logic of classes with unsharp boundaries. In the world of human cognition, classes with unsharp boundaries are the rule rather than exception. What this implies is that it is only a matter of time before the essentiality of fuzzy logic becomes apparent not only in science but in most domains of human thought and culture. In the history of fuzzy logic and its applications, Etienne's contributions will occupy a prominent place.

Berkeley (California), June 1, 2010 Lotfi A. Zadeh

# Preface

## About Etienne E. Kerre

Etienne E. Kerre: husband, father, grandfather, and ... pioneering researcher in fuzzy set theory and fuzzy logic. This book is dedicated to the long and exceptionally fruitful career of Etienne, and is published on the occasion of his retirement. Let us use this preface to provide a bird's eye view on the life of Etienne, both as a person and as a researcher, as a way to say "thank you" for the wonderful years that we worked together and enjoyed his guidance.

Etienne E. Kerre – 'E' stands for Emiel – was born in Zele, Belgium, on May 8, 1945, the very day World War II ended in Europe. He is the oldest son in a large family of 8 children. His father, who passed away earlier this year, worked for the Belgian railway company. His mother, who is now 90 years old, stayed at home to take care of the children. In the social context of the decades right after World War II, Etienne should be proud of having succeeded in starting what would prove to be a successful academic career. At that time, it was not obvious to complete a full cycle of secondary school. It were Etienne's primary school teachers who could convince his parents to allow him to go to secondary school, given the fact that Etienne was an excellent student. We are all very lucky that Etienne got and took this opportunity!

Etienne is married to Andrea De Kegel and they have one daughter Tessa. Andrea is director of the cultural department of the province Oost-Vlaanderen (together they share a passion for culture), Tessa is a hematologist and has a Ph.D. in medical sciences, and his son-in-law Bruno is a jazz photographer and reviewer. Whenever Etienne speaks about his family, it is always with love. We particularly remember his happiness when his grandson Henri was born on October 21, 1999. On many occasions Etienne told us entertaining anecdotes about this boy of whom he is so proud. Without any doubt, his retirement will allow him to spend even more time with his grandson.

After graduating from Sint-Lodewijkscollege Secondary School (Lokeren, Belgium) in 1963, Etienne studied mathematics at Ghent University where he obtained a Master's degree in 1967, and the certificate of teacher education one year later. After an unusually short period of less than three years, in 1970 he was able to complete his Ph.D. in Mathematics concerning "Low Energy Electron Diffraction", a topic on which he has published five papers in international scientific journals. It was the first year of what Etienne calls his "3 Fat Years": in 1970 he got his Ph.D., in 1971 he got married, and in 1972 he became a father. Etienne worked as an assistant of Ghent University from 1969 to 1978. In 1978 he became a work leader at Ghent University, and a lecturer in 1984. In 1991 he obtained a full professor position at Ghent University. In 2004 he furthermore obtained honorary professorships from Southwest Jiaotong University and Xihua University in Chengdu, China.

Etienne started his research on fuzzy set theory in 1976. In this year, he also founded the Fuzziness and Uncertainty Modelling Research Unit, which he guided until his retirement. Since then he has published, together with many colleagues from around the world (DBLP lists 91 different co-authors, ISI Web of Science more than 130), more than 390 research papers on fuzzy set theory (both theoretical and applied) in international scientific journals and international conference proceedings. His research interests cover Fuzzy Topology, Fuzzy Algebraic Structures, Fuzzy Relational Calculus, Fuzzy Reasoning and If-Then Rules, Possibility Theory and Reliability Theory, Fuzzy Image Processing and so on. He has also written 8 books on basic principles of Fuzzy Set Theory and its applications, and is editor of 17 other books. Besides these publications, Etienne has given 155 lectures at international conferences and institutes, and acted as organizer or chairman of sessions in 82 international conferences and workshops. He has been a member of the editorial boards of 24 international journals and has been a referee for more than 65 international journals. He has also been a member of the program committee of 127 international conferences and has been a referee for more than 83 international conferences. Already 29 students from Belgium, Bulgaria, China and Egypt obtained a Ph.D. degree under his guidance, and Etienne has been a member of the jury for another 15 Ph.D. theses.

These are all impressive numbers, and if in any way possible they quantify an impressive academic career. All editors of this book belong to the group of Etienne's 29 former Ph.D. students, and all enjoyed his guidance. We can rightly say that Etienne has always made great efforts to attract young people to academic research in fuzzy set theory. This seduction already started during his classes on fuzzy set theory, which he taught with great enthusiasm and dynamism. Quite logically, a lot of students made their master thesis in the field of fuzzy set theory, and this was a good recruiting area for young researchers. Etienne has obtained several large research projects that enabled him to employ these young people, and to train them into successful academics.

Etienne was also very active in teaching. He has given 7 courses about Mathematical Analysis and 11 courses about Fuzzy Set Theory and Fuzzy Logic to both undergraduates and postgraduate levels in Computer Science, Mathematics and Physics from 1969 until now. In fact, thanks to Etienne, Ghent University was one of the first universities worldwide where fuzzy set theory was taught in the basic curriculum. Students from Ghent University will certainly remember his white duster, which he wore whenever he was teaching. Etienne is also known as a firm but fair professor to his students (including his Ph.D. students): he demanded a lot of effort from his students and insisted that they formulated everything as precisely as possible. This was of course justified, from an academic point of view in general and from a mathematical point of view in particular. However, especially in the first year of an academic education many students lack the required level of formal rigour, and on occasion Etienne would come in our offices to complain about it. Teaching students this skill has always been one of his main missions. It is also interesting to discover how well Etienne is known by many current secondary school teachers in mathematics. If they took their university study in Ghent, chances are high that they had at least one course of Etienne, as an assistant in his beginning period and as a professor later on, and the teachers we encountered do remember him vividly. Teaching was also done outside of Ghent. Etienne has given 3 courses to undergraduates at the University Nebraska (Lincoln, U.S.A.) from 1987 to 1990, and 2 courses to postgraduates at Alcatel Bell Company (Antwerp, Belgium) in 1993 and 1994.

Beside his strong involvement in the academic world, Etienne always stayed a family man. He would not hesitate to offer his help if someone needed it. We often wondered how he found the time to do all these things. Etienne just has an extraordinary amount of energy, which is prefectly illustrated by the fact that he only needs a couple of hours of sleep every night. Chances are high that he is reading this book in the middle of the night!

When we are writing this preface, Etienne still has three months before his retirement. We know that he will retire with mixed feelings. On the one hand Etienne certainly deserves the rest of a retirement, which will allow him to spend more time with his family and to engage in other activities. On the other hand, Etienne is filled with passion for research and is not able to stop completely. Luckily, the university offers the possibility to stay active, of course at a slower pace, such that Etienne will be able to get a good mix between retirement and a continued involvement in the academic world. We wish him all the best with his retirement and his new or more intensified activities, and hope to keep meeting him, either in the context of research or just as friends to have a drink on a sunny day. Good luck Etienne, and thank you for being the wonderful person and researcher that you are!

## About This Volume

This volume contains contributions from researchers that have been close to Etienne in one way or another: former Ph.D. students, co-authors, and friends in general. The idea of compiling a volume dedicated to the retirement of Etienne originated in January 2010, and was carried out by the postdoctoral researchers in his group at that time. Several people have kindly offered their help in assisting us with the reviewing process: Kim Bauters, Martine De Cock, Jeroen Janssen, Bart Van Gasse, and Patricia Victor. We also thank the authors for the time and efforts they spent in preparing their contribution, within a tight time frame. We are grateful to Janusz Kacprzyk and the people from Springer-Verlag for their help, and for giving us the opportunity to publish this book in the series of Studies in Fuzziness and Soft Computing.

The chapters of this volume are divided in three parts: i) logics and connectives, ii) data analysis, and iii) media applications. The first part gathers contributions that are primarily of a theoretical nature, covering fuzzy logic, (interval-valued and intuitionistic) fuzzy logic connectives, and aggregation operators. The chapter by R. Mesiar and M. Komorníková introduces and discusses a general concept of aggregation on bounded partially ordered sets and on bounded lattices. The weakest and strongest aggregation functions on a poset are identified, and several construction methods are discussed. The properties of t-norms and related operators on a poset are discussed and several characterizations are given. The chapter ends with a discussion of aggregation functions on the underlying lattice of Atanassov's intuitionistic fuzzy set theory. Next, the chapter by K.T. Atanassov contains an overview of and new results on 23 of the 138 different intuitionistic fuzzy implications and 5 of the different intuitionistic fuzzy negations already introduced by the author. Some properties of these operators are discussed, and a partial solution to a problem by Baczynski and Jayaram relating the contrapositive property is given using intuitionistic fuzzy implications and negations. The axioms of intuitionistic logic are checked for the intuitionistic fuzzy implications and negations. The chapter by V. Novák and I. Perfilieva deals with the practicality of fuzzy logic, both in the narrow and in the broad sense. In the field of mathematical fuzzy logic in the narrow sense, traditional and evaluated syntax are reviewed, as well as fuzzy type theory. In the field of fuzzy logic in the broader sense, the formal theories of evaluative linguistic expressions, fuzzy if–then rules, perception based logical deduction and intermediate and generalized quantifiers are treated. Finally an example in commonsense human reasoning is given, and two theories related to fuzzy logic are discussed, namely fuzzy approximation and the fuzzy transform. The subsequent chapter by B. Van Gasse, C. Cornelis and G. Deschrijver studies propositional calculi for interval-valued fuzzy logics. Starting from the notion of the triangularization of a lattice, the authors focus on a specific case called interval-valued residuated lattices. Then the notion of a triangle algebra is introduced, and shown to be isomorphic to a particular class of interval-valued residuated lattices. Finally, triangle algebras are used to define several interval-valued fuzzy logics, for which soundness and completeness is analyzed. The last chapter of the first part by Y. Shi, B. Van Gasse and D. Ruan deals with fuzzy implications. The complete dependencies and independencies between eight commonly considered potential properties of implications are studied. Each independency is shown by a counterexample. Two of these counterexamples lead to a new class of implications which are only determined by a negation, and this new class is extensively investigated.

The second part of the book pertains to applications of fuzzy set theory to data analysis, covering classification, decision making, rough sets and formal concept analysis. The chapter by A.M. Radzikowska deals with $L$–fuzzy rough sets as a further generalization of rough sets. Rough sets were originally proposed as a formal tool for analyzing and processing incomplete information represented in data tables. Later on, fuzzy generalizations of rough sets were introduced and investigated to be able to deal with imprecision. Anja and Etienne cooperated on this topic for more than a decade. The next chapter is authored by Y. Djouadi, D. Dubois and H. Prade. One of Etienne Kerre's favorite topics is undoubtedly fuzzy relational calculus. Formal concept analysis (FCA) is known to be a successful and elegant application of relational calculus, revolving around a set of objects, a set of properties, and a binary relation between them. This chapter covers different graded extensions of formal concept analysis that allow to account for gradual properties, to handle uncertainty, and to acknowledge typicality of properties and importance of objects. The chapter offers insights in existing work on fuzzy FCA as well as interesting new results. The chapter by X. Wang, C. Wu and X. Wu gives a comprehensive overview of the research on fuzzy choice functions. First, it discusses Banerjee's framework, and then moves on to study various preferences derived from choice functions, where the focus is on rationality conditions and their interrelationships. At the end, Georgescu's framework, in which also the set of alternatives to choose from is fuzzified, is investigated. Finally, the chapter by G. Chen, Y. Xiong and Q. Wei discusses the use of fuzzy partitions to generate classification rules in domains with numerical attributes. In particular, they develop an extension of GARC, a classifier which is based on association rules. Experimental results show that the accuracy of GARC is maintained, while using a fewer number of rules.

The final part of the book comprises chapters in which fuzzy sets are used to represent or to manipulate different types of media objects, such as text documents and images. The first chapter is a contribution of the 5 former PhD students of Etienne that worked on fuzzy techniques in image processing (M. Nachtegael, T. Mélange, S. Schulte, V. De Witte and D. Van der Weken). In a prozaic style, the chapter gives an overview of the 12 years of research in this field, focussing on fuzzy mathematical morphology, similarity measures for images and noise reduction filters for both grayscale and colour images and video sequences. Next, the chapter by H. Bustince, M. Pagola, E. Barrenechea and J. Fernández deals with adapting a fuzzy thresholding algorithm to some extensions of fuzzy set theory, namely interval-valued fuzzy sets, type-2 fuzzy sets, interval type-2 fuzzy sets and intuitionistic fuzzy sets. The resulting fuzzy thresholding algorithms are used for image segmentation to solve the problem of accurate elicitation of membership functions. Experimental results suggest that in some cases, the adapted thresholding algorithm indeed outperforms the existing algorithm. The chapter by S. Schockaert, N. Makarytska, and M. De Cock provides an overview of fuzzy approaches and related techniques in the web intelligence field. In particular, the authors explain the need for, and discuss the potential of fuzzy concepts and fuzzy set theory for three classes of applications that are becoming increasingly important on the web: information retrieval methods, recommender systems and the semantic web. Finally, the chapter by S. Zadrozny, J. Kacprzyk and K. Nowacka is centered around the problem of text categorization. The authors propose a weighting scheme to represent text documents in which the weight of each term is a fuzzy set. An existing classification algorithm for binary attributes is generalized to cope with fuzzy set valued attributes, after which some preliminary experimental results are provided.

Ghent,
July 2010

Chris Cornelis
Glad Deschrijver
Mike Nachtegael
Steven Schockaert
Yun Shi

# Contents

# Part I
# Logics and Connectives

# Aggregation Functions on Bounded Posets

Radko Mesiar and Magda Komorníková

**Abstract.** A general concept of aggregation on bounded posets is introduced and discussed. Up to general results, several particular results due to E.E. Kerre and his research group are recalled in the light of our approach. A special attention is paid to triangular norms on bounded posets.

## 1 Introduction

Two basic attributes of aggregation functions introduced and discussed in [6, 28, 33, 34] are the monotonicity and the boundary conditions. Recall that an aggregation function $A : [0,1]^n \to [0,1]$ (an extended aggregation function $A : \bigcup_{n \in N} [0,1]^n \to [0,1]$) is supposed to be increasing, i.e., $A(\mathbf{x}) \leq A(\mathbf{y})$ whenever $\mathbf{x} \leq \mathbf{y}$, and $A(0,...,0) = 0$, $A(1,...,1) = 1$. The framework of aggregation functions on $[0,1]$ can be modified into any closed interval $[a,b] \subseteq [-\infty,\infty]$, or even to any interval $I \subseteq [-\infty,\infty]$ (with modified boundary conditions). Obviously, aggregation functions can be introduced to act on any (partially) ordered structure with bounds.

**Definition 1.** Let $(P, \leq, \mathbf{0}, \mathbf{1})$ be a bounded poset (partially ordered set). Let $n \in N$ be fixed. A mapping $A : P^n \to P$ is called an ($n$-ary) aggregation function on $P$ whenever it is increasing, i.e.,

$$A(\mathbf{x}) \leq A(\mathbf{y}) \text{ whenever } \mathbf{x} \leq \mathbf{y}, \text{ (i.e., } x_1 \leq y_1, \ldots, x_n \leq y_n) \tag{1}$$

and it satisfies boundary conditions

Radko Mesiar
Faculty of Civil Engineering, Slovak University of Technology, Radlinského 11, 813 68 Bratislava, Slovakia
e-mail: Radko.Mesiar@stuba.sk

Magda Komorníková
Faculty of Civil Engineering, Slovak University of Technology, Radlinského 11, 813 68 Bratislava, Slovakia
e-mail: Magdalena.Komornikova@stuba.sk

C. Cornelis et al. (Eds.): 35 Years of Fuzzy Set Theory, STUDFUZZ 261, pp. 3–17.
springerlink.com 

$$A(\mathbf{0},...,\mathbf{0}) = \mathbf{0},\ A(\mathbf{1},...,\mathbf{1}) = \mathbf{1}. \tag{2}$$

A mapping $B : \bigcup_{n\in N} P^n \to P$ is called an extended aggregation function on $P$ whenever $B/P^n$ is an $n$–ary aggregation function on $P$ for any $n \in N$.

Note that in the case $n = 1$, often a convention $A(x) = x$ is used for unary aggregation function on $[0,1]$, see [6, 28, 34]. We will not insist on this convention in our approach.

*Remark 1.* i) Note that for any poset $(P,\leq,\mathbf{0},\mathbf{1})$ a dual poset $(P^*,\leq^*,\mathbf{0}^*,\mathbf{1}^*)$ can be introduced, where $P^* = P$, $x \leq^* y$ if and only if $y \leq x$, and $\mathbf{0}^* = \mathbf{1}$, $\mathbf{1}^* = \mathbf{0}$. Evidently, any aggregation function $A : P^n \to P$ ($A : \bigcup_{n\in N} P^n \to P$) on $P$ can be considered also as an aggregation function on $P^*$. Several properties of $A$ on $P$ are the same as those of $A$ on $P^*$ (namely, all algebraic properties not linked to the orderings $\leq$ and $\leq^*$ ). However, properties based on the ordering should be modified by the above *–duality (for example, conjunctivity on $P$ is equivalent to the disjunctivity on $P^*$, see Definition 2).

ii) Extended aggregation functions acting on a poset $P$ with convention of identity for the unary aggregation were introduced in [10], where their categorical foundations were studied. Particularly, aggregation processes in probabilistic metric spaces were considered. As a typical example, convolution of distribution functions can be seen as a special aggregation function.

In this contribution, we discuss first aggregation functions which are defined on any poset, see Section 2. In Section 3, aggregation functions on general bounded lattices are studied. Special class of triangular norms and related aggregation functions are covered by Section 4. Section 5 brings an overview of some aggregation functions acting on special lattice $L^*$, with several results of E.E. Kerre's group. Finally, some concluding remarks are added.

## 2 Aggregation Functions on General Bounded Posets

Algebraic properties introduced for the classical aggregation functions acting on $[0,1]$ can be straightforwardly introduced for aggregation functions on general bounded posets. We recall some of these properties (for more details and discussion in the case $P = [0,1]$ we recommend [28]):

- symmetry
- associativity
- neutral element
- annihilator
- strict monotonicity
- decomposability (for extended aggregation functions)
- idempotency
- internality (i.e., $A(x_1,\ldots,x_n) \in \{x_1,\ldots,x_n\}$)
- bisymmetry.

Several other properties, like additivity, continuity, etc., can be introduced only in the case of particular posets equipped with additional properties, and thus we will not consider them. Note also that we can introduce conjunctive and disjunctive aggregation functions (even the concept of $k$–tolerantness and $k$–intolerantness due to Marichal [35]), while the concept of averaging aggregation functions is restricted to the class of lattices.

**Definition 2.** Let $A : P^n \longrightarrow P$ be an aggregation function on $P$. Then $A$ is called:

i) conjunctive whenever $A(x_1,\dots,x_n) \leq x_i$ for all $i \in \{1,\dots,n\}$,
ii) disjunctive whenever $A(x_1,\dots,x_n) \geq x_i$ for all $i \in \{1,\dots,n\}$,
iii) $k$–tolerant with $k \in \{1,\dots,n\}$ whenever $card(\{i|A(x_1,\dots,x_n) \geq x_i\}) \geq n-k+1$,
iv) $k$–intolerant with $k \in \{1,\dots,n\}$ whenever $card(\{i|A(x_1,\dots,x_n) \leq x_i\}) \geq n-k+1$.

Evidently, conjunctive aggregation functions coincide with *1*-intolerant, and disjunctive aggregation functions coincide with *1*-tolerant. The concepts of tolerant and intolerant aggregation functions allows to introduce a stronger form of averaging property.

**Definition 3.** Let $A : P^n \longrightarrow P$ be an aggregation function on $P$. Then $A$ is called strongly averaging whenever $A$ is simultaneously $n$-tolerant and $n$-intolerant, i.e., for each $(x_1,\dots,x_n) \in P^n$, there are $i, j \in \{1,\dots,n\}$ so that

$$x_i \leq A(x_1,\dots,x_n) \leq x_j.$$

Evidently, each internal aggregation function $A$ on $P$ is strongly averaging (but not vice–versa).

*Example 1.* Let $P = [0,1]^2$ be equipped with the standard partial order $\leq$. Observe that then $(P,\leq)$ is a lattice. Then $Min,\ Max : P^n \longrightarrow P$ defined by

$$Min((x_1,y_1),\dots,(x_n,y_n)) = (\min(x_1,\dots,x_n), \min(y_1,\dots,y_n))$$

and

$$Max((x_1,y_1),\dots,(x_n,y_n)) = (\max(x_1,\dots,x_n), \max(y_1,\dots,y_n))$$

are idempotent $n$-ary aggregation functions on $P$. Neither $Min$ nor $Max$ are strongly averaging. However, $Min$ is conjunctive and $Max$ is disjunctive.

On the other side, projection $Pr_i : P^n \longrightarrow P$ defined by

$$Pr_i((x_1,y_1),\dots,(x_n,y_n)) = (x_i,y_i)$$

are simultaneously $n$–tolerant and $n$–intolerant and thus strongly averaging (indeed, projections are internal aggregation functions).

We introduce some distinguished aggregation functions on $P$ (for any arity thus also as extended aggregation functions):

- the weakest aggregation function $A_w : \bigcup_{n\in N} P^n \to P$,

$$A_w(x_1,\dots,x_n) = \begin{cases} \mathbf{1} & \text{if } x_1 = \dots = x_n = \mathbf{1}, \\ \mathbf{0} & \text{else,} \end{cases}$$

- the strongest aggregation function $A_s : \bigcup_{n\in N} P^n \to P$,

$$A_s(x_1,\dots,x_n) = \begin{cases} \mathbf{0} & \text{if } x_1 = \dots = x_n = \mathbf{0}, \\ \mathbf{1} & \text{else,} \end{cases}$$

- the weakest t-norm (drastic product) $T_D : \bigcup_{n\in N} P^n \to P$,

$$T_D(x_1,\dots,x_n) = \begin{cases} x_i & \text{if } x_j = \mathbf{1} \text{ for each } j \neq i, \\ \mathbf{0} & \text{else,} \end{cases}$$

- the strongest t-conorm (drastic sum) $S_D : \bigcup_{n\in N} P^n \to P$,

$$S_D(x_1,\dots,x_n) = \begin{cases} x_i & \text{if } x_j = \mathbf{0} \text{ for each } j \neq i, \\ \mathbf{1} & \text{else.} \end{cases}$$

All these four aggregation functions are associative (i.e., their binary form is associative, and n-ary form is the genuine extension of that binary form).

Aggregation functions $A_w, T_D, A_s, S_D$ are symmetric. For any fixed $n \in N$, $n \geq 2$, projection functions $Pr_i : P^n \to P$, $Pr_i(x_1,\dots,x_n) = x_i$, are non–symmetric $n$-ary aggregation functions on $P$.

Observe that on the Boolean chain $P = \{0,1\}$, there are exactly four binary aggregation functions, namely $A_w = T_D, A_s = S_D, Pr_1$ and $Pr_2, Pr_1(x,y) = x$ and $Pr_2(x,y) = y$. Each of them is associative. This fact excludes the possibility to define a non-associative aggregation function $A : P^2 \to P$ by a unique formula independent of the underlying poset $P$. Obviously, in particular cases of posets $P$ with $card\, P > 2$, there are several non-associative aggregation functions. For example, the arithmetic mean $M : [0,1]^2 \to [0,1]$ given by $M(x,y) = \frac{x+y}{2}$ is not associative.

There are several construction methods appropriate for constructing aggregation functions on posets. First of all, for any fixed bounded poset $(P, \leq, \mathbf{0}, \mathbf{1})$, the composition of aggregation functions can be applied. It is not difficult to check that for any fixed $m \in N, n_1,\dots,n_m \in N$, and aggregation functions $A : P^m \to P$, $B_i : P^{n_i} \to P$, $i = 1,\dots,m$, the mapping $D : P^n \to P$, $n = \sum_{i=1}^m n_i$, $D(x_1,\dots,x_n) = A\left(B_1\left(x_1,\dots,x_{n_1}\right),\dots,B_m\left(x_{n-n_m+1},\dots,x_n\right)\right)$, is an $n$-ary aggregation function on $P$.

As a special case we can consider an $n$-ary aggregation function $A : P^n \to P$ and unary aggregation functions $f, g_1,\dots,g_n : P \to P$, yielding a transformed $n$-ary aggregation function $D : P^n \to P$ given by

$$D(x_1,\dots,x_n) = f\left(A\left(g_1(x_1),\dots,g_n(x_n)\right)\right). \tag{3}$$

Observe that (3) generalizes the standard isomorphic transformation of $A$, which corresponds to the case when $g_1 = \ldots = g_n$ and $f \circ g = id_P, f(g(x)) = x$.

Another construction method is linked to the ordinal sums of posets introduced by Birkhoff [3].

**Proposition 1.** *Let* $(K, \preceq, 0_K, 1_K)$ *be a bounded chain, and let* $((P_k, \leq_k, 0_k, 1_k))_{k \in K}$ *be a family of posets such that for* $k_1, k_2 \in K, k_1 \preceq k_2$, *either* $P_{k_1} \bigcap P_{k_2}$ *is empty, or* $1_{k_1} = 0_{k_2}$ *and* $P_{k_1} \bigcap P_{k_2} = \{1_{k_1}\}$. *For each* $k \in K$, *let* $A_k : P_k^n \to P_k$ $\left(A_k : \bigcup_{n \in N} P_k^n \to P_k\right)$ *be an aggregation function on* $P_k$. *Then:*

*(i)* $(P, \leq, \mathbf{0}, \mathbf{1})$ *is a poset, where* $P = \bigcup_{k \in K} P_k, \mathbf{0} = 0_{0_K}$, $\mathbf{1} = 1_{1_K}$, *and for any* $x, y \in P$, $x \leq y$ *if and only if* $x \in P_{k_1}, y \in P_{k_2}$ *and* $k_1 \prec k_2$ *or* $k_1 = k_2 = k$ *and* $x \leq_k y$.

*(ii)* *A mapping* $A : P^n \to P(A : \bigcup_{n \in N} P^n \to P)$ *given by* $A(x_1, \ldots, x_n) = A_k\left(x_1^k, \ldots, x_n^k\right)$, *where* $k \in K$ *satisfies* $x_i \in P_k$ *for some* $i \in \{1, \ldots, n\}$ *and* $\{x_1, \ldots, x_n\} \bigcap (\bigcup_{r \prec k} P_r) = \emptyset$, *and*

$$x_j^k = \begin{cases} x_j & \text{if } x_j \in P_k, \\ 1_k & \text{if } x_j > 1_k, \\ 0_k & \text{if } x_j < 0_k, \end{cases}$$

*is an aggregation function on* $P$. $A$ *is called conjunctive ordinal sum, and* $A_k, k \in K$, *are called summands with the notation* $A = C - (\langle P_k, A_k \rangle \mid k \in K)$.

*(iii)* *A mapping* $B : P^n \to P(B : \bigcup_{n \in N} P^n \to P)$ *given by* $B(x_1, \ldots, x_n) = A_k\left(x_1^k, \ldots, x_n^k\right)$, *where* $k \in K$ *satisfies* $x_i \in P_k$ *for some* $i \in \{1, \ldots, n\}$ *and* $\{x_1, \ldots, x_n\} \bigcap (\bigcup_{k \prec r} P_r) = \emptyset$, *is an aggregation function on* $P$. $B$ *is called a disjunctive ordinal sum with summands* $A_r$, *with the notation* $B = D - (\langle A_r, P_r \rangle \mid r \in K)$.

*Proof.* Item (i) is due to Birkhoff [3]. Items (ii) and (iii) can be checked directly. □

The next properties can be either checked directly, or they follow from Clifford's ordinal sum of semigroups construction [7].

**Corollary 1.** *Let* $A = C - (\langle P_k, A_k \rangle \mid k \in K)$ $(B = D - (\langle A_k, P_k \rangle \mid k \in K))$ *be a conjunctive (disjunctive) ordinal sum of aggregation functions. Then*

*(i)* *A is a conjunctive aggregation function (B is a disjunctive aggregation function) on P if and only if all summands* $A_k$, $k \in K$, *are conjunctive (all summands* $A_k$, $k \in K$, *are disjunctive); .*

*(ii)* *A (B) is symmetric if and only if all* $A_k$, $k \in K$, *are symmetric;*

*(iii)* *A is associative (B is associative) if and only if all* $A_k$, $k \in K$, *are associative, and if for* $k_1 \prec k_2$, $P_{k_1} \bigcap P_{k_2} \neq \emptyset$, *then* $1_{k_1}$ *is a neutral element of* $A_{k_1}$ *and* $0_{k_2} = 1_{k_1}$ *is the annihilator of* $A_{k_2}$ $(1_{k_1}$ *is an annihilator of* $A_{k_1}$ *and* $0_{k_2} = 1_{k_1}$ *is a neutral element of* $A_{k_2})$.

## 3 Aggregation Functions on Bounded Lattices

Two basic binary aggregation functions on a bounded lattice $(L, \leq, \mathbf{0}, \mathbf{1})$ are the lattice operations, i.e., the join $Max(x,y) = x \vee y$ and the meet $Min(x,y) = x \wedge y$. Due to their associativity, both $Max$ and $Min$ can be extended to $n$-ary aggregation functions, as well as to extended aggregation functions (keeping the original notation). Observe that $Min$ is the strongest conjunctive aggregation function on $L$, while $Max$ is the weakest disjunctive aggregation function on $L$.

For aggregation on lattices, the concept of averaging can be introduced as follows.

**Definition 4.** Let $(L, \leq, \mathbf{0}, \mathbf{1})$ be a lattice and let $A : L^n \to L$ $(A : \bigcup_{n \in N} L^n \to L)$ be an aggregation function on $L$. $A$ is called averaging whenever it satisfies $Min \leq A \leq Max$.

In general, each strongly averaging aggregation function on a lattice $L$ is averaging.

Note that if $L$ is a chain then the strong averaging and averaging properties coincide. In any other case, $Min$ and $Max$ are averaging aggregation functions which are not strongly averaging, compare also Example 1. Moreover, an aggregation function $A$ on a lattice $L$ is averaging if and only if it is idempotent.

Lattice operations of the join $\bigvee$ and of the meet $\bigwedge$ allows to introduce lattice polynomials $p : L^n \to L$ in the following way (see [27], section I.4):

(i) For any $i \in \{1, \dots, n\}$, $Pr_i : L^n \to L$ is a lattice polynomial.
(ii) If $A, B : L^n \to L$ are lattice polynomial functions, then also $A \vee B$ and $A \wedge B$ are lattice polynomial functions from $L^n$ to $L$.
(iii) Every lattice polynomial function $A : L^n \to L$ is formed by finitely many applications of rules (i) and (ii).

*Example 2.* Define $A, B : L^3 \to L$ by

$$A(x,y,z) = (x \vee y) \wedge z$$

and

$$B(x,y,z) = (x \wedge z) \vee (y \wedge z).$$

Then both $A$ and $B$ are lattice polynomials on $L$, and in general $A \neq B$. Obviously, if $L$ is a distributive lattice, then $A = B$.

Note that each lattice polynomial $A$ is an averaging aggregation function, which is strongly averaging if and only if $A$ is a projection function, $A = Pr_i$ for some $i$, or $L$ is a chain.

An interesting method how to construct $n$–ary aggregation functions on a lattice $L$ is linked to sections of aggregation functions on $L$ with higher dimensions (this method can be applied on posets, too):
consider an aggregation function $B : L^{k+n} \to L$, and choose elements $a_1, \dots, a_k \in L$ so that

$$B(a_1,\ldots,a_k,\mathbf{0},\ldots,\mathbf{0}) = \mathbf{0} \text{ and } B(a_1,\ldots,a_k,\mathbf{1},\ldots,\mathbf{1}) = \mathbf{1}$$

(observe that such elements need not exist, in general); then the mapping $A : L^n \to L$ given by

$$A(x_1,\ldots,x_n) = B(a_1,\ldots,a_k,x_1,\ldots,x_n)$$

is an $n$–ary aggregation function on $L$.

*Example 3.* (i) Let $Med : L^3 \to L$ be given by $Med(x,y,z) = (x \wedge y) \vee (x \wedge z) \vee (y \wedge z)$. Then $Med$ is a ternary lattice polynomial on $L$ ($Med$ is called median, following the standard terminology from the case when $L = [-\infty,\infty]$ is the extended real line). Obviously, for any $a \in L$, $Med_a : L^2 \to L$ given by $Med_a(x,y) = Med(a,x,y)$ is a binary aggregation function (so called a–median, compare [23, 26]). Note that $Med_{\mathbf{0}} = Min$ and $Med_{\mathbf{1}} = Max$.

(ii) Define $B : L^4 \to L$ by $B(x_1,x_2,x_3,x_4) = (x_1 \wedge x_3) \vee (x_2 \wedge x_4) \vee (x_3 \wedge x_4)$, i. e., $B$ is a quaternary lattice polynomial on $L$. For any fixed $a,b \in L, A : L^2 \to L$ given by $A(x,y) = B(a,b,x,y)$ is a binary aggregation function on $L$. Evidently, if $a = b$ then $A = Med_a$. Moreover, $A$ is always associative, and if $L = [0,1]$ then it coincides with the binary Sugeno integral, see [40].

On any lattice $L$, we can introduce also the following conjunctive aggregation function $Z : L^2 \to L$ and its dual $Z^* : L^2 \to L$ given by

$$Z(x,y) = \begin{cases} x \wedge y & \text{if } x \vee y = \mathbf{1}, \\ \mathbf{0} & \text{else,} \end{cases} \tag{4}$$

and

$$Z^*(x,y) = \begin{cases} x \vee y & \text{if } x \wedge y = \mathbf{0}, \\ \mathbf{1} & \text{else.} \end{cases} \tag{5}$$

Observe that the duality follows from Remark 1. Indeed, the formula for $Z$ on $(L, \leq, \mathbf{0}, \mathbf{1})$ is exactly the same as the formula for $Z^*$ on $(L^*, \leq^*, \mathbf{0}^*, \mathbf{1}^*)$ taking into account that $\wedge^* = \vee$ and $\vee^* = \wedge$.

Note that $Z = T_D$ ($Z^* = S_D$) if and only if $L$ has no unit multipliers, i.e., $x < \mathbf{1}$ and $y < \mathbf{1}$ imply $x \vee y < \mathbf{1}$ ($L$ has no zero divisors, i. e., $x > \mathbf{0}$ and $y > \mathbf{0}$ imply $x \wedge y > \mathbf{0}$). Both these constraints are satisfied if $L$ is a bounded chain. In remaining cases, $Z \geq T_D$ (and $Z \neq T_D$), and $Z^* \leq S_D$ (and $Z^* \neq S_D$).

The functions $T_D, S_D, Z, Z^*, Min$ and $Max$ are examples of triangular norms and related aggregation functions which we will discuss in the next section.

## 4 Triangular Norms and Related Aggregation Functions

E. E. Kerre with G. De Cooman [9] have introduced order norms on bounded posets, covering triangular norms, their weaking triangular seminorms, and related dual aggregation functions.

**Definition 5.** Let $A : P^2 \to P$ be a binary aggregation function on a fixed bounded poset $(P, \leq, \mathbf{0}, \mathbf{1})$. Then $A$ is called:

(i) a triangular seminorm ($t$–seminorm in short) whenever $\mathbf{1}$ is a neutral element of $A$, i. e., $A(x,\mathbf{1}) = A(\mathbf{1},x) = x$ for all $x \in P$;
(ii) a triangular semiconorm ($t$–semiconorm in short) whenever $\mathbf{0}$ is a neutral element of $A$.

Moreover:

(iii) a symmetric associative $t$–seminorm $A$ is called a triangular norm ($t$–norm) on $P$;
(iv) a symmetric associative $t$–semiconorm $A$ is called a triangular conorm ($t$–conorm) on $P$.

Finally, if $A$ is a $t$-norm, or a $t$-seminorm, or a $t$-conorm, or a $t$-semiconorm on $P$, then it is called an order norm on $P$.

Due to the associativity, triangular norms and triangular conorms can be extended to $n$–ary aggregation functions as well as to extended aggregation functions on $P$ (for $n = 1$, identity on $P$ is considered in both cases). Note that triangular norms are conjunctive, while triangular conorms are disjunctive aggregation functions. Moreover, it is possible to introduce $n$–ary $t$–seminorms ($t$–semiconorms) as aggregation functions $A : P^n \to P$ with a neutral element $\mathbf{1}$ ($\mathbf{0}$), i. e., satisfying $A(x_1,\ldots,x_n) = x_i$ whenever $x_j = \mathbf{1}$ for all $j \neq i$ ($A(x_1,\ldots,x_n) = x_j$ whenever $x_j = \mathbf{0}$ for all $j \neq i$).

As already mentioned in Section 2, $T_D$ is the weakest $t$–norm on $P$. Evidently, it is also the weakest $t$–seminorm on $P$. Similarly, $S_D$ is both the strongest $t$–conorm and the strongest $t$–semiconorm on $P$. In general, the strongest $t$–norm either $t$–seminorm need not exist. By duality, the weakest $t$–conorm ($t$–semiconorm) need not exist.

*Example 4.* Define a bounded poset $P$ by the next Hasse diagram:

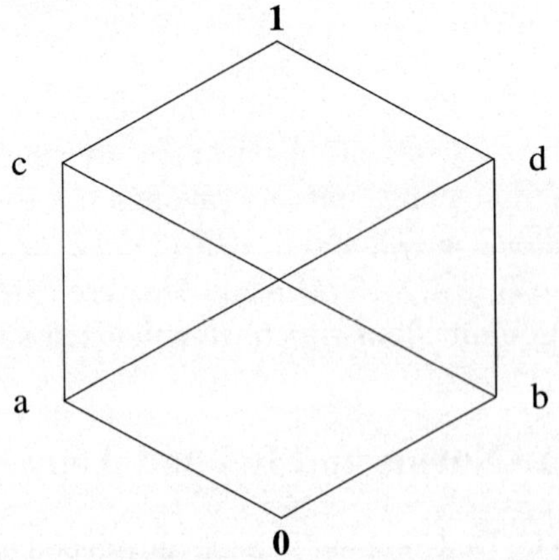


**Fig. 1** Hasse diagram of a poset $P$

and two mappings $A, B : P^2 \to P$ by the next tables 1 and 2:

**Table 1** The mapping $A$

| $A$ | **0** | a | b | c | d | **1** |
|---|---|---|---|---|---|---|
| **0** | **0** | **0** | **0** | **0** | **0** | **0** |
| a | **0** | a | **0** | a | a | a |
| b | **0** | **0** | b | b | b | b |
| c | **0** | a | b | c | a | c |
| d | **0** | a | b | a | d | d |
| **1** | **0** | a | b | c | d | **1** |

**Table 2** The mapping $B$

| $B$ | **0** | a | b | c | d | **1** |
|---|---|---|---|---|---|---|
| **0** | **0** | **0** | **0** | **0** | **0** | **0** |
| a | **0** | a | **0** | a | a | a |
| b | **0** | **0** | b | b | b | b |
| c | **0** | a | b | c | b | c |
| d | **0** | a | b | b | d | d |
| **1** | **0** | a | b | c | d | **1** |

Then both $A$ and $B$ are maximal $t$-seminorms on $P$. Moreover, both $A$ and $B$ are triangular norms on $P$.

On the other hand, if $P$ is a lattice, then $Min$ is the strongest $t$–norm as well as the strongest $t$–seminorm on $P$, and $Max$ is the weakest $t$–conorm as well as the weakest $t$–semiconorm on $P$.

Example 4 shows that there can be several idempotent $t$–norms ($t$–seminorms, $t$–conorms, $t$–semiconorms) on a general bounded poset $P$. However, if $P$ is a lattice, then the requirement of idempotency leads to the uniqueness of discussed order norms.

**Proposition 2.** *[9] Let $(L, \leq, \mathbf{0}, \mathbf{1})$ be a bounded lattice and let $A : L^2 \to L$ be an idempotent (i. e., averaging) aggregation function on L. Then:*

*(i) A is a t–norm if and only if A is a t–seminorm if and only if $A = Min$;*
*(ii) A is a t–conorm if and only if A is a t–semiconorm if and only if $A = Max$.*

Observe that due to the boundary conditions required for order norms, when considering a pair $A, B : P^2 \to P$ of a $t$–norm ($t$–seminorm) $A$ and a $t$–conorm ($t$–semiconorm) $B$ on $P$, the distributivity

$$A(x,B(y,z)) = B(A(x,y),A(x,z))$$

implies the absorption

$$B(A(x,y),x) = x$$

(see [9], Proposition 3.5), and this absorption implies the idempotency of $B$ (see [9], Proposition 3.6).

Similarly, the distributivity

$$A(B(x,y),z) = B(A(x,z),A(y,z))$$

implies the absorption

$$B(x,A(y,x)) = x$$

which implies the idempotency of $B$.

Note that the role of $A$ and $B$ can be reversed in the above claims.

Several other properties (and their relationships) of order norms are introduced and discussed in [9], having in mind in particular the role of $t$-norms and $t$-conorms as conjunctions and disjunctions in many valued logics with truth values domain $P$.

Order norms (in particular triangular norms) on special kinds of posets and/or possessing some special properties were discussed in several papers. For example, triangular norms on product lattices were studied, among others, in [8]. Evidently, for any product poset

$$(P, \leq, \mathbf{0}, \mathbf{1}) = \prod_{i \in I} (P_i, \leq_i, 0_i, 1_i),$$

and any system $(T_i)_{i \in I}$ of $t$–norms $T_i : P_i^2 \to P_i$, $i \in I$, the mapping $T : P^2 \to P$ given by

$$T\left((x_i)_{i \in I}, (y_i)_{i \in I}\right) = \left(T_i(x_i, y_i)\right)_{i \in I}$$

is a triangular norm on $P$. A similar result is valid also for the remaining types of order norms, i. e., for $t$–seminorms, for $t$–conorms, and for $t$–semiconorms. Interesting is the opposite claim, which is valid only under some special constraints.

**Proposition 3.** *[8] For a fixed chain* $(L, \leq, \mathbf{0}, \mathbf{1})$ *and* $n > 1$, *let* $T : (L^n)^2 \to L^n$ *be a t–norm on the product lattice* $L^n$. *Then* $T$ *is the product of t–norms* $T_1, \ldots, T_n$ *on* $L$, *i. e.,* $T((x_1, \ldots, x_n), (y_1, \ldots, y_n)) = (T_1(x_1, y_1), \ldots, T_n(x_n, y_n))$ *if and only if* $T(\mathbf{x} \vee \mathbf{y}, \mathbf{z}) = T(\mathbf{x}, \mathbf{z}) \vee T(\mathbf{y}, \mathbf{z})$ *for all* $\mathbf{x}, \mathbf{y}, \mathbf{z} \in L^n$ *(or* $T(\mathbf{x} \wedge \mathbf{y}, \mathbf{z}) = T(\mathbf{x}, \mathbf{z}) \wedge T(\mathbf{y}, \mathbf{z})$ *for all* $\mathbf{x}, \mathbf{y}, \mathbf{z} \in L^n$*).*

Not all $t$–norms are direct products of $t$–norms on $L$. For example, in [30] a method for constructing $t$–norms on product lattices which are not direct products is presented.

Many particular results on $t$–norms on lattices (thus by duality also on $t$–conorms) can be found in the domain of integral monoids, see, for example, [29].

Problems concerning ordinal sums of $t$–norms on bounded lattices were investigated in [36], and [38]. We recall only the next result from [38].

**Proposition 4.** *[38] Let* $(L, \leq, \mathbf{0}, \mathbf{1})$ *be a bounded lattice,* $(K, \leq, a, b)$ *its complete bounded sublattice, and* $H : K^2 \to K$ *a t–norm on* $K$. *Then* $T : L^2 \to L$ *given by*

$$T(x,y) = \begin{cases} x \wedge y & \textit{if } \mathbf{1} \in \{x,y\}, \\ \mathbf{0} & \textit{if } \mathbf{0} \in \{x,y\}, \\ H(x^*,y^*) & \textit{else,} \end{cases}$$

*where* $x^* = \sup\{z \in K \cup \{\mathbf{0},\mathbf{1}\} | z \leq x\}$, *is a t–norm on L extending H, i. e.,* $T|K^2 = H$.

For interested readers we recall some works on order norms on posets (lattices) which may be an inspiration for a deeper study of this topic. Some particular methods to construct triangular norms on special types of posets can be found for example in [42]. $\bigvee$-distributivity and infinitely $\bigvee$-distributivity of triangular norms on complete bounded lattices is studied in [31]. An interesting relation between triangular norms and triangular seminorms on the standard lattice $L = [0,1]$ (i. e., classical triangular norms are considered, as they were introduced in [39] and studied in [1, 32]) was shown in [24] (observe that triangular seminorms on $[0,1]$ are often called semicopulas [25]).

**Proposition 5.** *The class of all symmetric t–seminorms on* $[0,1]$ *is the* $\bigvee$*-closure (*$\bigwedge$*-closure) of the class of all t–norms on* $[0,1]$.

**Open problem**: Let $(L, \leq, \mathbf{0}, \mathbf{1})$ be a complete lattice. Is it possible to rewrite Proposition 5 for this case, i. e., is the class of all symmetric *t*–seminorms on *L* the $\bigvee$-closure ($\bigwedge$-closure) of the class of all *t*–norms on *L*?

## 5 Aggregation Functions on Lattice $L^*$

Consider the set $L^* = \{(x_1,x_2)|(x_1,x_2) \in [0,1]^2, x_1 + x_2 \leq 1\}$ and define a relation $\leq$ on $L^*$ by

$$(x_1,x_2) \leq (y_1,y_2) \text{ whenever } x_1 \leq x_2 \text{ and } x_2 \geq y_2.$$

Then $(L^*, \leq, \mathbf{0}, \mathbf{1})$ is a bounded complete lattice, where $\mathbf{0} = (0,1)$ and $\mathbf{1} = (1,0)$. Note that then the lattice operations on $L^*$ are defined as follows:

$$(x_1,x_2) \wedge (y_1,y_2) = (x_1 \wedge y_1, x_2 \vee y_2),$$

$$(x_1,x_2) \vee (y_1,y_2) = (x_1 \vee y_1, x_2 \wedge y_2).$$

Recall that the lattice $L^*$ is a background of intuitionistic fuzzy set theory introduced by [2] and it is isomorphic to the standard lattice of all closed subintervals of $[0,1]$.

The major contribution to the theory of aggregation functions on $L^*$ is due to the school of E. E. Kerre [13, 14, 15, 16, 17, 18, 19, 20, 21] and [4, 5, 22, 41]. In this section, we recall some of results presented in these manuscripts.

Note first of all that due to [12] , a mapping $N : L^* \to L^*$ is decreasing involution (i. e., a negator), if and only if there is decreasing involution $n : [0,1] \to [0,1]$ such that $N((x_1,x_2)) = (n(1-x_2), 1-n(x_1))$. It is not difficult to see that for any aggregation function $A : (L^*)^n \to L^*$ ($A : \bigcup_{n \in N}(L^*)^n \to L^*$), also the mapping $A^N : (L^*)^n \to L^*$ ($A^N : \bigcup_{n\in N}(L^*)^n \to L^*$) given by $A^N((x_1,y_1),\dots,(x_n,y_n)) = N(A(N(x_1,y_1),\dots,N(x_n,y_n)))$ is an aggregation function on $L^*$ ($A^N$ is then called $N$–dual of $A$). Recall that $N$–dual to a $t$–norm, $t$–seminorm, $t$–conorm, $t$–semiconorm on $L^*$ is a a $t$–conorm, $t$–semiconorm, $t$–norm, $t$–seminorm on $L^*$, respectively. A simple method how to construct aggregation functions on $L^*$ is based on aggregation functions on $[0,1]$.

**Lemma 1.** *Let $A : [0,1]^n \to [0,1]$ be an aggregation function on $[0,1]$. Then $A^* : (L^*)^n \to L^*$ given by*

$$A^*((x_1,y_1),\dots,(x_n,y_n)) = (A(x_1,\dots,x_n), 1 - A(1-y_1,\dots,1-y_n))$$

*is an aggregation function on $L^*$.*

This method can be further generalized to so called representable aggregation functions on $L^*$ [11, 12].

**Proposition 6.** *Let $A,B : [0,1]^n \to [0,1]$ ($A,B : \bigcup_{n\in N}[0,1]^n \to [0,1]$) be two aggregation functions of $[0,1]$ such that for all $x_1,\dots,x_n \in [0,1]$, the inequality*

$$A(x_1,\dots,x_n) + B(1-x_1,\dots,1-x_n) \leq 1$$

*holds. Then the mapping $C : (L^*)^n \to L^*$ ($C : \bigcup_{n\in N}(L^*)^n \to L^*$) given by*

$$C((x_1,y_1),\dots,(x_n,y_n)) = (A(x_1,\dots,x_n), B(y_1,\dots,y_n))$$

*is an aggregation function on $L^*$. $C$ is called a representable aggregation function.*

For arbitrary aggregation functions $A,B$ on $[0,1]$ such that $A \leq B^d$ ($B^d$ is the dual aggregation function to $B$, based on duality w.r.t. $n$, $n(x) = 1-x$) $A$ and $B$ satisfy requirements of Proposition 6. Thus Lemma 1 reflects the fact $A \leq A$ (observe that $(A^d)^d = A$). Note that not all aggregation functions on $L^*$ are representable.

*Example 5.* (i) Let $T : [0,1]^2 \to [0,1]$ be a $t$–norm on $[0,1]$. Then $T^* : (L^*)^2 \to L^*$ given by

$$T^*((x_1,y_1),(x_2,y_2)) = (T(x_1,x_2), S(y_1,y_2)),$$

where $S : [0,1]^2 \to [0,1]$ is a $t$–conorm on $[0,1]$ dual to $T$ (based on duality w.r.t. $n$, $n(x) = 1-x$). Obviously, $T^*$ is a representable $t$–norm on $L^*$.
Also $F : (L^*)^2 \to L^*$ given by

$$F((x_1,y_1),(x_2,y_2)) = (x_1x_2, y_1 \vee y_2)$$

is a representable $t$–norm on $L^*$ (linked to $A = T_P$, product $t$–norm on $[0,1]$, and $B = S_M$, maximum $t$–conorm on $[0,1]$).

(ii) For a fixed $t$–norm $T : [0,1]^2 \to [0,1]$, define a mapping $D : (L^*)^2 \to L^*$ by

$$D((x_1,y_1),(x_2,y_2)) = (T(x_1,x_2), S(1-x_1,y_2) \wedge S(x_1,1-x_2)).$$

Evidently, $D$ is a symmetric $t$–semiconorm on $L^*$ which is not representable. Observe that $D$ is associative and thus a $t$–norm on $L^*$. For more details see [12].

(iii) Define $C : (L^*)^3 \to L^*$ by

$$C((x_1,y_1),(x_2,y_2),(x_3,y_3)) = \left(\frac{x_1+2x_2+x_3}{4}, (y_1y_2^2y_3)^{\frac{1}{4}}\right).$$

Then $C$ is an idempotent representable aggregation function on $L^*$ which is bisymmetric but not symmetric ($C$ is linked to a weighted arithmetic mean and to a weighted geometric mean).

Several interesting details, including deep results on Archimedean $t$–norms and $t$–norms making $L^*$ a residuated monoid can be found in [11, 17].

## 6 Concluding Remarks

We have introduced and discussed aggregation functions on bounded partially ordered sets and on bounded lattices, stressing among others the pioneering work of E. E. Kerre and his research group in this domain. Our work opens new doors to the investigation of aggregation functions. For example, the class of strongly idempotent aggregation functions deserves a deeper look not only in the case of general bounded posets, but especially in the case of particular bounded lattices, such as $L^*$, $[0,1]^n$, etc. Our approach covers also the aggregation on cardinal and ordinal scales, on linguistic scales, as well as on special systems such as distributive functions.

**Acknowledgement.** The research summarized in this paper was supported by the Grants APVV–0012–07 and VEGA 1/0080/10.

## References

1. Alsina, C., Frank, M.J., Schweizer, B.: Associative functions. World Scientific Publishing Co. Pte. Ltd., Hackensack (2006)
2. Atanassov, K.: Intuitionistic fuzzy sets. Fuzzy Sets and Systems 20, 87–96 (1986)
3. Birkhoff, G.: Lattice theory, 3rd edn. American Mathematical Society Colloquium Publications. American Mathematical Society, Providence (1967)
4. Bustince, H., Barrenechea, E., Pagola, M.: Generation of interval-valued fuzzy and Atanassov's intuitionistic fuzzy connectives from fuzzy connectives and from K– operators: Laws for conjunctions and disjunctions, amplitude. International Journal of Intelligent Systems 23(6), 680–714 (2008)
5. Bustince, H., Montero, J., Pagola, M., Barrenechea, E., Gómez, D.: A survey on interval-valued fuzzy sets. In: Pedrycz, W., Skowron, A., Kreinovich, V. (eds.) Handbook of granular computing, pp. 489–515. Wiley, New York (2008)

6. Calvo, T., Kolesárová, A., Komorníková, M., Mesiar, R.: Aggregation operators: Properties, classes and construction methods. In: Calvo, T., Mayor, G., Mesiar, R. (eds.) Aggregation operators: Properties, Classes and Construction Methods, Aggregation Operators: New Trends and Applications, pp. 1–104. Physica-Verlag, Heilderberg (2002)
7. Clifford, A.H.: Naturally totally ordered commutative semigroups. Amer. J. Math. 76, 631–646 (1954)
8. De Baets, B., Mesiar, R.: Triangular norms on product lattices. Fuzzy Sets and Systems 104, 61–76 (1999)
9. De Cooman, G., Kerre, E.E.: Order Norms On Bounded Partially Ordered Sets. The Journal of Fuzzy Mathematics 2, 281–310 (1994)
10. Demirci, M.: Aggregation operators on partially ordered sets and their categorical foundations. Kybernetika 42, 261–277 (2006)
11. Deschrijver, G.: Additive and multiplicative generators in interval-valued fuzzy set theory. IEEE Transactions on Fuzzy Systems 15(2), 222–237 (2007)
12. Deschrijver, G.: Arithmetic operators in interval-valued fuzzy set theory. Information Sciences 177(14), 2906–2924 (2007)
13. Deschrijver, G.: Characterizations of (weakly) Archimedean t-norms in interval-valued fuzzy set theory. Fuzzy Sets and Systems 160(6), 778–801 (2009)
14. Deschrijver, G.: Generalized arithmetic operators and their relationship to t-norms in interval-valued fuzzy set theory. Fuzzy Sets and Systems 160(21), 3080–3102 (2009)
15. Deschrijver, G., Cornelis, C., Kerre, E.E.: Intuitionistic fuzzy connectives revisited. In: Proc. 9th Internat. Conf. on Information Processing and Management of Uncertainty in Knowledge-Based Systems, Annecy, France, pp. 1839–1844 (2002)
16. Deschrijver, G., Cornelis, C., Kerre, E.E.: On the representation of intuitionistic fuzzy t-norms and t-conorms. IEEE Transactions on Fuzzy Systems 12(1), 45–61 (2004)
17. Deschrijver, G., Kerre, E.E.: Classes of intuitionistic fuzzy t-norms satisfying the residuation principle. International Journal of Uncertainty, Fuzziness and Knowledge-Based Systems 11(6), 691–709 (2003)
18. Deschrijver, G., Kerre, E.E.: On the Relationship Between some Extensions of Fuzzy Set Theory. Fuzzy Sets and Systems 133, 227–235 (2003)
19. Deschrijver, G., Kerre, E.E.: Uninorms in $L^*$–fuzzy set theory. Fuzzy Sets and Systems 148, 243–262 (2004)
20. Deschrijver, G., Kerre, E.E.: Triangular norms and related operators in $L^*$-fuzzy set theory. In: Klement, E.P., Mesiar, R. (eds.) Logical, Algebraic, Analytic and Probabilistic Aspects of Triangular Norms, pp. 231–259. Elsevier, Amsterdam (2005)
21. Deschrijver, G., Kerre, E.: Aggregation operators in interval-valued fuzzy and Atanassov's intuitionistic fuzzy set theory. Studies in Fuzziness and Soft Computing 220, 183–203 (2008)
22. Dimuro, G.P., Bedregal, B.C., Reiser, R.H.S., Santiago, R.H.N.: Interval additive generators of interval t-norms. In: Hodges, W., de Queiroz, R. (eds.) Logic, Language, Information and Computation. LNCS (LNAI), vol. 5110, pp. 123–135. Springer, Heidelberg (2008)
23. Dubois, D., Prade, H.: Criteria aggregation and ranking of alternatives in the framework of fuzzy set theory. In: Fuzzy sets and decision analysis 20 of Stud. Management Sci., pp. 209–240. North-Holland, Amsterdam (1984)
24. Durante, F., Mesiar, R., Papini, P.-L.: The lattice-theoretic structure of the sets of triangular norms and semi-copulas. Nonlinear Anal. 69(1), 46–52 (2008)
25. Durante, F., Sempi, C.: Semicopula. Kybernetika 41(3), 315–328 (2005)

26. Fung, L.W., Fu, K.S.: An axiomatic approach to rational decision making in a fuzzy environment. In: Fuzzy sets and their applications to cognitive and decision processes (Proc. U. S.-Japan Sem., Univ. Calif., Berkeley, Calif., 1974), pp. 227–256. Academic Press, New York (1975)
27. Grätzer, G.: General lattice theory, 2nd edn. Birkhäuser Verlag, Basel (1998)
28. Grabisch, M., Marichal, J.–L., Mesiar, R., Pap, E.: Aggregation functions. Cambridge University Press, Cambridge (2009)
29. Höhle, U.: Commutative, residuated l-monoids. In: Höhle, U., Klement, E.P. (eds.) Non-classical Logics and their Applications to Fuzzy Subsets, pp. 53–106. Kluwer Academic Publishers, Dordrecht (1995)
30. Jenei, S., De Baets, B.: On the direct decomposability of t-norms on product lattices. Fuzzy Sets and Systems 139(3), 699–707 (2003)
31. Karacal, F., Khadjiev, D.: $\bigvee$–distributive and infinitely $\bigvee$–distributive t-norms on complete lattices. Fuzzy Sets and Systems 151(2), 341–352 (2005)
32. Klement, E.P., Mesiar, R., Pap, E.: Triangular norms. Trends in Logic, Studia Logica Library, vol. 8. Kluwer Academic Publishers, Dordrecht (2000)
33. Klir, G., Folger, T.: Fuzzy Sets, Uncertainty and Information. Prentice-Hall, Englewood Cliffs (1988)
34. Kolesárová, A., Komorníková, M.: Triangular norm-based iterative compensatory operators. Fuzzy Sets and Systems 104, 109–119 (1999)
35. Marichal, J.L.: k–intolerant capacities and Choquet integrals. European J. Oper. Res. 177(3), 1453–1468 (2007)
36. Saminger-Platz, S.: On ordinal sums of triangular norms on bounded lattices. Fuzzy Sets and Systems 157(10), 1403–1416 (2006)
37. Saminger-Platz, S.: The dominance relation in some families of continuous Archimedean t–norms and copulas. Fuzzy Sets and Systems 160(14), 2017–2031 (2009)
38. Saminger-Platz, S., Klement, E.P., Mesiar, R.: On extensions of triangular norms on bounded lattices. Indagationes Mathematicae 19(1), 135–150 (2008)
39. Schweizer, B., Sklar, A.: Statistical metric spaces. Pacific J. Math. 10, 313–330 (1960)
40. Sugeno, M.: Theory of fuzzy integrals and its applications. PhD thesis, Tokyo Institute of Technology (1974)
41. Turksen, I.B.: Interval-valued fuzzy sets and 'compensatory AND'. Fuzzy Sets and Systems 51(3), 295–307 (1992)
42. Zhang, D.: Triangular norms on partially ordered sets. Fuzzy Sets and Systems 153(2), 195–209 (2005)

# On the Intuitionistic Fuzzy Implications and Negations. Part 1

Krassimir T. Atanassov

To my friend Prof. Etienne Kerre

**Abstract.** The paper contains review and new results related to the first 23 of the 138 already introduced by the author different intuitionistic fuzzy implications and the first 5 of the 36 different intuitionistic fuzzy negations. Some of their basic properties are studied. The validity and non-validity of the Law of Excluded Middle and its modifications, and of De Morgan's Laws and their modifications are discussed. The axioms of Intuitionistic Logic are checked for the intuitionistic fuzzy implications and negations.

## 1 Introduction

During the last years, a discussion started related to the name "intuitionistic" given to the extension of the fuzzy set, introduced by the author in 1983 in [1]. He hopes that now, after [16], it is clear that the concept of "intuitionistic fuzzy set" introduced by Takeuti and Titani a year and a half later in [28] is a particular case of the ordinary fuzzy sets, while the author's concept of an Intuitionistic Fuzzy Set (IFS) essentially extends the fuzzy sets. Also, the author hopes that it is already clear that the construction of the IFSs contains Brouwer's idea for intuitionism [20]. In every way, the discussion was very useful for IFS theory.

First, and it is possible, that the most important author's mistake was that more than 20 years the IFS theory used only the classical negation (denoted below by $\neg_1$), that generates a classical implication (denoted below by $\rightarrow_4$). Really, another implication (denoted below by $\rightarrow_{11}$) that generates

Krassimir T. Atanassov
Department of Bioinformatics and Mathematical Modelling,
Institure of Biophysics and Biomedical Engineering
Bulgarian Academy of Sciences, Acad. G. Bonchev Str., Bl. 105,
Sofia-1113, Bulgaria
e-mail: krat@bas.bg

C. Cornelis et al. (Eds.): 35 Years of Fuzzy Set Theory, STUDFUZZ 261, pp. 19–38.
springerlink.com 

an intuitionistic fuzzy negation, was introduced in IFS theory in 1988 [2], but for a long period of time the author had not estimated their merits as an intuitionistic implication and negation and he had not used them because they are more complex than the standard ones. Now, the author realizes his mistake (better late than never) and hence the properties of the different versions of both these operations will be discussed below.

Second, searching arguments to defend the name of IFS, the author found a lot of new implications and negations. One part of them exhibits classical properties, another – intuitionistic properties and a third group exhibits more nonstandard properties. These new operations will be an object of discussion of the present and next author's research.

We must mention that another approach for introducing intuitionistic fuzzy implications and negations is demonstrated by a series of papers by Chris Cornelis, Glad Deschrijver and Etienne Kerre [21, 22, 23, 24].

## 2 Definitions and Properties of 23 Intuitionistic Fuzzy Implications and 5 Intuitionistic Fuzzy Negations

### 2.1 Definitions of 23 Intuitionistic Fuzzy Implications

The first ten variants of intuitionistic fuzzy implications are discussed in [4], using the book [26] by George Klir and Bo Yuan as a basis, where the conventional fuzzy implications are given. Other five implications, defined by the author and his colleagues Boyan Kolev and Trifon Trifonov, are introduced in [2, 3, 7, 8, 15, 17]. These fifteen implications generated five negations by the formulae

$$\neg x = x \to 0. \tag{1}$$

On the other hand, these negations are the basis for new eight implications introduced in [9] through the formulae

$$x \to y = \neg x \vee y \tag{2}$$

and

$$x \to y = \neg x \vee \neg\neg y. \tag{3}$$

The new (eight) implications generate negations that coincide with the respective negations generated by the corresponding implications, i.e., the generating process finishes. Therefore, using this scheme, we finally obtain 23 implications.

Below, we shall describe some properties of these implications and negations.

Let us denote below each of these implications by $I(x, y)$ and the negations by $N(x)$.

In intuitionistic fuzzy logic, if $x$ is a variable, then its truth-value is represented by the ordered pair

$$V(x) = \langle a, b \rangle,$$

such that $a, b, a + b \in [0, 1]$, where $a$ and $b$ are degrees of validity and of non-validity of $x$.

Assume that for the three variables $x, y$ and $z$ there hold the equalities: $V(x) = \langle a, b \rangle, V(y) = \langle c, d \rangle, V(z) = \langle e, f \rangle$ $(a, b, c, d, e, f, a + b, c + d, e + f \in [0, 1])$.

For the needs of the discussion below, we define the notion of Intuitionistic Fuzzy Tautology (IFT, see [2, 3]) by,

$$x \text{ is an IFT if and only if } a \geq b,$$

while $x$ will be a tautology iff $a = 1$ and $b = 0$.

We define the following relation:

$$V(x) \leq V(y) \text{ iff } a \leq c \text{ and } b \geq d.$$

In some definitions, we use functions sg and $\overline{\text{sg}}$, defined by

$$\text{sg}(x) = \begin{cases} 1 & \text{if } x > 0 \\ 0 & \text{if } x \leq 0 \end{cases}, \quad \overline{\text{sg}}(x) = \begin{cases} 0 & \text{if } x > 0 \\ 1 & \text{if } x \leq 0 \end{cases}$$

For two variables $x$ and $y$ operations "conjunction" (&) and "disjunction" ($\vee$) are defined (see [2, 3]) by:

$$V(x \wedge y) = \langle \min(a, c), \max(b, d) \rangle,$$

$$V(x \vee y) = \langle \max(a, c), \min(b, d) \rangle.$$

In Table 1, we include the implications from [4], but also the implications, introduced by the author in [2, 7, 8, 15, 17] with coauthors Boyan Kolev [15] and Trifon Trifonov [17].

The correctness of the above definitions of implications is directly checked. For example, to check the validity of the definition of $\rightarrow_{15}$, we check that for every $a, b, c, d \in [0, 1]$ such that $a + b \leq 1$ and $c + d \leq 1$ for the expression,

$$X \equiv 1 - (1 - \min(b, c)).\text{sg}(\text{sg}(a - c) + \text{sg}(d - b)) - \min(b, c).\text{sg}(a - c).\text{sg}(d - b)$$

$$+1 - (1 - \max(a, d)).\text{sg}(\overline{\text{sg}}(a - c) + \overline{\text{sg}}(d - b)) - \max(a, d).\overline{\text{sg}}(a - c).\overline{\text{sg}}(d - b)$$

we obtain,

If $a \leq c$ and $b \geq d$, then,

$$X = 1 - (1 - \min(b, c)).\text{sg}(0 + 0) - \min(b, c).0 + 1$$

$$-(1 - \max(a, d)).\text{sg}(1 + 1) - \max(a, d).1$$

$$= 1 + 1 - (1 - \max(a, d)) - \max(a, d) = 1.$$

**Table 1** List of intuitionistic fuzzy implications

| Notation | Name | Form of implication |
|---|---|---|
| $\rightarrow_1$ | Zadeh | $\langle \max(b, \min(a, c)), \min(a, d) \rangle$ |
| $\rightarrow_2$ | Gaines-Rescher | $\langle 1 - \mathrm{sg}(a - c), d.\mathrm{sg}(a - c) \rangle$ |
| $\rightarrow_3$ | Gödel | $\langle 1 - (1 - c).\mathrm{sg}(a - c), d.\mathrm{sg}(a - c) \rangle$ |
| $\rightarrow_4$ | Kleene-Dienes | $\langle \max(b, c), \min(a, d) \rangle$ |
| $\rightarrow_5$ | Lukasiewicz | $\langle \min(1, b + c), \max(0, a + d - 1) \rangle$ |
| $\rightarrow_6$ | Reichenbach | $\langle b + ac, ad \rangle$ |
| $\rightarrow_7$ | Willmott | $\langle \min(\max(b, c), \max(a, b), \max(c, d)),$ $\max(\min(a, d), \min(a, b), \min(c, d)) \rangle$ |
| Notation | Name | Form of implication |
| $\rightarrow_8$ | Wu | $\langle 1 - (1 - \min(b, c)).\mathrm{sg}(a - c),$ $\max(a, d).\mathrm{sg}(a - c).\mathrm{sg}(d - b) \rangle$ |
| $\rightarrow_9$ | Klir and Yuan 1 | $\langle b + a^2 c, ab + a^2 d \rangle$ |
| $\rightarrow_{10}$ | Klir and Yuan 2 | $\langle c.\overline{\mathrm{sg}}(1 - a) + \mathrm{sg}(1 - a)$ $.(\overline{\mathrm{sg}}(1 - c) + b.\mathrm{sg}(1 - c)),$ $d.\overline{\mathrm{sg}}(1 - a) + a.\mathrm{sg}(1 - a).\mathrm{sg}(1 - c) \rangle$ |
| $\rightarrow_{11}$ | Atanassov 1 | $\langle 1 - (1 - c).\mathrm{sg}(a - c),$ $d.\mathrm{sg}(a - c).\mathrm{sg}(d - b) \rangle$ |
| $\rightarrow_{12}$ | Atanassov 2 | $\langle \max(b, c), 1 - \max(b, c) \rangle$ |
| $\rightarrow_{13}$ | Atanassov and Kolev | $\langle b + c - b.c, a.d \rangle$ |
| $\rightarrow_{14}$ | Atanassov and Trifonov | $\langle 1 - (1 - c).\mathrm{sg}(a - c) - d.\overline{\mathrm{sg}}(a - c).\mathrm{sg}(d - b),$ $d.\mathrm{sg}(d - b) \rangle$ |
| $\rightarrow_{15}$ | Atanassov 3 | $\langle 1 - (1 - \min(b, c)).\mathrm{sg}(\mathrm{sg}(a - c) + \mathrm{sg}(d - b))$ $- \min(b, c).\mathrm{sg}(a - c).\mathrm{sg}(d - b),$ $1 - (1 - \max(a, d)).\mathrm{sg}(\overline{\mathrm{sg}}(a - c) + \overline{\mathrm{sg}}(d - b))$ $- \max(a, d).\overline{\mathrm{sg}}(a - c).\overline{\mathrm{sg}}(d - b) \rangle$ |
| $\rightarrow_{16}$ | | $\langle \max(1 - \mathrm{sg}(a), c), \min(\mathrm{sg}(a), d) \rangle$ |
| $\rightarrow_{17}$ | | $\langle \max(b, c), \min(a.b + a^2, d) \rangle$ |
| $\rightarrow_{18}$ | | $\langle \max(b, c), \min(1 - b, d) \rangle$ |
| $\rightarrow_{19}$ | | $\langle \max(1 - \mathrm{sg}(\mathrm{sg}(a) + \mathrm{sg}(1 - b)), c),$ $\min(\mathrm{sg}(1 - b), d) \rangle$ |
| $\rightarrow_{20}$ | | $\langle \max(1 - \mathrm{sg}(a), 1 - \mathrm{sg}(1 - \mathrm{sg}(c))), \min(\mathrm{sg}(a),$ $\mathrm{sg}(1 - sg(c))) \rangle$ |
| $\rightarrow_{21}$ | | $\langle \max(b, c(c + d)), \min(a(a + b),$ $d(c^2 + d + cd)) \rangle$ |
| $\rightarrow_{22}$ | | $\langle \max(b, 1 - d), \min(1 - b, d) \rangle$ |
| $\rightarrow_{23}$ | | $\langle 1 - \min(\mathrm{sg}(1 - b), \mathrm{sg}(1 - \mathrm{sg}(1 - d))),$ $\min(\mathrm{sg}(1 - b), \mathrm{sg}(1 - \mathrm{sg}(1 - d))) \rangle$ |

If $a \leq c$ and $b < d$, then,

$$X = 1 - (1 - \min(b,c)).\mathrm{sg}(0+1) - \min(b,c).0.1$$

$$+1 - (1 - \max(a,d)).\mathrm{sg}(1+0) - \max(a,d).1.0$$

$$= 1 - 1 + \min(b,c) + 1 - 1 + \max(a,d)$$

$$= \min(b,c) + \max(a,d) \leq 1.$$

If $a > c$ and $b \geq d$, then,

$$X = 1 - (1 - \min(b,c)).\mathrm{sg}(1+0) - \min(b,c).1.0$$

$$+1 - (1 - \max(a,d)).\mathrm{sg}(0+1) - \max(a,d).0.1$$

$$= 1 - 1 + \min(b,c) + 1 - 1 + \max(a,d)$$

$$= \min(b,c) + \max(a,d) \leq 1.$$

If $a > c$ and $b < d$ then,

$$X = 1 - (1 - \min(b,c)).\mathrm{sg}(1+1) - \min(b,c).1.1$$

$$+1 - (1 - \max(a,d)).\mathrm{sg}(0+0) - \max(a,d).0.0$$

$$= 1 - (1 - \min(b,c)).1 - \min(b,c) + 1 - (1 - \max(a,d)).0$$

$$= 1 - 1 + \min(b,c)) - \min(b,c) + 1 = 1.$$

Therefore, implication $\rightarrow_{15}$ is valid.

**Theorem 1**. If $I$ is any of the 23 implications, then

$$I(\langle 0,1\rangle, \langle 0,1\rangle) = \langle 1,0\rangle,$$

$$I(\langle 0,1\rangle, \langle 1,0\rangle) = \langle 1,0\rangle,$$

$$I(\langle 1,0\rangle, \langle 1,0\rangle) = \langle 1,0\rangle,$$

$$I(\langle 1,0\rangle, \langle 0,1\rangle) = \langle 0,1\rangle.$$

Therefore, the restriction of each of these implications coincides over the constants *false* and *true* with the implication from the ordinary propositional calculus.

Now, we introduce an implication, inspired by G. Takeuti and S. Titani's paper [28] and its answer by T. Trifonov and the author [16].

In [28] G. Takeuti and S.Titani introduced the following implication for $p, q \in [0,1]$

$$p \rightarrow q = \bigvee\{r \in [0,1] \mid p \wedge r \leq q\} = \begin{cases} 1, & \text{if } p \leq q \\ q, & \text{if } p > q \end{cases}$$

In [16], its intuitionistic fuzzy extension is given in the form

$$\langle a,b\rangle \to \langle c,d\rangle = \langle \max(c,\overline{\text{sg}}(a-c)), \min(d, \text{sg}(a-c))\rangle$$

$$= \begin{cases} \langle 1,0\rangle, & \text{if } a \le c \text{ and } b \ge d \\ \langle 1,0\rangle, & \text{if } a \le c \text{ and } b < d \\ \langle c,d\rangle, & \text{if } a > c \text{ and } b \ge d \\ \langle c,d\rangle, & \text{if } a > c \text{ and } b < d \end{cases}.$$

There, it is proved that the latter implication coincides with Gödel's implication ($\to_3$), because if

$$X \equiv 1-(1-c).\text{sg}(a-c) - \max(c, \overline{\text{sg}}(a-c)),$$

$$Y \equiv d.\text{sg}(a-c) - \min(d, \text{sg}(a-c)).$$

then we obtain the following. If $a > c$

$$X = 1-(1-c) - \max(c,0) = c - c = 0,$$

$$Y = d.1 - \min(d,1) = d - d = 0.$$

If $a \le c$

$$X = 1-(1-c).0 - \max(c,1) = 1-1 = 0,$$

$$Y = d.0 - \min(d,0) = 0 - 0 = 0,$$

i.e.

$$\langle 1-(1-c).\text{sg}(a-c), d.\text{sg}(a-c)\rangle = \langle \max(c,\overline{\text{sg}}(a-c)), \min(d, \text{sg}(a-c))\rangle.$$

All constructions in [28] are re-written in [16] for intuitionistic fuzzy case with the aim to show that Takeuti and Titani's sets are a particular case of intuitionistic fuzzy sets in the sense of [3]. Practically, the set constructed by them is an ordinary fuzzy set with elements, satisfying intuitionistic logic axioms. In [16], we show that there are *intuitionistic fuzzy sets* with elements, satisfying intuitionistic logic axioms.

Extending the research from [8], we shall study some properties of the 23 implications.

Let us introduce the expression

$$I_i = x \to_i y,$$

where $1 \le i \le 23$. We say that $I_i$ is more powerful than $I_j$ for $1 \le i,j \le 23$, if

$$V(I_i) \ge V(I_j) \text{ for all } x \text{ and } y.$$

We can construct Table 2, in which the lack of relation between two implications is denoted by " $*$ ".

**Table 2** Relations between elements of set $\{I_i \mid 1 \leq i \leq 23\}$

| | $I_1$ | $I_2$ | $I_3$ | $I_4$ | $I_5$ | $I_6$ | $I_7$ | $I_8$ | $I_9$ | $I_{10}$ | $I_{11}$ | $I_{12}$ | $I_{13}$ | $I_{14}$ | $I_{15}$ | $I_{16}$ | $I_{17}$ | $I_{18}$ | $I_{19}$ | $I_{20}$ | $I_{21}$ | $I_{22}$ | $I_{23}$ |
|---|---|---|---|---|---|---|---|---|---|---|---|---|---|---|---|---|---|---|---|---|---|---|---|
| $I_1$ | * | * | * | ≤ | ≤ | * | * | * | * | * | * | * | ≤ | * | * | * | ≤ | * | * | * | * | * | * |
| $I_2$ | * | * | ≤ | * | * | * | * | * | * | * | ≤ | * | * | * | * | * | * | * | * | * | * | * | * |
| $I_3$ | * | * | * | * | * | * | * | * | * | * | ≤ | * | * | * | * | * | * | * | * | * | * | * | * |
| $I_4$ | * | * | * | * | ≤ | * | * | * | * | * | * | * | ≤ | * | * | * | ≤ | * | * | * | * | * | * |
| $I_5$ | * | * | * | * | * | * | * | * | * | * | * | * | * | * | * | * | * | * | * | * | * | * | * |
| $I_6$ | * | * | * | * | ≤ | * | * | * | * | * | * | * | ≤ | * | * | * | * | * | * | * | * | * | * |
| $I_7$ | ≤ | * | * | ≤ | ≤ | * | * | * | * | * | * | * | ≤ | * | * | * | ≤ | * | * | * | * | * | * |
| $I_8$ | * | * | * | * | * | * | * | * | * | * | ≤ | * | * | * | * | * | * | * | * | * | * | * | * |
| $I_9$ | * | * | * | * | * | * | * | * | * | * | * | * | * | * | * | * | * | * | * | * | * | * | * |
| $I_{10}$ | * | * | * | ≤ | ≤ | * | * | * | * | * | * | * | ≤ | * | * | * | ≤ | * | * | * | * | * | * |
| $I_{11}$ | * | * | * | * | * | * | * | * | * | * | * | * | * | * | * | * | * | * | * | * | * | * | * |
| $I_{12}$ | * | * | * | ≤ | ≤ | * | * | * | * | * | * | * | ≤ | * | * | * | ≤ | ≤ | * | * | * | ≤ | * |
| $I_{13}$ | * | * | * | * | ≤ | * | * | * | * | * | * | * | * | * | * | * | * | * | * | * | * | * | * |
| $I_{14}$ | * | * | * | * | * | * | * | * | * | * | ≤ | * | * | * | * | * | * | * | * | * | * | * | ≤ |
| $I_{15}$ | * | * | ≤ | * | * | * | * | ≤ | * | * | ≤ | * | * | ≤ | * | * | * | * | * | * | * | * | ≤ |
| $I_{16}$ | * | * | ≤ | * | * | * | * | * | * | * | ≤ | * | * | * | * | * | * | * | * | * | * | * | * |
| $I_{17}$ | * | * | * | * | * | * | * | * | * | * | * | * | * | * | * | * | * | * | * | * | * | * | * |
| $I_{18}$ | * | * | * | ≤ | ≤ | * | * | * | * | * | * | * | ≤ | * | * | * | ≤ | * | * | * | * | ≤ | * |
| $I_{19}$ | * | * | ≤ | ≤ | ≤ | * | * | * | * | * | ≤ | * | ≤ | ≤ | * | ≤ | ≤ | ≤ | * | * | * | ≤ | ≤ |
| $I_{20}$ | * | * | * | * | * | * | * | * | * | * | * | * | * | * | * | * | * | * | * | * | * | * | * |
| $I_{21}$ | * | * | * | * | * | * | * | * | * | * | * | * | * | * | * | * | * | * | * | * | * | * | * |
| $I_{22}$ | * | * | * | * | * | * | * | * | * | * | * | * | * | * | * | * | * | * | * | * | * | * | * |
| $I_{23}$ | * | * | * | * | * | * | * | * | * | * | * | * | * | * | * | * | * | * | * | * | * | * | * |

There are many other implications, defined by Lilija Atanassova, Dimiter Dimitrov and the author, that will be discussed in a next author's research.

## *2.2 Definitions of 5 Intuitionistic Fuzzy Negations*

Now, following [5] and using, as a basis the equality (1) that has the intuitionistic fuzzy form

$$\neg\langle a, b\rangle = \langle a, b\rangle \to \langle 0, 1\rangle,$$

we shall construct negations, corresponding to the above implications. These negations are introduced in Table 3, but we must note that some of them have better (more compact) forms than those published already.

The following three properties are checked in [6, 7] for the separate negations:

**Property P1**: $A \to \neg\neg A$,
**Property P2**: $\neg\neg A \to A$,
**Property P3**: $\neg\neg\neg A = \neg A$.

Each of the negations from Table 3 satisfies Property 1.

**Table 3** List of the different intuitionistic fuzzy negations

| Notation | Form of negation |
|---|---|
| $\neg_1$ | $\langle b, a\rangle$ |
| $\neg_2$ | $\langle 1-\mathrm{sg}(a), \mathrm{sg}(a)\rangle$ |
| $\neg_3$ | $\langle b, a.b + a^2\rangle$ |
| $\neg_4$ | $\langle b, 1-b\rangle$ |
| $\neg_5$ | $\langle \overline{\mathrm{sg}}(1-b), \mathrm{sg}(1-b)\rangle$ |

Only negation $\neg_1$ satisfies Property 2, while the other negations do not satisfy it.

Each of the negations from Table 3 satisfies Property 3.

As an illustration of how the checks are performed, we shall prove the validity of the latest assertion for the case of negation $\neg_5$.

$$\neg\neg\neg\langle a, b\rangle = \neg\neg\langle \overline{\mathrm{sg}}(1-b), \mathrm{sg}(1-b)\rangle$$
$$= \neg\langle \overline{\mathrm{sg}}(1-\mathrm{sg}(1-b)), \mathrm{sg}(1-\mathrm{sg}(1-b))$$
$$= \langle \overline{\mathrm{sg}}(1-\mathrm{sg}(1-\mathrm{sg}(1-b))), \mathrm{sg}(1-\mathrm{sg}(1-\mathrm{sg}(1-b)))\rangle.$$

Let

$$X \equiv \overline{\mathrm{sg}}(1-\mathrm{sg}(1-\mathrm{sg}(1-b))) - \overline{\mathrm{sg}}(1-b).$$

If $b = 1$, then

$$X = \overline{\mathrm{sg}}(1-\mathrm{sg}(1)) - \overline{\mathrm{sg}}(0) = \overline{\mathrm{sg}}(0) - 1 = 0.$$

If $b < 1$, then

$$X = \overline{\mathrm{sg}}(1-\mathrm{sg}(1-1)) - 0 = \overline{\mathrm{sg}}(1) = 0.$$

Therefore Property 3 is valid for $\neg_5$.

Following [6], we show the relations between the different negations. By direct checks we can see the validity of these relations in Table 4.

The lack of relation between two implications is denoted in Table 4 by " $*$ ".

**Table 4** List of the relations between the different intuitionistic fuzzy negations

| | $\neg_1$ | $\neg_2$ | $\neg_3$ | $\neg_4$ | $\neg_5$ |
|---|---|---|---|---|---|
| $\neg_1$ | $=$ | $*$ | $\leq$ | $\geq$ | $\geq$ |
| $\neg_2$ | $*$ | $=$ | $*$ | $*$ | $\geq$ |
| $\neg_3$ | $\geq$ | $*$ | $=$ | $\geq$ | $\geq$ |
| $\neg_4$ | $\leq$ | $*$ | $\leq$ | $=$ | $\geq$ |
| $\neg_5$ | $\leq$ | $\leq$ | $*$ | $\leq$ | $=$ |

Therefore, we can construct the ordered graph (see Fig. 1) with vertices, corresponding to negations $\neg_1, ..., \neg_5$.

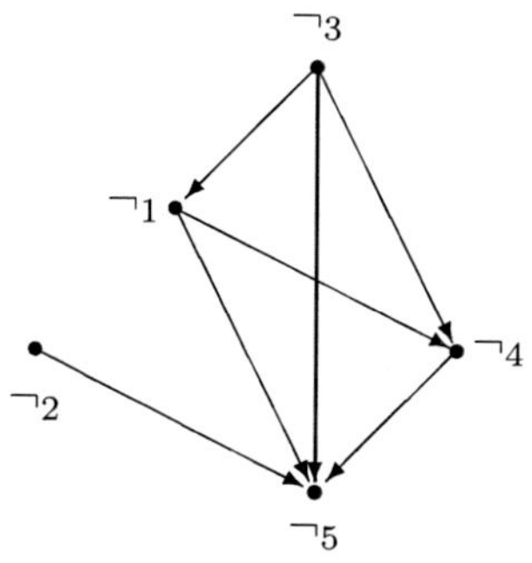


**Fig. 1**

George Klir and Bo Yuan discuss the following 9 axioms related to fuzzy implications in their book [26].

**Axiom 1:** $(\forall x, y)(x \leq y \rightarrow (\forall z)(I(x, z) \geq I(y, z))$.
**Axiom 2:** $(\forall x, y)(x \leq y \rightarrow (\forall z)(I(z, x) \leq I(z, y))$.
**Axiom 3:** $(\forall y)(I(0, y) = 1)$.
**Axiom 4:** $(\forall y)(I(1, y) = y)$.
**Axiom 5:** $(\forall x)(I(x, x) = 1)$.
**Axiom 6:** $(\forall x, y, z)(I(x, I(y, z)) = I(y, I(x, z)))$.
**Axiom 7:** $(\forall x, y)(I(x, y) = 1$ iff $x \leq y)$.
**Axiom 8:** $(\forall x, y)(I(x, y) = I(N(y), N(x)))$, where $N$ is an operation for a negation.
**Axiom 9:** $I$ is a continuous function.

Following and extending [29] we shall note that Table 5 summarizes the list of axioms of Klir and Yuan which are satisfied by the 23 implications.

If some axiom is valid as an IFT, its number in Table 5 is marked by an asterisk (*). We should note that Axiom 8 is checked using the classical intuitionistic fuzzy negation ($\neg_1$); if it is valid using the respective generated by implication $I$ negation $N(x)$, then the axiom is marked as $8^N$. We should also note that the validity of Axiom 7 does not imply the validity of Axiom 7*. The forms of the axioms with asterisks are

**Axiom 3*:** $(\forall y)(I(0, y)$ is an IFT).
**Axiom 5*:** $(\forall x)(I(x, x)$ is an IFT).
**Axiom 7*:** $(\forall x, y)(I(x, y)$ is an IFT iff $x \leq y)$.

The validity of each of these assertions can be checked directly. In [14] all 36 IF negations are given.

**Table 5** List of axioms of Klir and Yuan that are satisfied by the 23 intuitionistic fuzzy implications

| Notation | Axioms |
|---|---|
| $\rightarrow_1$ | 2,3,4,5*,9 |
| $\rightarrow_2$ | 1,2,3,5 |
| $\rightarrow_3$ | 1,2,3,4,5,6 |
| $\rightarrow_4$ | 1,2,3,4,5*,6,8,9 |
| $\rightarrow_5$ | 1,2,3,4,5*,6,8,9 |
| $\rightarrow_6$ | 2,3,4,5*,9 |
| $\rightarrow_7$ | 3*,4,5*,8,9 |
| $\rightarrow_8$ | 1,2,3,5 |
| $\rightarrow_9$ | 2,3,4,5* |
| $\rightarrow_{10}$ | 2,3,4 |
| $\rightarrow_{11}$ | 1,2,3,4,5,6 |
| $\rightarrow_{12}$ | 1,2,3,6,8,9 |
| $\rightarrow_{13}$ | 1,2,3,4,5*,6,8,9 |
| $\rightarrow_{14}$ | 1,2,3,4,5,6,7 |
| $\rightarrow_{15}$ | 1,2,3,5,7,7*,8 |
| $\rightarrow_{16}$ | 1,2,3,4,6 |
| $\rightarrow_{17}$ | 2,3,4,5*,6,9 |
| $\rightarrow_{18}$ | 1,2,3,4,5*,6,9 |
| $\rightarrow_{19}$ | 1,2,3,4,6 |
| $\rightarrow_{20}$ | 1,2,3,5,6,$8^N$ |
| $\rightarrow_{21}$ | 1,2,3,5*,9 |
| $\rightarrow_{22}$ | 1,2,3,5*,6,$8^N$,9 |
| $\rightarrow_{23}$ | 1,2,3,5,6,$8^N$ |

## 2.3 *On the Law of Excluded Middle and Its Modifications*

Following [6], we shall study the validity of the Law of Excluded Middle (LEM) in the forms:

$$\langle a, b\rangle \vee \neg_i \langle a, b\rangle = \langle 1, 0\rangle$$

(tautology-form) and

$$\langle a, b\rangle \vee \neg_i \langle a, b\rangle = \langle p, q\rangle$$

(IFT-form), where $1 \geq p \geq q \geq 0$ and $i = 1, 2, ..., 5$.

We will also study a Modified LEM in the forms:

$$\neg_i \neg_i \langle a, b\rangle \vee \neg_i \langle a, b\rangle = \langle 1, 0\rangle$$

(tautology-form) and

$$\neg_i \neg_i \langle a, b\rangle \vee \neg_i \langle a, b\rangle = \langle p, q\rangle,$$

(IFT-form), where $1 \geq p \geq q \geq 0$ and $i = 1, 2, ..., 5$.

None of our negations satisfies the LEM in the tautological form.

Negations $\neg_1, \neg_3$ and $\neg_4$ satisfy the LEM in the IFT-form.

Only $\neg_2$ and $\neg_5$ satisfy the Modified LEM in the tautological form.

All negations satisfy the Modified LEM in the IFT-form.

As illustration, we shall calculate

$$\neg_5\neg_5\langle a, b\rangle \vee \neg_5\langle a, b\rangle$$

$$= \langle \overline{\text{sg}}(1 - \text{sg}(1 - b)), \text{sg}(1 - \text{sg}(1 - b)) \vee \langle \overline{\text{sg}}(1 - b), \text{sg}(1 - b)\rangle$$

$$= \langle \max(\overline{\text{sg}}(1 - \text{sg}(1 - b)), \overline{\text{sg}}(1 - b)), \min(\text{sg}(1 - \text{sg}(1 - b)), \text{sg}(1 - b))\rangle.$$

If $b = 1$, then

$$\max(\overline{\text{sg}}(1 - \text{sg}(1 - b)), \overline{\text{sg}}(1 - b)) = \max(\overline{\text{sg}}(1), 1) = \max(0, 1) = 1.$$

and

$$\min(\text{sg}(1 - \text{sg}(1 - b)), \text{sg}(1 - b)) = \min(\text{sg}(1), 0) = \min(1, 0) = 0.$$

If $b < 1$, then

$$\max(\overline{\text{sg}}(1 - \text{sg}(1 - b)), \overline{\text{sg}}(1 - b)) = \max(\overline{\text{sg}}(0), 0) = \max(1, 0) = 1.$$

and

$$\min(\text{sg}(1 - \text{sg}(1 - b)), \text{sg}(1 - b)) = \min(\text{sg}(0), 1) = \min(0, 1) = 0.$$

Therefore, negation $\neg_5$ satisfies the Modified LEM in the IFT-form. On the other hand, in all cases the evaluation of the expression is equal to $\langle 1, 0\rangle$, i.e., this negation satisfies the Modified LEM in the tautological form.

## *2.4 On De Morgan's Laws and Their Modifications*

Below, following [11], we shall discuss some forms of De Morgan's Laws.

Usually, De Morgan's Laws have the forms:

$$\neg x \wedge \neg y = \neg(x \vee y),$$

$$\neg x \vee \neg y = \neg(x \wedge y).$$

In [11] it is proved that for every two propositional forms $x$ and $y$:

$$\neg_i x \wedge \neg_i y = \neg_i(x \vee y),$$

$$\neg_i x \vee \neg_i y = \neg_i(x \wedge y)$$

for $i = 1, 2, 4, 5$, while negation $\neg_3$ does not satisfy these equalities.

We shall illustrate only the fact that the De Morgan's Laws are not valid for $i = 3$. For example, if $a = b = 0.5, c = 0.1, d = 0$, then

$$V(\neg_3 x \wedge \neg_3 y) = \langle 0, 0.5 \rangle,$$

$$V(\neg_3(x \vee y)) = \langle 0, 0.25 \rangle.$$

The above mentioned change of the LEM inspired the idea to study the validity of De Morgan's Laws, which the classical negation $\neg$ (here it is negation $\neg_1$) satisfies. Really, it can be easily proved that the expressions

$$\neg_1(\neg_1 x \vee \neg_1 y) = x \wedge y$$

and

$$\neg_1(\neg_1 x \wedge \neg_1 y) = x \vee y$$

are IFTs, but the other negations do not satisfy these equalities. For them the following assertion is proved in [11].

For every two propositional forms $x$ and $y$ it holds that

$$\neg_i(\neg_i x \vee \neg_i y) = \neg_i \neg_i x \wedge \neg_i \neg_i y$$

and

$$\neg_i(\neg_i x \wedge \neg_i y) = \neg_i \neg_i x \vee \neg_i \neg_i y$$

for $i = 2, 4, 5$, while negation $\neg_3$ does not satisfy these equalities.

Now, we shall discuss another form of the LEM, also known as Law of Contradiction, that in propositional calculus is equivalent with the standard LEM, if De Morgan's Laws were valid. It is the following:

$$\neg(x \wedge \neg x).$$

Negations $\neg_2$ and $\neg_5$ satisfy it as tautologies, negations $\neg_1$, $\neg_3$ and $\neg_4$ satisfy it as IFTs.

## 2.5 On One Baczynski and Jayaram Problem and Its Extension

In [18] Michal Baczynski and Balasubramaniam Jayaram formulated some problems related to fuzzy implications $I$ and negations $N$. Here we give a solution to one of them:

**Problem 1.7.1.** Give examples of fuzzy implications $I$ such that
(i) $I$ satisfies only property

$$(CP) \qquad I(x, y) = I(N(y), N(x))$$

(ii) $I$ satisfies only property

$$(L-CP) \qquad I(N(x),y) = I(N(y),x)$$

(iii) $I$ satisfies both (CP) and (L-CP), but not

$$(R-CP) \qquad I(x,N(y)) = I(y,N(x))$$

with some fuzzy negation $N$, where $x, y \in [0,1]$.

We must note that in [18] no example is given.

Here we shall give examples of pairs of implications and negations that satisfy Problem 1.7.1 (ii) and other problems. Let the pair $(m,n)$ denote the expression with $m$-th implication and $n$-th negation.

Following [12], first, we shall formulate the following

**Theorem 2.** The pairs $(4,1)$, $(5,1)$, $(7,1)$, $(12,1)$, $(13,1)$, $(15,1)$, $(20,2)$, $(22,4)$, $(23,5)$, satisfy the three axioms.

We must note that no one pair for $1 \leq m \leq 23$ and $1 \leq n \leq 5$ satisfy exactly two axioms and more precisely, axioms $(L-CP)$ and $(R-CP)$. In [12], solutions are given for $1 \leq m \leq 138$ and $1 \leq n \leq 34$.

We had not found any pair of implication and negation that are solution of Problem 1.7.1 (iii).

Also, we had not found any pair of implication and negation that satisfy only the first axiom, i.e., we cannot give examples for the case of Problem 1.7.1 (i).

Another result of our search [12] is

**Theorem 3.** The pairs $(2,2)$, $(3,2)$, $(8,2)$, $(11,2)$, $(16,2)$, $(12,3)$, $(17,3)$, $(12,4)$, $(18,4)$, $(14,5)$, $(15,5)$, $(19,5)$ satisfy only the axiom $(R-CP)$.

The most interesting is the following

**Theorem 4.** Only the pair $(21,3)$ satisfies only Axiom $(L-CP)$.

We must note that this assertion has 135 solutions, described in [12] in the case when $23 < m \leq 138$ or $5 < n \leq 34$.

Now, we shall extend the Baczynski and Jayaram problem and shall give solutions to its new form.

Using the idea for Modified LEM, we can replace the equalities $(CP)$, $(L-CP)$ and $(R-CP)$ with the equalities:

$$I(N(N(x)),N(N(y))) = I(N(y),N(x)), \qquad (CP')$$

$$I(N(x),N(N(y))) = I(N(y),N(N(x))), \qquad (L-CP')$$

$$I(N(N(x)),N(y)) = I(N(N(y)),N(x)). \qquad (R-CP')$$

Here, we give examples of pairs of implications and negations that satisfy an extended form of Problem 1.7.1 (ii), where the assertions are related to $(CP')$, $(L-CP')$ and $(R-CP')$.

In [13], it is shown that there are 1322 pairs $(m,n)$ satisfying the three equalities, when $1 \le m \le 138$ and $1 \le n \le 34$. But, no one pair $(m,n)$ for $1 \le m \le 23$ and $1 \le n \le 5$ satisfies exactly two equalities, namely – $(L-CP')$ and $(R-CP')$.

We had not found any pair of implication and negation for $1 < m \le 138$ and $1 < n \le 34$, that solve the extended Problem 1.7.1 (iii).

Also, we had not found any pair of implication and negation that satisfy the first equality only, i.e., we cannot give examples for the case of Problem 1.7.1 (i).

Another result of our search is

**Theorem 5**. Only the pair $(21,3)$ satisfies only equality $(R-CP')$.

The most interesting is the following

**Theorem 6**. The pairs $(12,3)$, $(17,3)$ satisfy only equality $(L-CP')$.

We must note that this assertion has 375 solutions, described in [13] in the case when $23 < m \le 138$ or $5 < n \le 34$.

## *2.6 On the Axioms of Propositional Intuitionistic Logic*

The next and more important question is which of the introduced implications satisfy all the axioms of Propositional Intuitionistic Logic (IL) (see for example [27]).

The validity of the IL axioms was already checked for some implications in [15, 25, 29]. Here, we give a full list of valid axioms for each of the 23 implications. We will again verify the validity axioms in two variants - tautological validity (Table 6) and IFT validity (Table 7).

We use the following list of axioms for propositional intuitionistic logic:

(a) $A \to A$,
(b) $A \to (B \to A)$,
(c) $A \to (B \to (A \wedge B))$,
(d) $(A \to (B \to C)) \to (B \to (A \to C))$,
(e) $(A \to (B \to C)) \to ((A \to B) \to (A \to C))$,
(f) $A \to \neg\neg A$,
(g) $\neg(A \wedge \neg A)$,
(h) $(\neg A \vee B) \to (A \to B)$,
(i) $\neg(A \vee B) \to (\neg A \wedge \neg B)$,
(j) $(\neg A \wedge \neg B) \to \neg(A \vee B)$,
(k) $(\neg A \vee \neg B) \to \neg(A \wedge B)$,
(l) $(A \to B) \to (\neg B \to \neg A)$,
(m) $(A \to \neg B) \to (B \to \neg A)$,

(n) $\neg\neg\neg A \rightarrow \neg A$,
(o) $\neg A \rightarrow \neg\neg\neg A$,
(p) $\neg\neg(A \rightarrow B) \rightarrow (A \rightarrow \neg\neg B)$,
(q) $(C \rightarrow A) \rightarrow ((C \rightarrow (A \rightarrow B)) \rightarrow (C \rightarrow B))$.

**Table 6** List of axioms of the intuitionistic logic that are satisfied by intuitionistic fuzzy implications as tautologies

| | a | b | c | d | e | f | g | h | i | j | k | l | m | n | o | p | q |
|---|---|---|---|---|---|---|---|---|---|---|---|---|---|---|---|---|---|
| $\rightarrow_1$ | - | - | - | - | - | - | - | - | - | - | - | - | - | - | - | - | - |
| $\rightarrow_2$ | + | - | - | - | + | + | + | - | + | + | + | + | + | + | + | + | + |
| $\rightarrow_3$ | + | + | + | + | + | + | + | + | + | + | + | + | + | + | + | + | + |
| $\rightarrow_4$ | - | - | - | - | - | - | - | - | - | - | - | - | - | - | - | - | - |
| $\rightarrow_5$ | - | - | - | - | - | - | - | - | - | - | - | - | - | - | - | - | - |
| $\rightarrow_6$ | - | - | - | - | - | - | - | - | - | - | - | - | - | - | - | - | - |
| $\rightarrow_7$ | - | - | - | - | - | - | - | - | - | - | - | - | - | - | - | - | - |
| $\rightarrow_8$ | + | - | - | - | - | + | + | - | + | + | + | + | + | + | + | + | - |
| $\rightarrow_9$ | - | - | - | - | - | - | - | - | - | - | - | - | - | - | - | - | - |
| $\rightarrow_{10}$ | - | - | - | - | - | - | - | - | - | - | - | - | - | - | - | - | - |
| $\rightarrow_{11}$ | + | + | + | + | + | + | + | + | + | + | + | + | + | + | + | + | + |
| $\rightarrow_{12}$ | - | - | - | - | - | - | - | - | - | - | - | - | - | - | - | - | - |
| $\rightarrow_{13}$ | - | - | - | - | - | - | - | - | - | - | - | - | - | - | - | - | - |
| $\rightarrow_{14}$ | + | + | + | + | + | + | + | + | + | + | + | + | + | + | + | + | + |
| $\rightarrow_{15}$ | + | - | - | - | - | + | + | - | + | + | + | + | + | + | + | + | - |
| $\rightarrow_{16}$ | - | - | - | - | - | + | + | - | + | + | + | + | + | + | + | + | - |
| $\rightarrow_{17}$ | - | - | - | - | - | - | - | - | - | - | - | - | - | - | - | - | - |
| $\rightarrow_{18}$ | - | - | - | - | - | - | - | - | - | - | - | - | - | - | - | - | - |
| $\rightarrow_{19}$ | - | - | - | - | - | + | + | - | + | + | + | + | + | + | + | + | - |
| $\rightarrow_{20}$ | + | + | + | + | + | + | + | + | + | + | + | + | + | + | + | + | + |
| $\rightarrow_{21}$ | ? | ? | ? | ? | ? | ? | ? | ? | ? | ? | ? | ? | ? | ? | ? | ? | ? |
| $\rightarrow_{22}$ | - | - | - | - | - | - | - | - | - | - | - | - | - | - | - | - | - |
| $\rightarrow_{23}$ | + | + | + | + | + | + | + | + | + | + | + | + | + | + | + | + | + |

The most important of the results collected in Table 6 can be formulated as the following

**Theorem 7.** Implications $\rightarrow_3$, $\rightarrow_{11}$, $\rightarrow_{14}$, $\rightarrow_{20}$, $\rightarrow_{23}$ satisfy all intuitionistic logic axioms as tautologies.

The validity of axioms for cells marked by a question mark (?) in Tables 6 and 7 is not yet clear and it is an open problem.

The most important of the results collected in Table 7 can be formulated as follows:

**Theorem 8.** Implications $\rightarrow_1$, $\rightarrow_3$, $\rightarrow_4$, $\rightarrow_5$, $\rightarrow_{11}$, $\rightarrow_{14}$, $\rightarrow_{18}$, $\rightarrow_{20}$, $\rightarrow_{22}$, $\rightarrow_{23}$ satisfy all intuitionistic logic axioms as IFTs.

**Table 7** List of axioms of the intuitionistic logic that are satisfied by intuitionistic fuzzy implications as IFTs

| | a | b | c | d | e | f | g | h | i | j | k | l | m | n | o | p | q |
|---|---|---|---|---|---|---|---|---|---|---|---|---|---|---|---|---|---|
| $\rightarrow_1$ | + | + | + | + | + | + | + | + | + | + | + | + | + | + | + | + | + |
| $\rightarrow_2$ | + | - | - | - | + | + | + | - | + | + | + | + | + | + | + | + | + |
| $\rightarrow_3$ | + | + | + | + | + | + | + | + | + | + | + | + | + | + | + | + | + |
| $\rightarrow_4$ | + | + | + | + | + | + | + | + | + | + | + | + | + | + | + | + | + |
| $\rightarrow_5$ | + | + | + | + | + | + | + | + | + | + | + | + | + | + | + | + | + |
| $\rightarrow_6$ | + | + | + | ? | - | + | + | + | + | + | + | + | + | + | + | + | ? |
| $\rightarrow_7$ | + | - | - | - | - | + | + | - | + | + | + | ? | ? | + | + | ? | - |
| $\rightarrow_8$ | + | - | - | ? | ? | + | + | - | + | + | + | + | + | + | + | + | + |
| $\rightarrow_9$ | + | + | + | ? | ? | + | + | ? | ? | ? | ? | ? | ? | + | + | ? | ? |
| $\rightarrow_{10}$ | - | - | - | - | - | - | + | - | - | - | - | - | - | - | - | - | - |
| $\rightarrow_{11}$ | + | + | + | + | + | + | + | + | + | + | + | + | + | + | + | + | + |
| $\rightarrow_{12}$ | - | - | - | + | + | + | + | - | + | + | + | + | + | + | + | + | + |
| $\rightarrow_{13}$ | + | + | + | ? | ? | + | + | ? | + | + | + | + | + | + | + | + | ? |
| $\rightarrow_{14}$ | + | + | + | + | + | + | + | + | + | + | + | + | + | + | + | + | + |
| $\rightarrow_{15}$ | + | - | - | - | - | + | + | - | + | + | + | + | + | + | + | + | - |
| $\rightarrow_{16}$ | - | - | - | - | - | + | + | - | + | + | + | + | + | + | + | + | - |
| $\rightarrow_{17}$ | ? | ? | ? | ? | ? | ? | ? | ? | ? | ? | ? | ? | ? | ? | ? | ? | ? |
| $\rightarrow_{18}$ | + | + | + | + | + | + | + | + | + | + | + | + | + | + | + | + | + |
| $\rightarrow_{19}$ | - | - | - | - | - | + | + | - | + | + | + | + | + | + | + | + | - |
| $\rightarrow_{20}$ | + | + | + | + | + | + | + | + | + | + | + | + | + | + | + | + | + |
| $\rightarrow_{21}$ | ? | ? | ? | ? | ? | ? | ? | ? | ? | ? | ? | ? | ? | ? | ? | ? | ? |
| $\rightarrow_{22}$ | + | + | + | + | + | + | + | + | + | + | + | + | + | + | + | + | + |
| $\rightarrow_{23}$ | + | + | + | + | + | + | + | + | + | + | + | + | + | + | + | + | + |

## 3 A New Argument that the Intuitionistic Fuzzy Sets Are 'Intuitionistic' in Nature

The above assertions show that all negations, except the first one, satisfy conditions of the intuitionistic logic, but not of the classical logic. A part of these negations were generated by implications, that were generated by fuzzy implications. Now, let us return from the intuitionistic fuzzy negations to ordinary fuzzy negations. The result is shown on Table 8, where $b = 1 - a$.

Another approach to obtaining fuzzy negations from IF negations is to use the ideas from [19], where a procedure for de-intuitionistic-fuzzification is discussed. It will give the negations from Table 9.

Now, if we put $b = 1 - a$, we will again obtain the values from Table 8.

Therefore, from the intuitionistic fuzzy negations we can generate fuzzy negations, so that two of them ($\neg_3$ and $\neg_4$) coincide with the standard fuzzy negation ($\neg_1$). Therefore, there are intuitionistic fuzzy negations that lose

**Table 8** List of the fuzzy negations, generated by intuitionistic fuzzy negations

| Notation | Form of the intuitionistic fuzzy negation | Form of the fuzzy negation |
|---|---|---|
| $\neg_1$ | $\langle b, a\rangle$ | $1-a$ |
| $\neg_2$ | $\langle 1-\mathrm{sg}(a), \mathrm{sg}(a)\rangle$ | $1-\mathrm{sg}(a)$ |
| $\neg_3$ | $\langle b, a.b+a^2\rangle$ | $1-a$ |
| $\neg_4$ | $\langle b, 1-b\rangle$ | $1-a$ |
| $\neg_5$ | $\langle \overline{\mathrm{sg}}(1-b), \mathrm{sg}(1-b)\rangle$ | $1-\mathrm{sg}(a)$ |

**Table 9** List of the fuzzy negations, generated by intuitionistic fuzzy negations using the procedure for de-intuitionistic-fuzzification

| Notation | Form of the intuitionistic fuzzy negation | Form of the fuzzy negation |
|---|---|---|
| $\neg_1$ | $\langle b, a\rangle$ | $\frac{1-a+b}{2}$ |
| $\neg_2$ | $\langle 1-\mathrm{sg}(a), \mathrm{sg}(a)\rangle$ | $1-\mathrm{sg}(a)$ |
| $\neg_3$ | $\langle b, a.b+a^2\rangle$ | $\frac{1+b-a.b-a^2}{2}$ |
| $\neg_4$ | $\langle b, 1-b\rangle$ | $b$ |
| $\neg_5$ | $\langle \overline{\mathrm{sg}}(1-b), \mathrm{sg}(1-b)\rangle$ | $\overline{\mathrm{sg}}(1-b)$ |

their properties when they are restricted to ordinary fuzzy case. In other words, the construction of the intuitionistic fuzzy estimation

$$\langle \text{degree of membership/validity, degree of non-membership/non-validity}\rangle$$

which is specific for the intuitionistic fuzzy sets, is the reason for the intuitionistic behaviour of these sets. Over them we can define both intuitionistic fuzzy, and classical negations.

In the fuzzy case, negations $\neg_2$ and $\neg_5$ coincide, generating a fuzzy negation that satisfies properties P1 and P3 and does not satisfy Property P2 from Section 2.2, i.e., it is 'intuitionistic' in nature. In these cases, the implication from properties P1 and P2 may be any that generates negation $\neg_2$.

## 4 Etienne Kerre's Group and Intuitionistic Fuzzy Implications

In the beginning of the century, Prof. Etienne Kerre established in Ghent University one of the most active research groups in the world in the area of IFS theory. The author regrets that he does not dispose of all their publications, but among these, that he has got, there are very important ones. Papers [21, 22, 23, 24] are some of them. Their authors are Etienne Kerre and two of his then PhD-students – Chris Cornelis and Glad Deschijver, who nowadays are active and prospective scientists.

In a relation to our research, the results of Kerre's group on intuitionistic fuzzy implications generate some interesting problems. Below, we will mention two of them:

1. By analogy with our research on axioms of Klir and Yuan that are satisfied by the 23 intuitionistic fuzzy implications, to check which of these implications satisfy the axioms discussed in [21, 24].
2. To check the properties of the implications, discussed in [21, 22, 23, 24] for the introduced here 23 implications and for the rest implications, that already are defined in IFS theory.

## 5 Conclusion

The new operations open a very large field for future research. We hope that soon many other properties will be clarified and new results will be obtained. There are some possible directions that will be an object of an author's future research:

1. Study the relationships between the different implications and order them as vertices of an oriented graph with respect to the ordering "$\leq$". Similarly to the relationships between the different negations, discussed above.
2. Study all pairs $(\rightarrow_i, \neg_j)$ and determine which of them have nice properties, e.g. satisfy all axioms of intuitionistic logic.
3. Construct new sets of the existing implications and study their properties.

The most interesting problem is related to the theory of T- and S-norms. The results, related to De Morgan Laws show that the present theory of T- and S-norms must be revised totally, because it is based on De Morgan Laws and therefore, on the classical negation. In future, it must be modified, based on the different types of non-classical negations.

**Acknowledgements.** This paper is partially supported by grant BIn-2/09 "Design and development of intuitionistic fuzzy logic tools in information technologies" of the Bulgarian National Science Fund.

The author wishes to express his gratitude to Glad Deschrijver and Peter Vassilev for the valuable advice and corrections.

## References

1. Atanassov, K.: Intuitionistic fuzzy sets, VII ITKR's Session, Sofia (June 1983) (Deposed in Central Sci. - Techn. Library of Bulg. Acad. of Sci., 1697/84) (in Bulg.)
2. Atanassov, K.: Two variants of intuitonistc fuzzy propositional calculus. Preprint IM-MFAIS-5-88, Sofia (1988)
3. Atanassov, K.: Intuitionistic Fuzzy Sets. Springer Physica-Verlag, Berlin (1999)

4. Atanassov, K.: Intuitionistic fuzzy implications and Modus Ponens. Notes on Intuitionistic Fuzzy Sets 11(1), 1–5 (2005)
5. Atanassov, K.: On some types of intuitionistic fuzzy negations. Notes on Intuitionistic Fuzzy Sets 11(4), 170–172 (2005)
6. Atanassov, K.: On some intuitionistic fuzzy negations. Notes on Intuitionistic Fuzzy Sets 11(6), 13–20 (2005)
7. Atanassov, K.: A new intuitionistic fuzzy implication from a modal type. Advanced Studies in Contemporary Mathematics 12(1), 117–122 (2006)
8. Atanassov, K.: On some intuitionistic fuzzy implications. Comptes Rendus de l'Academie bulgare des Sciences, Tome 59(1), 19–24 (2006)
9. Atanassov, K.: On eight new intuitionistic fuzzy implications. In: Proc. of 3rd Int. IEEE Conf. "Intelligent Systems" IS06, London, September 4-6, pp. 741–746 (2006)
10. Atanassov, K.: On intuitionistic fuzzy negations. In: Computational Intelligence, Theory and Applications, pp. 159–167. Springer, Berlin (2006)
11. Atanassov, K.: On intuitionistic fuzzy negations and De Morgan's Laws. In: Proc. of Eleventh International Conf. IPMU 2006, Paris, July 2-7, pp. 2399–2404 (2006)
12. Atanassov, K., Dimitrov, D.: On one of Baczyncki-Jayaram's problems. Cybernetics and Information Technologies 9(2), 14–20 (2009)
13. Atanassov, K., Dimitrov, D.: Extension of one of Baczyncki-Jayaram's problems. Comptes Rendus de l'Academie bulgare des Sciences, Tome 62(11), 1377–1386 (2009)
14. Atanassov, K., Dimitrov, D.: On the negations over intuitionistic fuzzy sets. Part 1 Annual of "Informatics" Section Union of Scientists in Bulgaria 1, 49–58 (2008)
15. Atanassov, K., Kolev, B.: On an intuitionistic fuzzy implication from a possibilistic type. Advanced Studies in Contemporary Mathematics 12(1), 111–116 (2006)
16. Atanassov, K., Trifonov, T.: Towards combining two kinds of intuitionistic fuzzy sets. Notes on Intuitionistic Fuzzy Sets 11(2), 1–11 (2005)
17. Atanassov, K., Trifonov, T.: On a new intuitionistic fuzzy implication from Gödel's type. In: Proceedings of the Jangjeon Mathematical Society, vol. 8(2), pp. 147–152 (2005)
18. Baczynski, M., Jayaram, B.: Fuzzy Implications. Springer, Berlin (2008)
19. Ban, A., Kacprzyk, J., Atanassov, K.: On de-I-fuzzification of intuitionistic fuzzy sets. Comptes Rendus de l'Academie bulgare des Sciences, Tome 61(12), 1535–1540 (2008)
20. Brouwer, L.E.J.: Collected Works, vol. 1. North Holland, Amsterdam (1975)
21. Cornelis, C., Deschrijver, G., Kerre, E.: Implication on intuitionistic fuzzy and interval-valued fuzzy set theory: construction, classification, Application. Int. Journal of Approximate Reasoning 35, 55–95 (2004)
22. Deschrijver, G., Cornelis, C., Kerre, E.: On the representation of intuitionistic fuzzy t-norms and t-conorms. IEEE Transactions on Fuzzy Systems 12(1), 45–61 (2004)
23. Deschrijver, G., Kerre, E.: Classes of intuitionistic fuzzy T-norms satisfying the residuation principle. Int. Journal of Uncertainty, Fuzziness and Knowledge-Based Systems 11(6), 691–709 (2003)
24. Deschrijver, G., Kerre, E.: Smets-Magrez axioms for R-implicators in interval-values and intuitionistic fuzzy set theory. Int. Journal of Uncertainty, Fuzzyness and Knowledge-Based Systems 13(4), 453–464 (2005)

25. Dimitrov, D., Atanassov, K.: On intuitionistic logic axioms and intuitionistic fuzzy implications and negations. Developments in Fuzzy Sets, Intuitionistic Fuzzy Sets, Generalized Nets and Related Topics 1, 61–68 (2008)
26. Klir, G., Yuan, B.: Fuzzy Sets and Fuzzy Logic. Prentice Hall, New Jersey (1995)
27. Rasiova, H., Sikorski, R.: The Mathematics of Metamathematics. Warszawa, Pol. Acad. of Sci. (1963)
28. Takeuti, G., Titani, S.: Intuitionistic fuzzy logic and intuitionistic fuzzy set theory. The Journal of Symbolic Logic 49(3), 851–866 (1984)
29. Trifonov, T., Atanassov, K.: On some intuitionistic properties of intuitionistic fuzzy implications and negations. In: Computational Intelligence, Theory and Applications, pp. 151–158. Springer, Berlin (2006)

# Mathematical Fuzzy Logic: A Good Theory for Practice*

Vilém Novák and Irina Perfilieva

*The paper is dedicated to Prof. Etienne Kerre*

**Abstract.** We discuss the present state of mathematical fuzzy logic in narrow sense, its extension — fuzzy logic in broader sense (FLb) as a logic of natural human reasoning and also some related theories, e.g., the fuzzy transform. We argue that these are good theories with potential to be very practical.

**Keywords:** Soft computing, Fuzzy logic in broader sense, Fuzzy decision-making, Evaluative linguistic expressions, Fuzzy transform.

## 1 Introduction

"There is nothing more practical than a good theory" — this sentence, which has probably been first proclaimed by the psychologist K. Lewin in [27], characterizes in a a very succinct form essentially important balance between basic and applied research. We are convinced that only such balance can lead to really new innovative solutions which can drive our society forward and bring also tangible economical benefit. This can be especially visible in physics where only good theory brought us to the atomic age (with all its positive and also negative impacts), but also many other mathematical theories such as the theory of differential equations, elasticity theory, control theory, statistics, etc.

One can ask whether also fuzzy logic is such a practical theory. We argue for the positive: the *mathematical fuzzy logic* (MFL), its extension — *fuzzy logic in broader*

Vilém Novák · Irina Perfilieva
University of Ostrava
Institute for Research and Applications of Fuzzy Modeling
30. dubna 22, 701 03 Ostrava 1, Czech Republic
e-mail: {Vilem.Novak, Irina.Perfilieva}@osu.cz

* The research was supported by the project MSM 6198898701 of the MŠMT ČR.

C. Cornelis et al. (Eds.): 35 Years of Fuzzy Set Theory, STUDFUZZ 261, pp. 39–55.
springerlink.com 

*sense* (FLb) and some related theories such as *fuzzy approximation of functions* or *fuzzy transform* are very practical.

Recall that MFL is a special mathematical theory the goal of which is to provide a working mathematical model of the vagueness phenomenon and to become a well established sound formal system that makes its applications well justified. The mathematization of vagueness is based on introduction of degrees of truth taken from some scale. Many people have significantly contributed to its development, for example P. Hájek, F. Esteva, L. Godo, S. Gottwald, F. Montagna, P. Cintula, D. Mundici, G. Gerla, V. Novák, and others. There are several books [3, 16, 18, 19, 46, 48] and many papers published on this topic, which are hard to be listed. Perhaps the most comprehensive list can be found on the WEB page www.mathfuzzlog.org/. In this paper, we will provide a brief overview of the present state of mathematical fuzzy logic and the mentioned related topics.

## 2 Mathematical Fuzzy Logic

### 2.1 Fuzzy Logic in Narrow Sense

Let us remember that the seminal paper [67] on fuzzy set theory written by L. A. Zadeh has been followed by the paper on fuzzy logic [17] written by J. A. Goguen. The latter can be taken as the seminal paper on MFL because the residuated lattice has been proposed there as a convenient structure of truth values. The paper contains a lot of solid material accompanied by the detailed philosophical discussion but it is not sufficiently formal. The first, mathematically highly sophisticated paper is that of J. Pavelka [50]. Since then, MFL has been significantly developed.

Fuzzy logic with traditional syntax

This branch of MFL has been elaborated in a deep and seminal way by P. Hájek in [19]. It is specific for it that formulas are dealt with classically, i.e. the language of such fuzzy logic differs from that of classical logic especially by having more connectives. On the other hand, the concepts of inference rule, theory, proof and many other ones remain classical.

The semantics is many valued and based on the concept of a commutative, integral, bounded residuated lattice $\mathcal{L}$ (shortly, residuated lattice) which is an algebra of type (2, 2, 2, 2, 0, 0)

$$\mathcal{L} = \langle L, \vee, \wedge, \otimes, \rightarrow, \mathbf{0}, \mathbf{1} \rangle, \tag{1}$$

where $\langle L, \vee, \wedge, \mathbf{0}, \mathbf{1} \rangle$ is a lattice with the bottom element $\mathbf{0}$ and the top element $\mathbf{1}$. The operation $\otimes$ is called *multiplication* and $\langle L, \otimes, \mathbf{1} \rangle$ is a commutative monoid. The operation $\rightarrow$ is the *residuation* operation tied with the multiplication by adjointness: for all $a, b, c \in L$ it holds that

$$a \otimes b \leq c \quad \text{iff} \quad a \leq b \rightarrow c. \tag{2}$$

A BL-algebra is a residuated lattice fulfilling, moreover, *prelinearity* ($(a \to b) \vee (b \to a) = \mathbf{1}$) and *divisibility* ($a \otimes (a \to b) = a \wedge b$).

The most prominent is the standard Łukasiewicz MV-algebra (taken as the residuated lattice)

$$\mathcal{L}_Ł = \langle [0,1], \vee, \wedge, \otimes, \to, 0, 1\rangle \tag{3}$$

where

$$a \otimes b = \max(0, a + b - 1), \qquad \text{(Łukasiewicz conjunction)}$$
$$a \to b = \min(1, 1 - a + b). \qquad \text{(Łukasiewicz implication)}$$

Quite recently, a new algebra of truth values called EQ-algebra has been introduced in [44]. It is not identical with residuated lattice. It is specific by taking fuzzy equality $\sim$ as the basic operation so that the implication operation is derived:

$$a \to b = (a \wedge b) \sim a.$$

The study of residuated lattices is closely connected with the study of t-norms (cf. [26] and also [22, 24]) and it is important for the study of fuzzy logics (with traditional syntax). One of the first logics in which conjunction is interpreted by a t-norm has been proposed in [6]. At present, there are many formal systems of MFL, both propositional as well as predicate. The prominent fuzzy logic calculi are MTL, IMTL, BL, product, or Łukasiewicz (cf. [40]). All of them are embraced in a wide class of the, so called, *core fuzzy logics* [4, 20].

All core fuzzy logics enjoy completeness in the form

$$\vdash A \quad \text{iff} \quad \models A \tag{4}$$

for every formula $A$ where $\models A$ means that $A$ is true in the degree $\mathbf{1}$ ($\mathcal{M}(A) = \mathbf{1}$) in every interpretation $\mathcal{M}$ of the language of the given core fuzzy logic.

MFL has been studied in detail especially from the algebraic point of view. One can hardly estimate, how far this work can still continue. It seems that MFL is in the position of a theory promising a deeply justified technique for modeling of various manifestations of the vagueness phenomenon and for various applications. There are not so many results in the latter, however. One of the exceptions is the program of formalization of fuzzy set theory — *fuzzy mathematics* — using fuzzy logic as a metatheory that was announced by P. Cintula and L. Běhounek in [1, 2]. Several papers have already been published and the program seems to be very promising because it encompasses and quite often even generalizes most (if not all) of the results in fuzzy set theory obtained so far. Another promising program is development of *fuzzy logic in broader sense* which we will discuss below.

### Fuzzy type theory

It turns out that a fully-fledged treatment of vagueness cannot be made using first-order fuzzy logic only, especially when the meaning of natural language is at play. Thus, the need for higher-order fuzzy logic raised up. This has been developed in

[36] and called the *fuzzy type theory* (FTT). It generalizes classical type theory initiated by B. Russel, A. Church and L. Henkin. Let us mention that another version of higher-order fuzzy logic has been developed in [1] which focuses especially on construction of special classes and which thus aims at becoming the general theory of "fuzzy mathematics" mentioned above.

The structure of truth values of FTT is generally supposed to form one of the following: a complete $\mathrm{IMTL}_\Delta$-algebra (see [13]), standard Łukasiewicz algebra, ŁΠ-algebra, or BL-algebra. The most important for applications in linguistics is the standard Łukasiewicz algebra (3) extended by the operation

$$\Delta(a) = \begin{cases} 1 & \text{if } a = 1, \\ 0 & \text{otherwise.} \end{cases}$$

which raises boolean structure from the corresponding algebra.

Important concept in FTT is that of a *fuzzy equality*. This is a fuzzy relation $\doteq: M \times M \longrightarrow L$ which has the following properties:

(i) reflexivity $[m \doteq m] = \mathbf{1}$,
(ii) symmetry $[m \doteq m'] = [m' \doteq m]$,
(iii) $\otimes$-transitivity $[m \doteq m'] \otimes [m' \doteq m''] \leq [m \doteq m'']$

for all $m, m', m'' \in M$ where $[m \doteq m']$ denotes a *truth value* of $m \doteq m'$.

A special case of fuzzy equality on the algebra of truth values is *biresiduation*. Example of a fuzzy equality on $M = \mathbb{R}$ with respect to the standard Łukasiewicz algebra is

$$[m \doteq n] = 1 - (1 \wedge |m - n|), \qquad m, n \in \mathbb{R}.$$

Syntax of FTT is a generalization of the lambda-calculus that is constructed in a classical way. The main difference from classical type theory is thus in definition of additional special connectives. Note that all essential syntactical elements of FTT are formulas (alternatively, they can be called lambda-terms as is usual in classical type theory).

As usual, each formula $A$ has a certain type. The basic types are $o$ (truth values) and $\varepsilon$ (elements). These can be then iterated to more complex types.

Formulas of type $o$ (truth value) can be joined by the following connectives: $\equiv$ (equivalence), $\vee$ (disjunction), $\wedge$ (conjunction), $\&$ (strong conjunction), $\nabla$ (strong disjunction), $\Rightarrow$ (implication). General ($\forall$) and existential ($\exists$) quantifiers are defined as special formulas. For the details about their definition and semantics — see [36].

If $A \in Form_{o\alpha}$, then $A$ represents a fuzzy set of elements. It can also be understood as a first-order property of elements of the type $\alpha$. Similarly, $A_{(o\alpha)\alpha}$ is a fuzzy relation (between elements of type $\alpha$).

There are over 15 *logical axioms* in FTT depending on the chosen structure of truth values. FTT has two inference rules and classical concept of provability. The rules of modus ponens and generalization are derived rules.

A theory $T$ is a set of formulas of type $o$. A specific formula of type $o$ is a formula $\dagger$ representing the most indefinite truth value for which $\vdash \neg\dagger \equiv \dagger$ holds. Note that such a truth value, in general, needs not be present in the given algebra of truth values. In the standard Łukasiewicz algebra is $\dagger$ interpreted by 0.5.

A formula $\mathbf{\Delta} A_o$ is crisp, i.e., its interpretation is either $\mathbf{0}$ or $\mathbf{1}$. There are formulas which are not crisp.

Quite useful are the following special crisp formulas:

$$\Upsilon_{oo} \equiv \lambda z_o \cdot \neg\mathbf{\Delta}(\neg z_o),$$
$$\hat{\Upsilon}_{oo} \equiv \lambda z_o \cdot \neg\mathbf{\Delta}(z_o \vee \neg z_o).$$

Thus, the formula $\Upsilon_{oo} z_o$ says that $z_o$ is non-zero truth value and $\hat{\Upsilon}_{oo} z_o$ says that $z_o$ is a general truth value between $\mathbf{0}$ and $\mathbf{1}$.

Note that $\mathbf{\Delta}$ corresponds to $D$-operator of supervaluation theory (e.g., $A \vdash C$ implies $\vdash \mathbf{\Delta} A \Rightarrow C$ as well as $\neg C \vdash \neg \mathbf{\Delta} A$) and $\hat{\Upsilon}$ corresponds to $I$-operator *indefinitely*.

The semantics of FTT is defined using generalization of the concept of a frame that is a system $\langle (M_\alpha, =_\alpha)_{\alpha \in Types}, \mathcal{L} \rangle$ where $=_\alpha$ is a special fuzzy equality in each set $M_\alpha$. Interpretation of each formula $A_{\beta\alpha}$ is a function assigning to every $m \in M_\alpha$ an element from $M_\beta$.

**Theorem 1 (completeness, [36])**

*(a) A theory $T$ of FTT is consistent iff $T$ has a general model.*
*(b) $T \vdash A_o$ iff $T \models A_o$ holds for every theory $T$ and a formula $A_o$.*

We claim that all essential properties of vague predicates are formally expressible in FTT and so, they have a many-valued model.

Fuzzy logic with evaluated syntax (Ev$_{Ł}$)

This logic is specific by considering formulas that are evaluated also on syntactic level. Namely, we deal with *evaluated formulas* of the form $a/A$ where $A$ is a formula and $a \in L$ is its syntactic evaluation. This has nice interpretation since it allows to consider axioms which need not be fully satisfactory and so, their initial truth value can be lower than $\mathbf{1}$.

This logic has been in propositional version established by J. Pavelka in [50] and extended to predicate version by V. Novák in [33] and especially in the book [48]. Ev$_{Ł}$ has been established as a generalization of classical mathematical logic with clearly distinguished syntax and semantics. The syntax consists of precise definitions of evaluated formula, evaluated proof, fuzzy theory, model, evaluated provability, etc. and the semantics is formed by the standard Łukasiewicz algebra only. The reason for choosing the latter follows from Theorem 1.7 in [50], Part III, which states that completeness in the sense of (7) below requires the following four equations to hold in the given algebra of truth values:

$$\bigvee_{i\in I}(a \to b_i) = a \to \bigvee_{i\in I} b_i, \qquad \bigwedge_{i\in I}(a \to b_i) = a \to \bigwedge_{i\in I}(b_i) \tag{5}$$

$$\bigvee_{i\in I}(a_i \to b) = \bigwedge_{i\in I} a_i \to b, \qquad \bigwedge_{i\in I}(a_i \to b) = \bigvee_{i\in I} a_i \to b \tag{6}$$

These equations are in $[0,1]$ equivalent to continuity of $\rightarrow$, which is fulfilled only by the Łukasiewicz implication and its isomorphs. The resulting logic is quite strong since it is complete with respect to the generalized syntax, in which all formulas are evaluated by elements from the underlying algebra. A fuzzy theory $T$ is determined by a *fuzzy set* of axioms. Hence, besides the concept of *truth of a formula A in a theory T*, $T \models_a A$, which is infimum of truth values of $A$ in all models of $T$, also the *provability degree of A in T* is introduced: $T \vdash_a A$, which is supremum of values of all proofs of $A$ in $T$. Then we obtain generalization of the Gödel-Henkin completeness theorem:

$$T \vdash_a A \quad \text{iff} \quad T \models_a A, \qquad a \in L, \tag{7}$$

for all all formulas $A \in F_{J(T)}$ and all *fuzzy theories* $T$. The system of Ev$_{Ł}$is open to extension by new connectives.

## 2.2 Fuzzy Logic in Broader Sense

The paradigm of fuzzy logic in broader sense[1] was proposed by V. Novák in 1995 in [34] as a program for extension of FLn, which aims at developing a *formal theory of natural human reasoning that would include mathematical models of special expressions in natural language with regard to their vagueness*. This program overlaps with two other paradigms proposed in the literature, namely *commonsense reasoning* and *precisiated natural language* (PNL).

The idea of commonsense reasoning has been proposed by J. McCarthy in [31] as a part of the program of logic-based artificial intelligence. Its paradigm is to develop formal commonsense theories and systems using mathematical logic that exhibit commonsense behavior. The reason is that commonsense reasoning is a central part of human behavior and no real intelligence is possible without it.

The main drawback of the up-to-date formalizations of commonsense reasoning, in our opinion, is that it neglects vagueness present in the meaning of natural language expressions (cf. [5] and the citations therein).

The concept of PNL, which has been proposed by L. A. Zadeh in [70], is based on two main premises:

(a) Much of the world knowledge is perception based,
(b) perception based information is intrinsically fuzzy.

It is important to stress that the term *precisiated natural language* means "a reasonable working formalization of semantics of natural language without pretension to capture it in detail and fineness". Its goal is to provide an acceptable and applicable technical solution. It should also be noted that the term *perception* is not considered here as a psychological term but rather as a result of human, intrinsically imprecise measurement.

[1] It should be noted that FLb should be distinguished from the very general concept of fuzzy logic in wide sense announced informally by L. A. Zadeh in the nineties.

PNL methodology requires the presence of *World Knowledge Database* and *Multiagent, Modular Deduction Database*. The former contains all the necessary information including perception based propositions describing the knowledge acquired by direct human experience, which can be used in the deduction process. No exact formalization of PNL, however, has been developed until recently, and so it should be taken mainly as a reasonable methodology.

The concept of FLb is a glue between both paradigms that should consider the best of each. During the years, it has been slowly developed and so far, it consists of the following theories:

(a) Formal theory of evaluative linguistic expressions, which is in detail explained in [42].
(b) Formal theory of fuzzy IF-THEN rules which is presented in [11, 45].
(c) Formal theory of perception-based logical deduction, which is presented in [9, 37, 38, 47],
(d) Formal theory of intermediate and generalized quantifiers presented in [10, 21, 39, 43].

The most convenient formal system as the basis for further development of FLb seems to be the *fuzzy type theory*. One of the reasons supporting the latter is argumentation of many logicians and linguists (cf. [29, 30, 66]) that the first order logic is not sufficient for capturing semantics of natural language.

Evaluative linguistic expressions

This theory is one of the most essential constituents of FLb. Recall that evaluative expressions are expressions of natural language, for example, *small, medium, big, about twenty five, roughly one hundred, very short, more or less deep, not very tall, roughly warm or medium hot, quite roughly strong, roughly medium size*, and many others. They form a small but very important part of natural language and they are present in its everyday use any time. The reason is that people very often need to evaluate phenomena around them. Moreover, they often make important decisions based on them, learn how to control, and many other activities. Note that from the linguistic point of view, they also include the, so called, gradable adjectives (cf. [23]). We argue that the meaning of evaluative expressions is a fundamental bearer of the vagueness phenomenon and namely, that its vagueness is a consequence of the indiscernibility between objects.

All the details about formal theory of evaluative linguistic expressions can be found in [42]. We distinguish *intension* (a property), and *extension* in a given *context* of use (i.e., a *possible world*; see [14]). Mathematical representation of *intension is a function defined on a set of contexts which assigns to each context a fuzzy set of elements*. Intension is invariant with respect to change of context. *Extension* is a class of elements (i.e., a fuzzy set) determined by intension when setting a *specific context*. It depends on the particular context of use and changes whenever the context is changed. For example, the expression "large town" is a name of an intension being a property of some feature of objects, i.e. number of people in a town. Its meaning can be, e.g., 100 thousand people in the Czech Republic, 1 mil. people in France, or 10 mil. people in Asia.

The *pure evaluative expressions* are expressions of the form

$$\langle\text{linguistic hedge}\rangle\langle\text{atomic evaluative expression}\rangle$$

where linguistic hedges are e.g. *very*, *more or less* and atomic evaluative expressions are gradable adjectives *small*, *medium* and *big*, and many others. We must also distinguish the meaning of *evaluative predications* that are linguistic expressions of the form

$$X \text{ is } \langle\text{linguistic hedge}\rangle\langle\text{atomic expression}\rangle \tag{8}$$

where $X$ is a variable for values of some specific *feature* of objects (e.g., temperature, pressure, height, depth, etc.).

There are good means in FTT using which notions of context, horizon, and many other general characteristics of the meaning of evaluative expressions can be modeled. The result is a formal logical theory of evaluative linguistic expressions $T^{\mathrm{Ev}}$ which is determined by 11 special axioms.

Interpretation of extensions of evaluative expressions in a model is depicted in Fig. 1. The context is determined a by the triple $\langle v_L, v_S, v_R\rangle$ of points (left bound, middle point, and right bound). The *LH*, *MH* and *RH* are fuzzy sets interpreting left, medium and right horizon, respectively which are determined by a special fuzzy equality $\approx_w$ constructed for each context $w$ from one universal one. The $\nu_{a,b,c}$ is a function specific for each linguistic hedge $\boldsymbol{\nu}$ which represents the corresponding horizon shift. The composition of horizon and its shift provides extension of the evaluative predication (8) being a fuzzy set of elements. Note that there are various mathematical models of linguistic hedges which, however, in most cases do not meet all the necessary requirements. A nice overview of them can be found in [25].

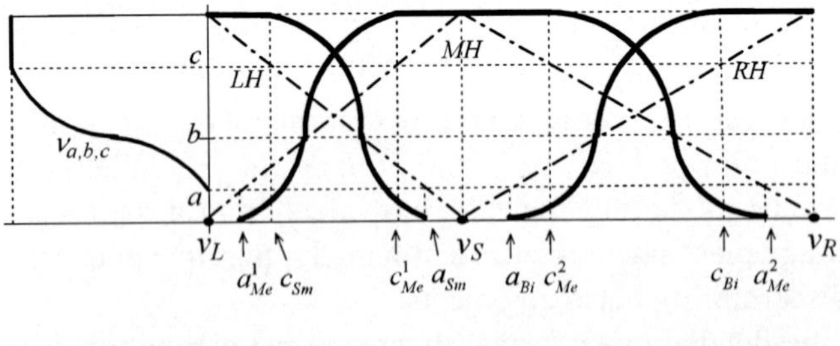


**Fig. 1** Scheme of the construction of extensions of evaluative expressions

It is possible to construct a model of the theory $T^{\mathrm{Ev}}$ of evaluative expression. Hence, using the completeness theorem, we can prove the following:

**Theorem 2.** *The theory of evaluative linguistic expressions is consistent.*

This theorem is a strong theoretical support for the functionality of our theory.

Fuzzy IF-THEN rules and intermediate quantifiers

The theory of evaluative expressions is the point of departure for more advanced theories of FLb. The first of them is the theory of fuzzy IF-THEN rules that are conditional expressions of natural language having the form

$$\textsf{IF } X \text{ is } \mathcal{A} \textsf{ THEN } Y \text{ is } \mathcal{B}. \tag{9}$$

The $\mathcal{A}, \mathcal{B}$ in (9) are just evaluative linguistic expressions, or more precisely, the fuzzy IF-THEN rule consists of two (or more) evaluative linguistic predications joined by an implication. Sets (finite) of fuzzy IF-THEN rules are called *linguistic descriptions* and they are the fundamental tool, on the basis of which are can describe various decision, control or other situations which can give us more information about the reality.

The principal method for derivation of a conclusion on the basis of a linguistic description in *perception-based logical deduction* (see [37, 47]). We will explain the main idea on the following example.

*Example 1.* Given a linguistic description $LD$

$$\mathcal{R}_1 := \textsf{IF } X \text{ is } \textit{small} \textsf{ THEN } Y \text{ is } \textit{very big},$$
$$\mathcal{R}_2 := \textsf{IF } X \text{ is } \textit{very big} \textsf{ THEN } Y \text{ is } \textit{small}.$$

Each rule provides us with a certain knowledge (related to the concrete application). We are able to *distinguish* between the rules despite the fact that their meaning is vague.

Let us now consider specific linguistic contexts: $w \in W$ for values of $X$, say, $w = \langle 150, 330, 600\rangle$ (for example, temperature in some oven) and $w' \in W$ for values of $Y$, $w' = \langle 0, 36, 90\rangle$ (for example, a turncock position in degrees). Then "small $X$" are values of $X$ around 180–210 (and smaller) and "very big $X$" are values around, at least, 550 or higher. Similarly, "small $Y$" are values of $Y$ around 8 (and smaller) and "very big $Y$" values around, at least, 80–85 (and higher).

Let the value $X = 180$ be given as an observation of $X$ (this may be, e.g., a result of some measurement). To derive a conclusion on the basis of the given linguistic description, we must first test whether this value falls in its topic. Since the value of 180 is in the context $w = \langle 150, 330, 600\rangle$ apparently *small*, we may apply knowledge determined by $LD$ and because of rule $\mathcal{R}_1$, we expect a *very big* value of $Y$ (a value around 85). Similarly, $X = 560$ which is *very big* leads to a value of $Y$ around 8 due to the rule $\mathcal{R}_2$.

More details, formal explanation and description of some applications can be found also in [41].

An interesting formal theory elaborated in the frame of FLb is the theory of intermediate quantifiers. These are expressions such as *most, a lot of, many, a few, a great deal of, large part of, small part of* which have been in detail informally elaborated in [63]. The main idea is to introduce intermediate quantifiers as special formulas of fuzzy type theory which express that they are, in fact, classical general or existential quantifiers but limited to smaller (fuzzy) sets (for example, fuzzy sets with smaller support) whose size is measured using some measure. Formulas representing intermediate quantifiers are constructed in a certain extension of the formal theory $T^{\mathrm{Ev}}$ of evaluative linguistic expressions. The theory is sufficiently general to encompass a wide class of the generalized quantifiers and provides a unique definition for all of them.

We have proved in [32] validity of all 105 generalizations of classical Aristotlean syllogisms presented in [63]. Their form is, for example, as follows:

| | | | |
|---|---|---|---|
| | All *M* are *Y* | | All *M* are *Y* |
| **APK-I:** | Almost all *X* are *M* | **ATT-I:** | Most *X* are *M* |
| | Many *X* are *Y* | | Most *X* are *Y* |

There is a great application potential of the theory presented above. One of interesting possibilities is the commonsense human reasoning.

## *2.3 Model of Commonsense Human Reasoning*

We will very briefly mention an FLb-model of a detective story inspired by one episode of the famous TV series about Lt. Columbo. Note that detective stories are typical examples of human reasoning where crucial role is played also by natural language, knowledge of common things and situations, and standard logical deduction. FLb can provide a formal logical mechanism that can mimic Lt. Columbo's reasoning.

**The plot:** *Mr. John Smith has been shot dead in his house. He was found by his friend, Mr. Robert Brown. Lt. Columbo suspects Mr. Brown to be the murderer.*

**Mr. Brown's testimony:** *I have started from my home at about 6:30, arrived to John's house at about 7, found John dead and went immediately to the phone box to call police. They came immediately.*

**Lt. Columbo's evidence:** *Mr. Smith had high quality suit, broken wristwatch stopped at 5:45. No evidence of strong strike on his body. Lt. Columbo touched engine of Mr. Brown's car and found it to be more or less cold.*

Lt. Columbo concluded that Mr. Brown lies because:

(i) Mr. Brown's car engine is *more or less cold*, however he went *long* (more than about 30 minutes). He could not arrive and continue to call the police (which came *immediately*).

(ii) *High quality wristwatch* does not break after *not too strong strike*. A man having *high quality dress* and a *luxurious house* is supposed to have also *high quality wristwatch*. Wristwatch of John Smith is of *low quality* and so, it does not belong to him. It does not display time of Mr. Smith's death.

**The method:** combination of logical rules, world knowledge and evidence with the help of *non-monotonic reasoning* realized on syntactic level in FTT.

- **Context**

  (a) *Drive duration to heat the engine (minutes):* $w_D = \langle 0, 5, 30 \rangle$
  (b) *Temperature of engine (degrees Celsius):* $w_T = \langle 0, 45, 100 \rangle$
  (c) Abstract degrees: quality, state, strike strength: $\langle 0, 0.5, 1 \rangle$

- **Logical rules** – logical theorems of FTT and theorems given by some considered theory, for example

  IF $X$ is *Sm* THEN $X$ is *not Bi*
  IF $X$ is *Bi* THEN $X$ is *not Sm*

  *World knowledge*

- **Knowledge from physics**

  IF *drive duration* is *Bi* THEN *engine temperature* is *Bi*
  IF *drive duration* is *Sm* THEN *engine temperature* is *ML Sm*

  ..........................................................

- **Customs of people**

  IF quality of $x_\pi$'s suit is *Bi* AND quality of $x_\pi$'s house is *Ve Bi*
  THEN wealth of $x_\pi$ is *Bi*

  ..............................................

- etc.

(the following shorts for evaluative expressions have been used: *Sm*–small, *Bi*-big, *Ve*-very, *ML*-more or less)

We construct a specific model $\mathcal{M}$ given by the evidence of the contexts of wealthy people and perceptions of Lt. Columbo, for example:

- Touching Mr. Brown's car engine by hand does not burn; its temperature is *MLSm*,
- Quality of Mr. Smith's house is *VeBi*.

**Lt. Columbo's conclusion:** *The two special constructed theories are contradictory. Since my evidence is correct, Mr. Brown lies and had an opportunity to kill Mr. Smith.*

The vagueness is manifested in this reasoning by the fact that a slight change in the evidence may lead to disappearance of the contradiction, despite the fact that the basic knowledge has not changed. For example, the truth of "temperature of engine is low" can be higher, or vice versa, the truth of "temperature of engine is more or less big" can be nonzero, etc. so that the conclusion can be made more, or less, convincing. Consequently, further evidence might be necessary, or vice-versa. We conclude that vagueness plays an important role in reasoning and drawing conclusions. Classical boolean logic can provide solutions only at the price of imposing, quite often, improper precision and thus, making the model of reasoning unrealistic. Introduction of truth degrees may help to overcome such restrictions since we can balance them: the higher the truth values, the more convincing the conclusion — but the limit (i.e., full truth 1 or full falsity 0) is unnecessary.

We should see the above example of commonsense reasoning as a general methodology which can have various kinds of applications. For example, in [12] it has also been applied to analysis of economic texts.

## 3 Related Theories with Great Application Potential

One could conclude from the above discussion that fuzzy logic is indeed a good well established theory which can be very practical. But we can go farther. Besides the linguistic interpretation of the rules (9), they are also interpreted in the frame of predicate fuzzy logic as fuzzy relations (cf. [19, 48]).

### 3.1 *Fuzzy Approximation*

The linguistic description is assigned certain predicate formulas in BL- or Łukasiewicz fuzzy logic in one of two possible forms ([19, 48, 51, 52, 55]):

1. *Disjunctive normal form* (DNF) where each rule is interpreted by a *conjunction* of the antecedent and consequent and the fuzzy relations of all rules are joined by disjunction. This approach is largely known under the name *Mamdani-Assilian interpretation* [28], although it has its roots in [68, 69]. Interpretation of DNF provides the following fuzzy relation

$$R_{DNF}(x,y) = \bigvee_{i=1}^{n} (A_i(x) * B_i(y)), x \in X, y \in Y. \tag{10}$$

2. *Conjunctive normal form* (CNF) where the individual rules are interpreted as implications and joined by conjunction (see also [8]). Interpretation of CNF provides the following fuzzy relation

$$R_{CNF}(x,y) = \bigwedge_{i=1}^{n} (A_i(x) \to B_i(y)), x \in X, y \in Y. \tag{11}$$

The linguistic description characterizes a specific dependence between values of $X$ and $Y$, where the values of $X$ are the possible *inputs* and the values of $Y$ are the possible *outputs* of the given system described by $\mathcal{R}$. The constructed fuzzy relation $R_{DNF}$ or $R_{CNF}$ can then be used to compute an output for a given input. This input is either a crisp value $x \in X$ or a fuzzy set $A \in \mathcal{F}(X)$. The process how the output is computed using fuzzy relations is usually called the *Computational Rule of Inference* (CRI) first considered in [68]. The core of such an inference mechanism is the computation of an image of the input fuzzy set via a given fuzzy relation. The result is an output fuzzy set $B \in \mathcal{F}(Y)$. Logical justification of this process has been provided in [19, 48].

One of the practical problems is how the fuzzy sets $A_i, B_i$ in (10), (11) should be specified. This problem turned out to be closely related to solvability of fuzzy relation equations which has been initiated by E. Sanchez in [65]. The result says that the fuzzy relations $R_{DNF}$ or $R_{CNF}$ and, consequently, CRI rule are incorrectly defined if the corresponding system of fuzzy relation equations is not solvable (cf. [54, 59, 60, 61]). This also implies that evaluative linguistic expressions cannot be, in general, used with CRI (see [35]) and so, the latter is not convenient tool for logical reasoning inside natural language (i.e., to fit the paradigm of FLb).

Relational interpretation of linguistic descriptions has thousands of various kinds of applications. It can be considered as one of most practical theory based on fuzzy predicate logic.

## 3.2 Fuzzy (F)-Transform

The fuzzy transform (F-transform for short) introduced in detail by I. Perilieva in [56] is a special technique that can be applied to a continuous function, defined on a fixed real interval $[a,b] \subset \mathbb{R}$. The essential idea is to transform a given function defined in one space into another, usually simpler space, and then to transform it back. The simpler space consists of a finite vector of numbers obtained on the basis of the well established *fuzzy partition* of the domain of the given function. The reverse transform then leads to a function approximately reconstructing the original one. Thus, the first step, sometimes called the *direct F-transform*, results in a vector of averaged functional values. The second step, called the *inverse transform*, converts this vector into another continuous function, which approximates the original one.

The essential idea of F-transform is defined with respect to a *fuzzy partition* which is a special system of fuzzy sets defined on a set of nodes in $[a,b]$ which fulfils several conditions. Some examples of fuzzy partitions see Fig. 2.

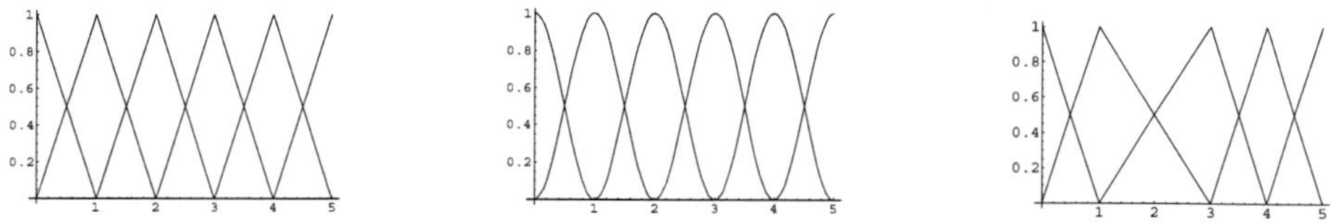

**Fig. 2** Graphical presentation of several fuzzy partitions.

Let a fuzzy partition of $[a,b]$ be given by $A_1,\dots,A_n, n \geq 2$, and let $f : [a,b] \longrightarrow \mathbb{R}$ be an arbitrary continuous function. The $n$-tuple of real numbers $[F_1,\dots,F_n]$ given by

$$F_i = \frac{\int_a^b f(x)A_i(x)dx}{\int_a^b A_i(x)dx}, \qquad i = 1,\dots,n, \tag{12}$$

is a *direct fuzzy transform (F-transform)* of $f$ with respect to the given fuzzy partition. The numbers $F_1,\dots,F_n$ are called the *components* of the F-transform of $f$.

In practice, the function $f$ is usually not given analytically, but we are at least provided some data, obtained, for example, by some measurements. In this case, the previous definition can be modified in such a way that the definite integrals in Formula (12) are replaced by finite summations. It has been proven [56] that the components of the F-transform are weighted mean values of the original function, where the weights are determined by the basic functions.

The original function $f$ can be approximately reconstructed from the direct F-transform of $f$ with respect to $A_1,\ldots,A_n \in \mathcal{F}([a,b])$ using the following inversion formula:

$$f_{F,n}(x) = \sum_{i=1}^{n} F_i A_i(x). \tag{13}$$

The function $f_{F,n}$ is called the *inverse F-transform* of $f$. It is a continuous function on $[a,b]$.

**Theorem 3.** *Let $f$ be a continuous function on $[a,b]$. Then for any $\varepsilon > 0$ there exist $n_\varepsilon$ and a fuzzy partition $A_1,\ldots,A_{n_\varepsilon}$ of $[a,b]$ such that for all $x \in [a,b]$*

$$|f(x) - f_{F,n_\varepsilon}(x)| \le \varepsilon \tag{14}$$

*where $f_{F,n_\varepsilon}$ is the inverse F-transform of $f$ with respect to the fuzzy partition $A_1,\ldots,A_{n_\varepsilon}$.*

It should be noted that this theorem holds true independently on shapes of the fuzzy sets $A_1,\ldots,A_{n_\varepsilon}$ (of course, provided that they fulfil the general required conditions). For various properties of the F-transform and detailed proofs — see [56]. There are numerous applications of the F-transform in image processing, data mining, time series analysis, solution of differential equations, some numerical methods and elsewhere (see, e.g. [7, 49, 53, 57, 58, 62, 64]). One can see that this theory is very practical.

## 4 Conclusion

As can be seen from the previous brief presentation, mathematical fuzzy logic is a well developed formal theory which gave rise to several more special theories. Its structure is now already known quite well. For example, in the paper [4], almost 60 various fuzzy logics are analyzed together. We still do not know exactly which logic is the most convenient to solve problems related to models of vagueness and their applications but one can guess from the discussion above that the leading position is played by MFL based on standard Łukasiewicz MV-algebra of truth values.

We think that there are good reasons to continue the development of FLb and move it further towards human (i.e., commonsense) reasoning, following the methodology of PNL and results obtained in the AI *theory of commonsense reasoning*. Another promising direction is to include also uncertainty in FLb in the sense that has been nicely established by T. Flaminio and F. Montagna in [15]. In their paper, fuzzy logic is joined with probability theory.

To summarize, we have presented several theories and mentioned their applications. We are convinced that this demonstrates that MFL and the related theories are very practical.

## References


1. Běhounek, L., Cintula, P.: Fuzzy class theory. Fuzzy Sets and Systems 154, 34–55 (2005)
2. Běhounek, L., Cintula, P.: From fuzzy logic to fuzzy mathematics: A methodological manifesto. Fuzzy Sets and Systems 157(5), 642–646 (2006)
3. Cignoli, R.L.O., D'ottaviano, I.M.L., Mundici, D.: Algebraic Foundations of Many-valued Reasoning. Kluwer, Dordrecht (2000)
4. Cintula, P., Hájek, P., Horčík, R.: Formal systems of fuzzy logic and their fragments. Annals of Pure and Applied Logic 150, 40–65 (2007)
5. Davis, E., Morgenstern, L.: Introduction:progress in formal commonsense reasoning. Artifical Intelligence 153, 1–12 (2004)
6. De Cooman, G., Kerre, E., Cappelle, B., Ruan, D., Vanmassenhove, F.: On the extension of classical propositional logic by means of a triangular norm. Int. J. of Intelligent Systems 5, 307–322 (1990)
7. Di Martino, F., Loia, V., Perfilieva, I., Sessa, S.: An image coding/decoding method based on direct and inverse fuzzy transforms. Int. Journ. of Appr. Reasoning 48, 110–131 (2008)
8. Dubois, D., Prade, H.: What are fuzzy rules and how to use them. Fuzzy Sets and Systems 84, 169–185 (1996)
9. Dvořák, A., Novák, V.: Fuzzy logic deduction with crisp observations. Soft Computing 8, 256–263 (2004)
10. Dvořák, A., Holčapek, M.: L-fuzzy quantifiers of the type $\langle 1 \rangle$ determined by measures. Fuzzy Sets and Systems 160, 3425–3452 (2009)
11. Dvořák, A., Novák, V.: Formal theories and linguistic descriptions. Fuzzy Sets and Systems 143, 169–188 (2004)
12. Dvořák, A., Novák, V.: Towards automatic modeling of economic texts. Mathware & Soft Computing XIV(3), 217–231 (2007)
13. Esteva, F., Godo, L.: Monoidal t-norm based logic: towards a logic for left-continuous t-norms. Fuzzy Sets and Systems 124, 271–288 (2001)
14. Fitting, M.: Intensional logic. In: Zalta, E.N. (ed.) The Stanford Encyclopedia of Philosophy (2006)
15. Flaminio, T., Montagna, F.: MV-algebras with internal states and probabilistic fuzzy logics. International Journal of Approximate Reasoning 50, 138–152 (2009)
16. Galatos, N., Jipsen, P., Kowalski, T., Ono, H.: Residuated Lattices. In: An Algebraic Glimpse At Substructural Logics. Elsevier, Amsterdam (2007)
17. Goguen, J.A.: The logic of inexact concepts. Synthese 19, 325–373 (1968-1969)
18. Gottwald, S.: A Treatise on Many-Valued Logics. Research Studies Press Ltd., Baldock (2001)
19. Hájek, P.: Metamathematics of Fuzzy Logic. Kluwer, Dordrecht (1998)
20. Hájek, P., Cintula, P.: On theories and models in fuzzy predicate logics. Journal of Symbolic Logic 71(3), 863–880 (2006)
21. Holčapek, M.: Monadic L-fuzzy quantifiers of the type $\langle 1^n, 1 \rangle$. Fuzzy Sets and Systems 159, 1811–1835 (2008)
22. Jenei, S., Kerre, E.: Convergence of residuated operators and connective stability. Fuzzy Sets and Systems 114, 411–415 (2000)

23. Kennedy, C.: Vagueness and grammar: The semantics of relative and absolute gradable adjectives. Linguistics and Philosophy 30(3), 1–45 (2007)
24. Kerre, E., De Cock, M.: A comparative study of the behaviour of some popular fuzzy implication operators on the generalized modus ponens. In: Martinez, J. (ed.) Fuzzy Logic for the Management of Uncertainty, pp. 281–296. John Wiley and Sons, New York (1992)
25. Kerre, E., De Cock, M.: Linguistic modifiers: an overview. In: Martinez, J. (ed.) Fuzzy Logic and Soft Computing, pp. 69–86. Kluwer Academic, Boston (1999)
26. Klement, E.P., Mesiar, R., Pap, E.: Triangular Norms. Kluwer, Dordrecht (2000)
27. Lewin, K.: Field theory in social science: Selected theoretical papers. Harper & Row, New York (1951)
28. Mamdani, E., Assilian, S.: An experiment in linguistic synthesis with a fuzzy logic controller. Int. J. of Man-Machine Studies 7, 1–13 (1975)
29. Materna, P.: Concepts and Objects, Acta Philosophica Fennica 63, Helsinki (1998)
30. Materna, P.: Conceptual Systems. Logos Verlag, Berlin (2004)
31. McCarthy, J.: Programs with common sense. In: Proc. of the Teddington Conf. on Mechanization of Thought Processes, Her Majesty's Stationary Office, London (1959)
32. Murinová, P., Novák, V.: A formal theory of generalized intermediate syllogisms. Fuzzy Sets and Systems
33. Novák, V.: On the syntactico-semantical completeness of first-order fuzzy logic I, II. Kybernetika 26, 47–66, 134–154 (1990)
34. Novák, V.: Towards formalized integrated theory of fuzzy logic. In: Bien, Z., Min, K. (eds.) Fuzzy Logic and Its Applications to Engineering, Information Sciences, and Intelligent Systems, pp. 353–363. Kluwer, Dordrecht (1995)
35. Novák, V.: Fuzzy relation equations with words. In: Nikravesh, M., Zadeh, L., Korotkikh, V. (eds.) Fuzzy Partial Differential Equations and Relational Equations, pp. 167–185. Springer, Berlin (2004)
36. Novák, V.: On fuzzy type theory. Fuzzy Sets and Systems 149, 235–273 (2005)
37. Novák, V.: Perception-based logical deduction. In: Reusch, B. (ed.) Computational Intelligence, Theory and Applications, pp. 237–250. Springer, Berlin (2005)
38. Novák, V.: Perception-based logical deduction as alternative approximate reasoning method. In: Proc. Int. Conf. FUZZ-IEEE 2005, Reno, USA (May 2005)
39. Novák, V.: Fuzzy logic theory of evaluating expressions and comparative quantifiers. In: Proc. $11^{th}$ Int. Conf. IPMU, Paris, Éditions EDK, Les Cordeliers, Paris, vol. 2 (July 2006)
40. Novák, V.: Which logic is the real fuzzy logic? Fuzzy Sets and Systems 157, 635–641 (2006)
41. Novák, V.: Mathematical fuzzy logic in modeling of natural language semantics. In: Wang, P., Ruan, D., Kerre, E. (eds.) Fuzzy Logic – A Spectrum of Theoretical & Practical Issues, pp. 145–182. Elsevier, Berlin (2007)
42. Novák, V.: A comprehensive theory of trichotomous evaluative linguistic expressions. Fuzzy Sets and Systems 159(22), 2939–2969 (2008)
43. Novák, V.: A formal theory of intermediate quantifiers. Fuzzy Sets and Systems 159(10), 1229–1246 (2008), doi:10.1016/j.fss.2007.12.008
44. Novák, V.: EQ-algebra-based fuzzy type theory and its extensions. Logic Journal of the IGPL (to appear, 2010)
45. Novák, V., Lehmke, S.: Logical structure of fuzzy IF-THEN rules. Fuzzy Sets and Systems 157, 2003–2029 (2006)
46. Novák, V., Perfilieva, I. (eds.): Discovering the World With Fuzzy Logic. Studies in Fuzziness and Soft Computing, vol. 57. Springer, Heidelberg (2000)

47. Novák, V., Perfilieva, I.: On the semantics of perception-based fuzzy logic deduction. International Journal of Intelligent Systems 19, 1007–1031 (2004)
48. Novák, V., Perfilieva, I., Močkoř, J.: Mathematical Principles of Fuzzy Logic. Kluwer, Boston (1999)
49. Novák, V., Štěpnička, M.U., Dvořák, A., Perfilieva, I., Pavliska, V., Vavříčková, L.: Analysis of seasonal time series using fuzzy approach. Int. Journal of General Systems 39, 305–328 (2010)
50. Pavelka, J.: On fuzzy logic I, II, III, Zeitschrift für Mathematische Logik und Grundlagen der Mathematik 25, 45–52, 119–134, 447–464 (1979)
51. Perfilieva, I.: Fuzzy logic normal forms for control law representation. In: Verbruggen, H., Zimmermann, H.-J., Babusha, R. (eds.) Fuzzy Algorithms for Control, pp. 111–125. Kluwer, Boston (1999)
52. Perfilieva, I.: Logical approximation. Soft Computing 7(2), 73–78 (2002)
53. Perfilieva, I.: Fuzzy transform: Application to reef growth problem. In: Demicco, R.B., Klir, G.J. (eds.) Fuzzy Logic in Geology, pp. 275–300. Academic Press, Amsterdam (2003)
54. Perfilieva, I.: Fuzzy function as an approximate solution to a system of fuzzy relation equations. Fuzzy Sets and Systems 147, 363–383 (2004)
55. Perfilieva, I.: Normal forms in BL-algebra of functions and their contribution to universal approximation. Fuzzy Sets and Systems 143, 111–127 (2004)
56. Perfilieva, I.: Fuzzy transforms: theory and applications. Fuzzy Sets and Systems 157, 993–1023 (2006)
57. Perfilieva, I.: Fuzzy transforms: A challenge to conventional transforms. In: Hawkes, P.W. (ed.) Advances in Images and Electron Physics, vol. 147, pp. 137–196. Elsevier Academic Press, San Diego (2007)
58. Perfilieva, I., De Meyer, H., De Baets, B., Plskova, D.: Cauchy problem with fuzzy initial condition and its approximate solution with the help of fuzzy transform. In: Proc. of WCCI 2008, IEEE Int. Conf. on Fuzzy Systems, Hong Kong (2008)
59. Perfilieva, I., Lemhke, S.: Correct models of fuzzy if–then rules are continuous. Fuzzy Sets and Systems 157, 3188–3197 (2006)
60. Perfilieva, I., Nosková, L.: System of fuzzy relation equations with inf $\rightarrow$ composition: complete set of solutions. Fuzzy Sets and Systems 159, 2256–2271 (2008)
61. Perfilieva, I., Novák, V.: System of fuzzy relation equations as a continuous model of IF-THEN rules. Information Sciences 177, 3218–3227 (2007)
62. Perfilieva, I., Novák, V., Dvořák, A.: Fuzzy transform in the analysis of data. Int. Journ. of Appr. Reasoning 48, 36–46 (2008)
63. Peterson, P.: Intermediate Quantifiers. Logic, linguistics, and Aristotelian semantics, Ashgate, Aldershot (2000)
64. Plšková, D.: Fuzzy transform in geological applications. Journal of Electrical Engineering 57, 43–46 (2006)
65. Sanchez, E.: Resolution of composite fuzzy relation equations. Information and Control 30, 38–48 (1976)
66. Sgall, P., Hajičová, E., Panevová, J.: The Meaning of the Sentence in Its Syntactic and Pragmatic Aspects. D. Reidel, Dordrecht (1986)
67. Zadeh, L.A.: Fuzzy sets. Information and Control 8, 338–353 (1965)
68. Zadeh, L.A.: Outline of a new approach to the analysis of complex systems and decision processes. IEEE Trans. on Systems, Man, and Cybernetics SMC 3, 28–44 (1973)
69. Zadeh, L.A.: A rationale for fuzzy control, Trans. ASME, Ser. G, J. Dynamic. Systems, Measurement and Control 94, 3–4 (1974)
70. Zadeh, L.A.: Precisiated natural language. AI Magazine 25, 74–91 (2004)

# Interval-Valued Algebras and Fuzzy Logics

Bart Van Gasse, Chris Cornelis, and Glad Deschrijver

**Abstract.** In this chapter, we present a propositional calculus for several interval-valued fuzzy logics, i.e., logics having intervals as truth values. More precisely, the truth values are preferably subintervals of the unit interval. The idea behind it is that such an interval can model imprecise information. To compute the truth values of 'p implies q' and 'p and q', given the truth values of p and q, we use operations from residuated lattices. This truth-functional approach is similar to the methods developed for the well-studied fuzzy logics. Although the interpretation of the intervals as truth values expressing some kind of imprecision is a bit problematic, the purely mathematical study of the properties of interval-valued fuzzy logics and their algebraic semantics can be done without any problem. This study is the focus of this chapter.

## 1 Introduction

Classical logic is a two-valued logic: propositions in this logic are either true or false. In the first case, the truth value 1 is attributed to the proposition, while in the second case the attributed truth value is 0. Given the truth values of two propositions

Bart Van Gasse
Department of Applied Mathematics and Computer Science, Ghent University, Krijgslaan 281 (S9), B-9000 Gent, Belgium
e-mail: Bart.VanGasse@UGent.be

Chris Cornelis
Department of Applied Mathematics and Computer Science, Ghent University, Krijgslaan 281 (S9), B-9000 Gent, Belgium
e-mail: Chris.Cornelis@UGent.be

Glad Deschrijver
Department of Applied Mathematics and Computer Science, Ghent University, Krijgslaan 281 (S9), B-9000 Gent, Belgium
e-mail: Glad.Deschrijver@UGent.be

C. Cornelis et al. (Eds.): 35 Years of Fuzzy Set Theory, STUDFUZZ 261, pp. 57–82.
springerlink.com 

$p$ and $q$, it is possible to derive the truth values of the negation 'not $p$' (and 'not $q$'), the conjunction '$p$ and $q$', the disjunction '$p$ or $q$' and the implication '$p$ implies $q$'. These formulas are denoted as $\neg p$, $p \& q$, $p \vee q$ and $p \rightarrow q$. The truth values are calculated using the operations[1] $\neg$, $*$, $\sqcup$ and $\Rightarrow$. The truth tables of these operations are given in Table 1.

**Table 1** Truth tables of the operations in classical logic

| $x$ | $y$ | $\neg x$ | $x * y$ | $x \sqcup y$ | $x \Rightarrow y$ |
|---|---|---|---|---|---|
| 0 | 0 | 1 | 0 | 0 | 1 |
| 0 | 1 | 1 | 0 | 1 | 1 |
| 1 | 0 | 0 | 0 | 1 | 0 |
| 1 | 1 | 0 | 1 | 1 | 1 |

For more complicated formulas the truth values can be computed in the same way. For example, if $p$ is true and $q$ is false, then the truth value of $(p \vee q) \rightarrow ((p \rightarrow q) \rightarrow q)$ is calculated as follows: the truth value of $p \rightarrow q$ is $1 \Rightarrow 0 = 0$, so the truth value of $(p \rightarrow q) \rightarrow q$ is $0 \Rightarrow 0 = 1$. The truth value of $p \vee q$ is $1 \sqcup 0 = 1$. So we conclude that the truth value of $(p \vee q) \rightarrow ((p \rightarrow q) \rightarrow q)$ is $1 \Rightarrow 1 = 1$. Interestingly, the truth value of this formula is always 1, even if other truth values are attributed to $p$ and $q$. Such formulas are called tautologies. If a formula $\phi$ is a tautology, this is denoted as $\models \phi$. More generally, for a set of formulas $\Gamma$, $\Gamma \models \phi$ means "no matter what truth values are attributed to the propositions, if the truth values of the formulas in $\Gamma$ are 1, then the truth value of $\phi$ is 1".

The two values 0 and 1, together with the defined operations, form a Boolean algebra. Therefore we say that this Boolean algebra is the semantics of classical logic. Saying that $(p \vee q) \rightarrow ((p \rightarrow q) \rightarrow q)$ is a tautology in classical logic, is the same as saying that $(x \sqcup y) \Rightarrow ((x \Rightarrow y) \Rightarrow y) = 1$ is an identity in this Boolean algebra (meaning "whatever value of the Boolean algebra we give to $x$ and $y$, the calculation of $(x \sqcup y) \Rightarrow ((x \Rightarrow y) \Rightarrow y)$ yields 1"). Now, identities in this Boolean algebra are also identities in every other Boolean algebra (we say that this Boolean algebra generates all Boolean algebras). Therefore classical logic does not only have the Boolean algebra with two elements as semantics, but also the whole variety of Boolean algebras: the general semantics of classical logic consist of all Boolean algebras.

Interestingly it is also possible to describe classical logic without using semantics. This is done with axioms and deduction rules, which allow to prove a formula from a set of formulas. When a formula $\phi$ is provable from a theory $\Gamma$, this is denoted as $\Gamma \vdash \phi$. Two important results in classical logic are soundness (if $\Gamma \vdash \phi$, then $\Gamma \models \phi$) and completeness (if $\Gamma \models \phi$, then $\Gamma \vdash \phi$). We write this shortly as $\Gamma \vdash \phi$ iff $\Gamma \models \phi$.

[1] Note that we use different symbols, to distinguish the logical connectives from the corresponding operations. Only for the negation we employ the same symbol.

Now, for the truth values of several propositions one might prefer more than the two options 0 (false) and 1 (true). Indeed, for vague propositions like 'it is raining hard', it would be useful if one could attribute an intermediate truth value, somewhere between 'false' and 'true'. This can be done using fuzzy set theory, in which every element of the unit interval $[0,1]$ serves as a truth value, instead of only 0 and 1. The operations for the negation, conjunction, disjunction and implication were generalized to this setting. Later the structure of the unit interval was generalized to an arbitrary bounded lattice to allow for incomparabilities among elements, and triangular norms and conorms are quite common nowadays as generalized representations of logical conjunction and disjunction, respectively. An interesting class of these generalizations, especially from the logical point of view, are MTL-algebras [16]. In these structures, the operations modelling (strong) conjunction and implication are connected by the residuation principle. These MTL-algebras form the general semantics of monoidal t-norm based logic (MTL, [16]), in the same way Boolean algebras form the general semantics of classical logic. Similarly as for classical logic, these general semantics can be restricted. Indeed, MTL is also sound and complete w.r.t. standard MTL-algebras, i.e., MTL-algebras on the unit interval. Therefore MTL is called a (formal) fuzzy logic. But it is definitely not the only fuzzy logic. Indeed, by adding more axioms and/or deduction rules to the axioms and deduction rule of MTL, we obtain other fuzzy logics. It is even possible to retrieve classical logic in this way. Semantically speaking, this means that Boolean algebras are special cases of MTL-algebras. Some well-known fuzzy logics, situated between MTL and classical logic, are Hájek's Basic Logic (BL) [24], Łukasiewicz Logic (Ł) [30] and Gödel Logic (GL) [14, 22]. Also Intuitionistic Logic (IL) [25] can be seen as a fuzzy logic. These logics are sound and complete w.r.t. BL-algebras, MV-algebras (or, equivalently, Wajsberg algebras [20]), G-algebras and Heyting algebras, respectively. We refer to [17] for a comprehensive overview of these and other logics. Other general references on fuzzy logics are [6, 23, 24].

In [43], Zadeh introduced type-2 fuzzy sets, a generalization of fuzzy sets. The idea behind these structures is that they provide a way to express incomplete as well as graded knowledge; as opposed to fuzzy sets, which only express gradedness, not incompleteness. Unfortunately, type-2 fuzzy sets are quite complicated to work with. Therefore often interval-valued fuzzy sets are used. These special cases of type-2 fuzzy sets are easier to handle. Indeed, truth values in this setting are closed subintervals of the unit interval, and such an interval is determined by just two values: its lower and upper bound. The aim of this chapter is to develop a logic that has intervals as truth values. The intended semantics are residuated lattices on the set of closed subintervals of the unit interval. We call this set the triangularization of the unit interval. A particular subset of this triangularization is its so-called diagonal, consisting of those intervals for which the lower and upper bound coincide. These intervals are called exact intervals and represent truth values of propositions about which the knowledge is complete. Intuitively, the truth values of formulas constructed with these propositions should be exact intervals as well (because in these cases, the situation is similar to working with formulas in fuzzy logics). The semantics of so-called interval-valued fuzzy logics have already been examined by different authors. Especially interval-valued triangular norms, triangular conorms and implicators have received ample attention. Most of these authors [1, 2, 15, 21, 26] only consider interval-valued operations that map the diagonal on the diagonal, although the most general definitions of triangular norms, triangular conorms and implicators allow other operations as well [7, 10, 11, 28]. Generally speaking, interval-valued operations do not satisfy as many properties as operations on the unit interval. For example, standard interval-valued residuated lattices can never satisfy prelinearity [8]. A lot of other properties can hold though. There are even interval-valued implicators that satisfy all the Smets-Magrez axioms [9].

The three main sections of this chapter are conceived as follows:

- In Section 2 we elaborate the theory of interval-valued residuated lattices, which include the intended semantics of interval-valued fuzzy logics.
- In Section 3 we give the definition of triangle algebras, which are algebraic structures describing interval-valued residuated lattices.
- In Section 4 we then introduce several interval-valued fuzzy logics and examine their properties, in particular the soundness and completeness w.r.t. the intended and the general semantics.

Before we continue, we recall some algebraic concepts that will be used in this chapter.

- An algebra of type $(n_1, n_2, \ldots, n_m)$, with $n_1$, $n_2$, ..., $n_m$ non-negative integers, is a structure $(A, f_1, \ldots, f_m)$ in which $A$ is a set, $f_1$ an $n_1$-ary operation on $A$, ... and $f_m$ an $n_m$-ary operation on $A$. If $n_i$ is 0, then $f_i$ is a constant.
- A reduct of an algebra is an algebra on the same set, but in which some of the operations are left out. An algebra $\mathscr{A}$ is an expansion of an algebra $\mathscr{B}$ if $\mathscr{B}$ is a reduct of $\mathscr{A}$.
- A subalgebra of an algebra $\mathscr{A} = (A, f_1, \ldots, f_m)$ is an algebra on a subset $A'$ of $A$ in which all operations of $\mathscr{A}$ are restricted to $A'$. Of course, this is only possible if $A'$ is closed under all these operations, i.e., if for every operation $f_i$ of $\mathscr{A}$, $f_i(a_1, \ldots, a_n) \in A'$ whenever the arguments $a_1$, ..., $a_n$ are in $A'$ (with $n$ the arity of $f_i$).
- A morphism from an algebra $\mathscr{A} = (A, f_1, \ldots, f_m)$ to an algebra $\mathscr{B} = (B, g_1, \ldots, g_m)$ of the same type, is a mapping $h$ from $A$ to $B$ such that $h(f_i(a_1, \ldots, a_n)) = g_i(h(a_1), \ldots, h(a_n))$ for all operations $f_i$ of $\mathscr{A}$ and all $a_1$, ..., $a_n$ in $A$ (with $n$ the arity of $f_i$).
- An embedding of an algebra $\mathscr{A} = (A, f_1, \ldots, f_m)$ in an algebra $\mathscr{B} = (B, g_1, \ldots, g_m)$ of the same type, is a morphism $h$ from $A$ to $B$ such that $h(a_1) \neq h(a_2)$ whenever $a_1 \neq a_2$.
- An isomorphism from an algebra $\mathscr{A} = (A, f_1, \ldots, f_m)$ to an algebra $\mathscr{B} = (B, g_1, \ldots, g_m)$ of the same type, is an embedding of $A$ in $B$ such that for every element $b$ of $B$, $b = h(a)$ for some $a$ in $A$.

## 2 Interval-Valued Structures

The most general semantics of fuzzy logics do not only contain algebraic structures on the unit interval, it consists of all residuated lattices. For interval-valued fuzzy logics, the situation is comparable: the most general semantics are interval-valued residuated lattices. In this section, we propose a definition of these structures.

### 2.1 *Triangularizations of Partially Ordered Sets*

**Definition 1.** [39] Given any partially ordered set (shortly: poset) $\mathscr{P} = (P, \leq)$, we can define its triangularization $\mathbb{T}(\mathscr{P}) = (Int(\mathscr{P}), \preceq)$ in the following way:

- $Int(\mathscr{P}) = \{[p_1, p_2] \mid (p_1, p_2) \in P^2 \text{ and } p_1 \leq p_2\}$,
- $[p_1, p_2] \preceq [q_1, q_2]$ iff $p_1 \leq q_1$ and $p_2 \leq q_2$, for all $[p_1, p_2]$ and $[q_1, q_2]$ in $Int(\mathscr{P})$.

The elements of $Int(\mathscr{P})$ are called the intervals of $\mathscr{P}$.
The first and the second projection $\mathrm{pr}_1$ and $\mathrm{pr}_2$ are the mappings from $Int(\mathscr{P})$ to $P$, defined by $\mathrm{pr}_1([x_1, x_2]) = x_1$ and $\mathrm{pr}_2([x_1, x_2]) = x_2$, for all $[x_1, x_2]$ in $Int(\mathscr{P})$.
The vertical and the horizontal projection $\mathrm{pr}_v$ and $\mathrm{pr}_h$ are the mappings from $Int(\mathscr{P})$ to $Int(\mathscr{P})$, defined by $\mathrm{pr}_v([x_1, x_2]) = [x_1, x_1]$ and $\mathrm{pr}_h([x_1, x_2]) = [x_2, x_2]$, for all $[x_1, x_2]$ in $Int(\mathscr{P})$.

It is straightforward to verify that for any poset $\mathscr{P}$, $\mathbb{T}(\mathscr{P})$ is also a poset. Moreover, the original poset $(P, \leq)$ is contained in $\mathbb{T}(\mathscr{P})$ in some way: indeed, the mapping $i : P \to Int(\mathscr{P})$ defined by $i(p) = [p, p]$ for all $p$ in $P$, is injective and preserves the ordering (if $p \leq q$, then $i(p) = [p, p] \preceq [q, q] = i(q)$). The image $i(P)$ consists of the intervals $[p_1, p_2]$ in $Int(\mathscr{P})$ for which $p_1 = p_2$. The elements of $i(P)$ are called exact intervals. The subset $i(P)$ of $Int(\mathscr{P})$ is often referred to as the diagonal of $\mathbb{T}(\mathscr{P})$. Note that $\mathrm{pr}_v = i \circ \mathrm{pr}_1$ and $\mathrm{pr}_h = i \circ \mathrm{pr}_2$, and that $i(P) = \mathrm{pr}_v(Int(\mathscr{P})) = \mathrm{pr}_h(Int(\mathscr{P}))$.

*Example 1.* The poset that will be of central interest in this chapter is $\mathbb{T}([0, 1], \leq)$: the closed subintervals of the unit interval. This poset is complete and its order is not linear. Its graphical representation as a triangle is shown in Figure 1. The diagonal is the hypothenuse of this triangle. Note that the shape of this representation is triangular. This holds for all triangularizations of bounded linear posets, hence the name 'triangularization'.

### 2.2 *Triangular Lattices*

Recall that a lattice is a poset in which the supremum and infimum of every two elements exist; on the other hand, often the following equivalent definition is also used.

**Definition 2.** A lattice is an algebra $(L, \sqcap, \sqcup)$ of type $(2, 2)$ such that $\sqcap$ ('meet') and $\sqcup$ ('join') are idempotent, commutative and associative operations satisfying the following absorption laws: for all $x$ and $y$ in $L$, $x \sqcup (x \sqcap y) = x$ and $x \sqcap (x \sqcup y) = x$. The lattice order $\leq$ is defined by $x \leq y$ iff $x \sqcap y = x$ (or, equivalently, iff $x \sqcup y = y$), for all $x$ and $y$ in $L$.

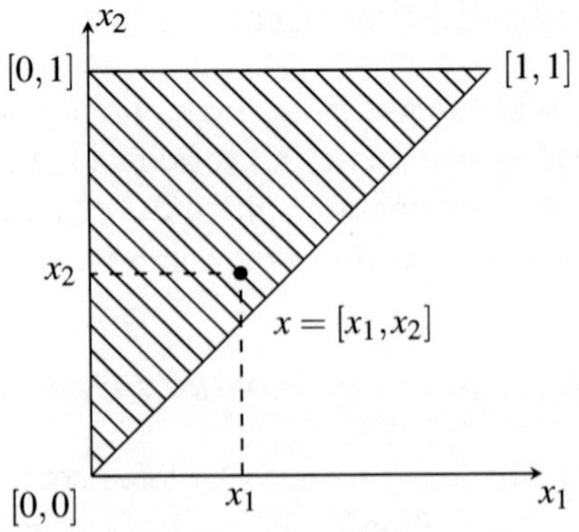


**Fig. 1** The lattice $\mathbb{T}([0,1],\leq)$

Because lattices can be seen as posets, we can consider their triangularizations. One easily observes that the infimum (resp. supremum) on the triangularization of a lattice is obtained by taking the infimum (resp. supremum) of the first and second projections. More precisely: for any lattice $\mathscr{L} = (L,\sqcap,\sqcup)$, the infimum $\bigsqcap$ and supremum $\bigsqcup$ on its triangularization $\mathbb{T}(\mathscr{L})$ are given by

- $[x_1,x_2]\bigsqcap[y_1,y_2] = [x_1 \sqcap y_1, x_2 \sqcap y_2]$,
- $[x_1,x_2]\bigsqcup[y_1,y_2] = [x_1 \sqcup y_1, x_2 \sqcup y_2]$,

for all $[x_1,x_2]$ and $[y_1,y_2]$ in $Int(\mathscr{L})$. Note that we use big $\bigsqcap$- and $\bigsqcup$-symbols for the intervals of the triangularization, and small $\sqcap$- and $\sqcup$-symbols for the elements of the original lattice.

It can be verified that $\mathbb{T}(\mathscr{L})$ is a lattice iff $\mathscr{L}$ is a lattice, that $i$ (as defined in Section 2.1) is a morphism from $(L,\sqcap,\sqcup)$ to $(Int(\mathscr{L}),\bigsqcap,\bigsqcup)$, and that the set $i(L)$ of exact intervals is therefore closed under $\bigsqcap$ and $\bigsqcup$ and forms a sublattice $\mathscr{E}(\mathscr{L}) = (i(L),\bigsqcap,\bigsqcup)$.

On the other hand, $\mathbb{T}(\mathscr{L})$ is bounded iff $\mathscr{L}$ is bounded. In this case, the smallest (resp. greatest) element of $\mathscr{L}$ is usually denoted by 0 (resp. 1), and the smallest (resp. greatest) element of $\mathbb{T}(\mathscr{L})$ by $[0,0]$ (resp. $[1,1]$). As we will see later on, the element $[0,1]$ will also play an important role, along with the projections $\mathrm{pr}_v$ and $\mathrm{pr}_h$.

For any triangularization $(Int(\mathscr{L}),\bigsqcap,\bigsqcup)$ of a bounded lattice $\mathscr{L} = (L,\sqcap,\sqcup)$ (with smallest element 0 and greatest element 1), we call $(Int(\mathscr{L}),\bigsqcap,\bigsqcup,\mathrm{pr}_v,\mathrm{pr}_h,$ $[0,0],[0,1],[1,1])$ the extended triangularization of $\mathscr{L}$. Below, we show that extended triangularizations can be captured by a class of algebraic structures defined only with identities: the variety of triangular lattices.

**Definition 3.** [36] A triangular lattice[2] is an algebra $(L,\sqcap,\sqcup,\nu,\mu,0,u,1)$ of type $(2,2,1,1,0,0,0)$ such that $(L,\sqcap,\sqcup)$ is a bounded lattice with smallest element 0 and greatest element 1 such that

[2] The reason we call the last two conditions (T.10) and (T.10') instead of (T.7) and (T.7') is that we would like to keep the same notations as in the papers [36, 37, 38, 40, 42].

$$\begin{array}{llll}
(T.1) & \nu x \leq x, & (T.1') & x \leq \mu x, \\
(T.2) & \nu x \leq \nu \nu x, & (T.2') & \mu \mu x \leq \mu x, \\
(T.3) & \nu(x \sqcap y) = \nu x \sqcap \nu y, & (T.3') & \mu(x \sqcap y) = \mu x \sqcap \mu y, \\
(T.4) & \nu(x \sqcup y) = \nu x \sqcup \nu y, & (T.4') & \mu(x \sqcup y) = \mu x \sqcup \mu y, \\
(T.5) & \nu u = 0, & (T.5') & \mu u = 1, \\
(T.6) & \nu \mu x = \mu x, & (T.6') & \mu \nu x = \nu x, \\
(T.10) & x = \nu x \sqcup (\mu x \sqcap u), & (T.10') & x = \mu x \sqcap (\nu x \sqcup u).
\end{array}$$

The unary operators $\nu$ and $\mu$ are called the necessity and possibility operator, respectively.

Note that from (T.10) it follows that for all $x$ and $y$ in $L$,

$$x = y \text{ whenever } \nu x = \nu y \text{ and } \mu x = \mu y. \qquad (1)$$

From (T.1) and (T.2), it is clear that in a triangular lattice, always $\nu\nu x = \nu x$. Similarly, $\mu\mu x = \mu x$. Each of (T.3) and (T.4) implies that $\nu$ is an increasing operator. In the same way, (T.3') or (T.4') force $\mu$ to be increasing too. Other properties that follow easily are $\nu 1 = 1$ and $\mu 0 = 0$: $\nu 1 = \nu \mu u = \mu u = 1$ and $\mu 0 = \mu \nu u = \nu u = 0$. Together with (T.1), (T.1'), (T.2), (T.2'), (T.3) and (T.4'), they mean that $\mu$ is a closure operator, and $\nu$ is an interior operator. Both are also lattice morphisms.

Note that (T.1')–(T.4') are conditions for $\mu$, which is similar to the modal possibility operator; they are dual to (T.1)–(T.4) for $\nu$, which is similar to the modal necessity operator. Only (T.4) and (T.3') are different: in the modal setting, they are in general not true; and one doesn't want them to be true either (see e.g. [44]). In general, we do not require dependency of $\mu$ on $\nu$. The conditions (T.5) and (T.5') express the complete lack of knowledge about $u$: its necessity is 0, but its possibility is 1. The conditions (T.6) and (T.6') are known in modal logics as the S5-principles [31, 32].

**Proposition 1.** *[36] Let* $(L, \sqcap, \sqcup, \nu, \mu, 0, u, 1)$ *be a triangular lattice. Then* $(L, \sqcap, \sqcup, \nu, \mu, 0, u, 1)$ *is isomorphic to the extended triangularization of a bounded lattice. Conversely, every extended triangularization of a bounded lattice is a triangular lattice.*

In the remainder, we will use the set of exact elements $E(\mathscr{L})$ of a triangular lattice $\mathscr{L} = (L, \sqcap, \sqcup, \nu, \mu, 0, u, 1)$ defined by $\{x \in L \mid \nu x = x\}$. It is closed under all the defined (unary and binary) operators, and therefore $\mathscr{E}(\mathscr{L}) = (E(\mathscr{L}), \sqcap, \sqcup)$ (in which the binary operators are restricted to $E(\mathscr{L})$) is a bounded lattice.

## 2.3 Interval-Valued Residuated Lattices

For most formal fuzzy logics the semantics require not only a partial order on the set of truth values, but also some extra operations that model 'AND' (the strong conjunction - the infimum being the weak conjunction) and 'IMPLIES' (the implication). A very commonly used structure – and also the basic structure that we will use in this work – is that of residuated lattices.

**Definition 4.** A residuated lattice[3] is a structure $\mathscr{L} = (L, \sqcap, \sqcup, *, \Rightarrow, 0, 1)$ in which $\sqcap$, $\sqcup$, $*$ and $\Rightarrow$ are binary operators on the set $L$ and

- $(L, \sqcap, \sqcup)$ is a bounded lattice with 0 as smallest and 1 as greatest element,
- $*$ is commutative and associative, with 1 as neutral element, and
- $x * y \leq z$ iff $x \leq y \Rightarrow z$ for all $x$, $y$ and $z$ in $L$ (residuation principle).

The binary operations $*$ and $\Rightarrow$ are called product and implication, respectively. We will use the notations $\neg x$ for $x \Rightarrow 0$ (negation), $x \Leftrightarrow y$ for $(x \Rightarrow y) \sqcap (y \Rightarrow x)$.

The following kinds of residuated lattices are used in this chapter.

**Definition 5**

- An MTL-algebra [16] is a prelinear residuated lattice, i.e., a residuated lattice in which $(x \Rightarrow y) \sqcup (y \Rightarrow x) = 1$ for all $x$ and $y$ in $L$.
- A BL-algebra [24] is a divisible MTL-algebra, i.e., an MTL-algebra in which $x \sqcap y = x * (x \Rightarrow y)$ for all $x$ and $y$ in $L$. The weaker property $x \sqcap y = (x * (x \Rightarrow y)) \sqcup (y * (y \Rightarrow x))$ is called weak divisibility.
- An MV-algebra [3, 4] is a BL-algebra in which the negation is an involution, i.e., $(x \Rightarrow 0) \Rightarrow 0 = x$ for all $x$ in $L$.
- A Boolean algebra [27] is an MV-algebra that is also a Heyting-algebra, i.e., in which $x * x = x$ for all $x$ in $L$, or, equivalently, in which $* = \sqcap$.

In a residuated lattice, the operator $*$ is always a residuated t-norm, with $\Rightarrow$ as its residual implicator[4]. Conversely, if $T$ is a residuated t-norm on a bounded lattice $(L, \sqcap, \sqcup)$, then $(L, \sqcap, \sqcup, T, I_T, 0, 1)$ is a residuated lattice. Note however that not all t-norms are residuated. In complete lattices $(L, \sqcap, \sqcup)$, a t-norm $T$ is residuated (and therefore induces a residuated lattice) iff it satisfies $T(x, \sup Y) = \sup\{T(x,y) \mid y \in Y\}$, for all $x$ in $L$ and $Y \subseteq L$ [35].

*Example 2.* Let $T$ be a t-norm on $([0,1], \min, \max)$. It is well-known (see, e.g., [17, 24] that

- $T$ is residuated iff $T$ is left-continuous,
- $([0,1], \min, \max, T, I_T, 0, 1)$ is an MTL-algebra[5] iff $T$ is left-continuous,
- $([0,1], \min, \max, T, I_T, 0, 1)$ is a BL-algebra iff $T$ is continuous,

---

[3] In the literature (e.g. in [26]), the name residuated lattice is sometimes used for structures more general than what we call residuated lattices. In the most general terminology, our structures would be called bounded integral commutative residuated lattices.

[4] Recall that a triangular norm (t-norm, for short) on a poset $(P, \leq)$ with largest element 1, is a binary, increasing, commutative and associative operator $T : P^2 \to P$ that satisfies $T(x, 1) = 1$, for all $x$ in $P$. If for every pair $(x, y)$ in $P^2$, $\sup\{z \in P \mid T(x,z) \leq y\}$ exists, then the map $I_T$ defined by $I_T(x,y) = \sup\{z \in P \mid T(x,z) \leq y\}$ is called the residual implicator of $T$. A t-norm $T$ is called residuated if it has a residual implicator satisfying $I_T(x,y) = \max\{z \in P \mid T(x,z) \leq y\}$, in other words if for any pair $(x, y)$ in $P^2$ the set $\{z \in P \mid T(x,z) \leq y\}$ has a maximum.

[5] Because $([0,1], \min, \max)$ is linear, every residuated lattice on this lattice is automatically an MTL-algebra.

- $([0,1],\min,\max,T,I_T,0,1)$ is an MV-algebra iff $T$ is conjugated to the Łukasiewicz t-norm $T_W$, i.e., iff there exists a strictly increasing bijection $\phi : [0,1] \to [0,1]$ such that $T(x,y) = \phi^{-1}(T_W(\phi(x),\phi(y)))$, where $T_W(x,y) = \max(0,x+y-1)$.

Several t-norms can be defined on triangularizations of bounded lattices. A class that will be of great importance later on, consists of t-norms constructed from a t-norm on and an element of the original lattice: if $T$ is a t-norm on a lattice $\mathscr{L} = (L,\sqcap,\sqcup)$ which has a greatest element 1, and $t$ is an element of $L$, then $\mathscr{T}_{T,t}$, defined by

$$\mathscr{T}_{T,t}([x_1,x_2],[y_1,y_2]) = [T(x_1,y_1),T(T(x_2,y_2),t) \sqcup T(x_1,y_2) \sqcup T(x_2,y_1)], \quad (2)$$

for all $[x_1,x_2]$ and $[y_1,y_2]$ in $Int(\mathscr{L})$, is a t-norm on $\mathbb{T}(\mathscr{L})$. These t-norms were introduced by Deschrijver and Kerre in [11].

Recall that we are working towards a variety of algebraic structures suitable as semantics for a logic with intervals as truth values. At this point it might seem a good idea to choose residuated lattices on triangularizations (or, equivalently, residuated lattices on triangular lattices). However, in these structures the set of exact intervals is not necessarily closed under the product and implication. This counters the intuition that the truth values of the propositions $p\&q$ and $p \to q$ should be exact if the truth values of $p$ and $q$ are exact. Therefore, residuated lattices on triangularizations are too general to serve as the desired semantics. This leads us to the definition of interval-valued residuated lattices.

**Definition 6.** [41]

- An interval-valued residuated lattice (IVRL) is a residuated lattice $(Int(\mathscr{L}),\bigsqcap,\bigsqcup,*,\Rightarrow,[0,0],[1,1])$ on the triangularization $\mathbb{T}(\mathscr{L})$ of a bounded lattice $\mathscr{L} = (L,\sqcap,\sqcup)$, in which the diagonal $i(L)$ is closed under $*$ and $\Rightarrow$, i.e., $[x_1,x_1] * [y_1,y_1] \in i(L)$ (with $i$ the injection defined in Section 2.1) and $[x_1,x_1] \Rightarrow [y_1,y_1] \in i(L)$ for all $x_1$ and $y_1$ in $L$.
- When we add $[0,1]$ as a constant, and $\mathrm{pr}_v$ and $\mathrm{pr}_h$ (as defined in Section 2.1) as unary operators, the structure $(Int(\mathscr{L}),\bigsqcap,\bigsqcup,*,\Rightarrow,\mathrm{pr}_v,\mathrm{pr}_h,[0,0],[0,1],[1,1])$ is called an extended IVRL.

An IVRL in which $\mathscr{L} = ([0,1],\min,\max)$ is called a *standard IVRL*. An extended IVRL in which $\mathscr{L} = ([0,1],\min,\max)$ is called a *standard extended IVRL*.

By Proposition 1, if $(Int(\mathscr{L}),\bigsqcap,\bigsqcup,*,\Rightarrow,\mathrm{pr}_v,\mathrm{pr}_h,[0,0],[0,1],[1,1])$ is an extended IVRL, then it is also a triangular lattice. Now we show that the extra operations $*$ and $\Rightarrow$ satisfy the following two properties, for all $x$ and $y$ in $Int(\mathscr{L})$:

- $\mathrm{pr}_v(x) * \mathrm{pr}_v(y) \le \mathrm{pr}_v(\mathrm{pr}_v(x) * \mathrm{pr}_v(y))$,
- $\mathrm{pr}_v(x) \Rightarrow \mathrm{pr}_v(y) \le \mathrm{pr}_v(\mathrm{pr}_v(x) \Rightarrow \mathrm{pr}_v(y))$.

Indeed, the first property is equivalent to $\mathrm{pr}_v(x) * \mathrm{pr}_v(y) = \mathrm{pr}_v(\mathrm{pr}_v(x) * \mathrm{pr}_v(y))$, which means that for any $a$ and $b$ in $i(L)$, $a * b = \mathrm{pr}_v(a * b)$, in other words $a * b \in i(L)$. So it tells us exactly that the diagonal $i(L)$ is closed under $*$. And similarly, the second property means the diagonal $i(L)$ is closed under $\Rightarrow$ too.

These two properties suggest a way to describe these IVRLs, which seem suitable as semantics for interval-valued fuzzy logic, using only identities. This leads us to the next section, where we will introduce this variety (called triangle algebras) and study its properties in detail.

## 3 Triangle Algebras

In this section, we introduce a variety of structures called triangle algebras, and show that they are isomorphic to extended IVRLs. Then, we investigate the product and implication of triangle algebras and show that they are determined by their action on the exact elements and by one specific product: $u * u$. This characterization is used to uncover the connections between properties on triangle algebras and properties on their subalgebras of exact elements.

### 3.1 Definition and Connection with IVRLs

In the definition of a triangle algebra we want to combine the structure of a residuated lattice and the structure of intervals (equipped with the order in Definition 1), plus the desired property that the subset of exact intervals is closed under all defined operations. This leads us to the following definition.

**Definition 7.** [41, 36] A triangle algebra is a structure $(A,\sqcap,\sqcup,*,\Rightarrow,\nu,\mu,0,u,1)$ of type $(2,2,2,2,1,1,0,0,0)$ such that $(A,\sqcap,\sqcup,\nu,\mu,0,u,1)$ is a triangular lattice, $(A,\sqcap,\sqcup,*,\Rightarrow,0,1)$ is a residuated lattice, and satisfying for all $x$ and $y$ in $A$,

$$(T.7')\quad \nu x * \nu y \leq \nu(\nu x * \nu y),$$
$$(T.9)\quad \nu x \Rightarrow \nu y \leq \nu(\nu x \Rightarrow \nu y).$$

In other words, a structure $(A,\sqcap,\sqcup,*,\Rightarrow,\nu,\mu,0,u,1)$ of type $(2,2,2,2,1,1,0,0,0)$ such that $(A,\sqcap,\sqcup,*,\Rightarrow,0,1)$ is a residuated lattice, and satisfying for all $x$ and $y$ in $A$,

$$
\begin{array}{llll}
(T.1) & \nu x \leq x, & (T.1') & x \leq \mu x,\\
(T.2) & \nu x \leq \nu\nu x, & (T.2') & \mu\mu x \leq \mu x,\\
(T.3) & \nu(x\sqcap y) = \nu x \sqcap \nu y, & (T.3') & \mu(x\sqcap y) = \mu x \sqcap \mu y,\\
(T.4) & \nu(x\sqcup y) = \nu x \sqcup \nu y, & (T.4') & \mu(x\sqcup y) = \mu x \sqcup \mu y,\\
(T.5) & \nu u = 0, & (T.5') & \mu u = 1,\\
(T.6) & \nu\mu x = \mu x, & (T.6') & \mu\nu x = \nu x,\\
(T.7\dagger) & \nu x * \nu y \leq \nu(\nu x * \nu y), & & \\
(T.9) & \nu x \Rightarrow \nu y \leq \nu(\nu x \Rightarrow \nu y), & & \\
(T.10) & x = \nu x \sqcup (\mu x \sqcap u), & (T.10') & x = \mu x \sqcap (\nu x \sqcup u).
\end{array}
$$

A triangle algebra $(A,\sqcap,\sqcup,*,\Rightarrow,\nu,\mu,0_A,u_A,1_A)$ is called a standard triangle algebra iff $(A,\sqcap,\sqcup) = \mathbb{T}([0,1],\min,\max)$.

In a standard triangle algebra $(A,\sqcap,\sqcup,*,\Rightarrow,\nu,\mu,0_A,u_A,1_A)$ it holds that $0_A = [0,0]$, $1_A = [1,1]$, $u = [0,1]$, $\nu[x_1,x_2] = [x_1,x_1]$ and $\mu[x_1,x_2] = [x_2,x_2]$ for all $[x_1,x_2]$ in $Int([0,1],\min,\max)$ [41]. Because a triangle algebra is an expansion of both a triangular lattice and a residuated lattice, the properties of these kinds of structures remain valid in triangle algebras. The connections between triangle algebras and several related algebraic structures from the literature are studied extensively in [39, 41, 36]. In this chapter, we focus on the relationship with IVRLs.

**Proposition 2.** *[41] Let $(Int(\mathscr{L}),\sqcap,\sqcup,*,\Rightarrow,\mathrm{pr}_v,\mathrm{pr}_h,[0,0],[0,1],[1,1])$ be an extended IVRL. Then $(Int(\mathscr{L}),\sqcap,\sqcup,*,\Rightarrow,\mathrm{pr}_v,\mathrm{pr}_h,[0,0],[0,1],[1,1])$ is a triangle algebra. Conversely, let $\mathscr{A} = (A,\sqcap,\sqcup,*,\Rightarrow,\nu,\mu,0,u,1)$ be a triangle algebra. Then $\mathscr{A}$ is isomorphic to an extended IVRL.*

The isomorphism $\chi$ that is used in the proof of Proposition 2 is depicted graphically in Figure 2.

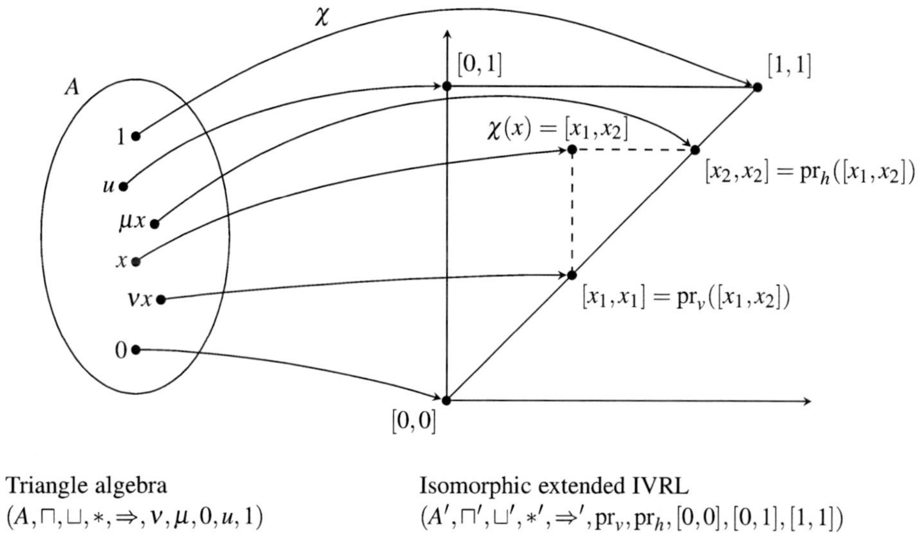


**Fig. 2** The isomorphism $\chi$ from a triangle algebra to an extended IVRL

## *3.2 Characterization of Product and Implication. Decomposition Theorem.*

The following important proposition reveals that the implication $\Rightarrow$ and the product $*$ are completely determined by their action on the diagonal and the value of $u * u$:

**Proposition 3.** *[40] In a triangle algebra* $\mathscr{A} = (A, \sqcap, \sqcup, *, \Rightarrow, \nu, \mu, 0, u, 1)$, *it holds that*

- $\nu(x \Rightarrow y) = (\nu x \Rightarrow \nu y) \sqcap (\mu x \Rightarrow \mu y)$,
- $\mu(x \Rightarrow y) = (\mu x \Rightarrow (\mu(u * u) \Rightarrow \mu y)) \sqcap (\nu x \Rightarrow \mu y)$,
- $\nu(x * y) = \nu x * \nu y$,
- $\mu(x * y) = (\nu x * \mu y) \sqcup (\mu x * \nu y) \sqcup (\mu x * \mu y * \mu(u * u))$

*and therefore (by (T.10) and (T.10'))*

$$\begin{aligned}
&x \Rightarrow y \\
&= \Big(\Big(\big(\mu x \Rightarrow (\mu(u * u) \Rightarrow \mu y)\big) \sqcap (\nu x \Rightarrow \mu y)\Big) \sqcap u\Big) \sqcup \Big((\mu x \Rightarrow \mu y) \sqcap (\nu x \Rightarrow \nu y)\Big) \\
&= \Big(((\mu x \Rightarrow \mu y) \sqcap (\nu x \Rightarrow \nu y)) \sqcup u\Big) \sqcap \Big(\big(\mu x \Rightarrow (\mu(u * u) \Rightarrow \mu y)\big) \sqcap (\nu x \Rightarrow \mu y)\Big)
\end{aligned}$$

*and*

$$\begin{aligned}
x * y &= \Big(\big((\nu x * \mu y) \sqcup (\mu x * \nu y) \sqcup (\mu x * \mu y * \mu(u * u))\big) \sqcap u\Big) \sqcup (\nu x * \nu y) \\
&= \Big((\nu x * \nu y) \sqcup u\Big) \sqcap \Big((\nu x * \mu y) \sqcup (\mu x * \nu y) \sqcup (\mu x * \mu y * \mu(u * u))\Big).
\end{aligned}$$

Because of Proposition 3, the product and implication in triangle algebras are always of a specific form, which implies that a triangle algebra is completely determined by its subalgebra of exact elements and the value $u * u$ (in triangle algebras, $u$ is a constant, playing the role of the interval $[0, 1]$ in IVRLs). Conversely, for a fixed residuated lattice $\mathscr{L}$ and element $\alpha$ in that lattice, we can construct a triangle algebra with $\mathscr{L}$ as subalgebra of exact elements and $u * u$ determined by $\mu(u * u) = \alpha$. So we can conclude that there is a one-to-one correspondence between triangle algebras and couples $(\mathscr{L}, \alpha)$, in which $\alpha$ is an element of the residuated lattice $\mathscr{L}$. This characterization implies that every property that can be imposed on triangle algebras, can be formulated in terms of such couples.

In particular, any property defined in Definition 5 can be weakened, by imposing it on $E(\mathscr{A})$ (instead of $\mathscr{A}$) only. We will denote this with the prefix 'pseudo'. For example, a triangle algebra is said to be pseudo-linear if its set of exact elements is linearly ordered (by the original ordering, restricted to the diagonal). Another example: a triangle algebra is pseudo-divisible if $\nu x \sqcap \nu y = \nu x * (\nu x \Rightarrow \nu y)$ for all $x$ and $y$ in $A$ ($E(\mathscr{A})$ consists exactly of the elements of the form $\nu x$).

It is well-known that MTL-algebras are isomorphic to subdirect products of linear residuated lattices [16, 24]. This is a very useful result, as it implies that identities valid in all linear residuated lattices are also valid in all MTL-algebras, which significantly simplifies several proofs, and which is also needed for the chain-completeness of the corresponding logic MTL (see Section 4). The 'interval-valued counterpart' of this result is given below:

**Theorem 1.** *[42] Every pseudo-prelinear triangle algebra $\mathscr{A}$ is isomorphic to a subalgebra of the direct product of a system of pseudo-linear triangle algebras.*

# 4 Interval-Valued Fuzzy Logics

In Section 2 we have given the definition of interval-valued residuated lattices (IVRLs), and in Section 3 we have introduced triangle algebras to capture their structure by means of identities and/or inequalities. Using this characterization, we give the definition of several (propositional) interval-valued fuzzy logics in Section 4.2. But first, in Section 4.1 we give an overview of the well-studied common fuzzy logics, on which our interval-valued fuzzy logics are based, and mention their most important properties. In Section 4.3 we then investigate which of these properties hold for interval-valued fuzzy logics as well. In particular, we prove the soundness and completeness with respect to the algebraic semantics and the deduction theorem.

## *4.1 Formal Fuzzy Logics*

Because in interval-valued fuzzy logics there will be more formulae, we make a distinction between formulae for fuzzy logics (FL-formulae) and formulae for interval-valued fuzzy logics (IVFL-formulae).

**Definition 8.** FL-formulae are built up from a countable set of propositional variables (denoted by $p$, $q$, $r$, $p_1$, $p_2$, $\ldots$) and the constant $\overline{0}$. These symbols are FL-formulae by definition. The other FL-formulae are defined recursively: if $\phi$ and $\psi$ are FL-formulae, then so are $(\phi \wedge \psi)$, $(\phi \vee \psi)$, $(\phi \& \psi)$ and $(\phi \rightarrow \psi)$.
The set of FL-formulae is denoted by $\mathscr{F}_{FL}$.

In order to avoid unnecessary brackets, we agree on the following priority rules:

- among the connectives, & has the highest priority; furthermore $\wedge$ and $\vee$ take precedence over $\rightarrow$,
- the outermost brackets are not written.

The following notations are used: $\overline{1}$ for $\overline{0} \rightarrow \overline{0}$, $\neg\phi$ for $\phi \rightarrow \overline{0}$, $\phi^2$ for $\phi\&\phi$, $\phi^n$ (with $n \in \{3,4,5,\ldots\}$) for $(\phi^{n-1})\&\phi$ (moreover, $\phi^0$ is $\overline{1}$ and $\phi^1$ is $\phi$), and $\phi \leftrightarrow \psi$ for $(\phi \rightarrow \psi) \wedge (\psi \rightarrow \phi)$, for FL-formulae $\phi$ and $\psi$.

The FL-formulae $\phi\&\psi$, $\phi \rightarrow \psi$ and $\neg\phi$ stand for what we understand intuitively by '$\phi$ and $\psi$' (strong conjunction), '$\phi$ implies $\psi$' (implication) and 'not $\phi$' (negation).

It is impossible to list all true FL-formulae of a specific fuzzy logic, because their number is not finite. Therefore axioms and deduction rules are used. In the logics we deal with, an FL-formula is true if it is provable from the axioms using the deduction rules. We will explain this in more detail. This method also allows to prove FL-formulae from a given set of FL-formulae (usually called a theory). This means that in the proof of an FL-formula not only axioms of the logic can be used but also formulae of the theory. If an FL-formula $\phi$ is provable from a theory $\Gamma$ in a fuzzy logic $L$, this is denoted as $\Gamma \vdash_L \phi$. The relation $\vdash$ is called provability relation or syntactic consequence.

It is often not very easy to find out if an FL-formula is true in a specific fuzzy logic. A proof might be difficult to find and such a proof can become very long. This is why soundness and completeness of (fuzzy) logics is so important. It provides a way to determine if a formula is true or provable from a theory in a purely algebraic way. Indeed, soundness and completeness of a fuzzy logic are two properties relative to a class of algebraic structures. We call such a class a semantics of the fuzzy logic. To explain this connection between formal logic and algebra in more detail, we need some terminology first.

**Definition 9.** Let $\mathscr{L} = (L, \sqcap, \sqcup, *, \Rightarrow, 0, 1)$ be a residuated lattice, $\Gamma$ a theory (i.e., a set of FL-formulae). An $\mathscr{L}$-evaluation is a mapping $e$ from the set of FL-formulae[6] to $L$ that satisfies, for each two formulae $\phi$ and $\psi$:

- $e(\phi \wedge \psi) = e(\phi) \sqcap e(\psi)$,
- $e(\phi \vee \psi) = e(\phi) \sqcup e(\psi)$,
- $e(\phi \& \psi) = e(\phi) * e(\psi)$,
- $e(\phi \rightarrow \psi) = e(\phi) \Rightarrow e(\psi)$ and
- $e(\overline{0}) = 0$.

If an $\mathscr{L}$-evaluation $e$ satisfies $e(\chi) = 1$ for every $\chi$ in $\Gamma$, it is called an $\mathscr{L}$-model[7] for $\Gamma$. We write $\Gamma \models_{\mathscr{L}} \phi$ if $e(\phi) = 1$ for all $\mathscr{L}$-models $e$ for $\Gamma$. If $\Gamma$ is empty, we simply write $\models_{\mathscr{L}} \phi$ instead of $\emptyset \models_{\mathscr{L}} \phi$. FL-formulae $\phi$ for which $\models_{\mathscr{L}} \phi$ are called $\mathscr{L}$-tautologies. The relation $\models$ is called semantic consequence.

Evaluations form a connection between the connectives of the logic and the algebraic operators in residuated lattices. Note that $e(\overline{1}) = e(\overline{0} \rightarrow \overline{0}) = e(\overline{0}) \Rightarrow e(\overline{0}) = 0 \Rightarrow 0 = 1$ and $e(\neg\phi) = e(\phi \rightarrow \overline{0}) = e(\phi) \Rightarrow e(\overline{0}) = \neg e(\phi)$.

Now let $\mathscr{C}$ be a class of residuated lattices and $L$ a fuzzy logic.

- We say $L$ is sound w.r.t. $\mathscr{C}$ if for all $\Gamma \subseteq \mathscr{F}_{FL}$ and $\phi \in \mathscr{F}_{FL}$, $\Gamma \vdash_L \phi$ implies $\Gamma \models_{\mathscr{L}} \phi$ for all $\mathscr{L}$ in $\mathscr{C}$.
- We say $L$ is complete[8] w.r.t. $\mathscr{C}$ if for all $\phi \in \mathscr{F}_{FL}$, ($\models_{\mathscr{L}} \phi$ for all $\mathscr{L}$ in $\mathscr{C}$) implies $\vdash_L \phi$.
- We say $L$ is strong complete w.r.t. $\mathscr{C}$ if for all $\Gamma \subseteq \mathscr{F}_{FL}$ and $\phi \in \mathscr{F}_{FL}$, ($\Gamma \models_{\mathscr{L}} \phi$ for all $\mathscr{L}$ in $\mathscr{C}$) implies $\Gamma \vdash_L \phi$.

We will illustrate these definitions in the following subsections.

---

[6] Note that an $\mathscr{L}$-evaluation is completely determined by its action on the propositional variables.

[7] Note that $\mathscr{L}$-models for the empty set are just $\mathscr{L}$-evaluations.

[8] For every logic $L$ appearing in this chapter, completeness w.r.t. a class $\mathscr{C}$ of residuated lattices (or, in the next sections, triangle algebras) implies that, for every finite theory $\Gamma \subseteq \mathscr{F}_{FL}$ and $\phi \in \mathscr{F}_{FL}$, $\Gamma \vdash_L \phi$ if $\Gamma \models_{\mathscr{L}} \phi$ for all $\mathscr{L}$ in $\mathscr{C}$. In other words, completeness implies 'strong completeness for finite theories'.

### 4.1.1 Monoidal Logic

Monoidal logic (ML) was introduced by Höhle in [26]. Its axioms[9] are:

$$
\begin{array}{ll}
(ML.1) & (\phi \rightarrow \psi) \rightarrow ((\psi \rightarrow \chi) \rightarrow (\phi \rightarrow \chi)),\\
(ML.2) & \phi \rightarrow (\phi \vee \psi),\\
(ML.3) & \psi \rightarrow (\phi \vee \psi),\\
(ML.4) & (\phi \rightarrow \chi) \rightarrow ((\psi \rightarrow \chi) \rightarrow ((\phi \vee \psi) \rightarrow \chi)),\\
(ML.5) & (\phi \wedge \psi) \rightarrow \phi,\\
(ML.6) & (\phi \wedge \psi) \rightarrow \psi,\\
(ML.7) & (\phi \& \psi) \rightarrow \phi,\\
(ML.8) & (\phi \& \psi) \rightarrow (\psi \& \phi),\\
(ML.9) & (\phi \rightarrow \psi) \rightarrow ((\phi \rightarrow \chi) \rightarrow (\phi \rightarrow (\psi \wedge \chi))),\\
(ML.10) & (\phi \rightarrow (\psi \rightarrow \chi)) \rightarrow ((\phi \& \psi) \rightarrow \chi),\\
(ML.11) & ((\phi \& \psi) \rightarrow \chi) \rightarrow (\phi \rightarrow (\psi \rightarrow \chi)),\\
(ML.12) & \overline{0} \rightarrow \phi.
\end{array}
$$

This means that for all possible choices of FL-formulae for $\phi$, $\psi$ and $\chi$, the above FL-formulae are provable in ML. For example, $(\overline{0} \rightarrow q) \rightarrow ((p\&q) \vee (\overline{0} \rightarrow q))$ is provable in ML, because it is an instance[10] of the third axiom, with $\psi = (\overline{0} \rightarrow q)$ and $\phi = p\&q$.

To show that other FL-formulae are provable in ML, ML has one deduction rule. This deduction rule is called modus ponens (MP) and states that if $\phi$ and $\phi \rightarrow \psi$ are provable then so is $\psi$. Now we can formally define what a proof in ML of an FL-formula $\phi$ from a theory $\Gamma$ is: it is a finite sequence of FL-formulae in which every FL-formula is either an instance of an axiom of ML, an element of $\Gamma$ or the result of an application of the modus ponens to two FL-formulae appearing earlier in the sequence. If such a proof exists, this is denoted as $\Gamma \vdash_{ML} \phi$. If $\Gamma$ is empty, we simply write $\vdash_{ML} \phi$ instead of $\emptyset \vdash_{ML} \phi$. An important result about ML is its soundness and completeness w.r.t. residuated lattices.

**Theorem 2.** *[26] Monoidal logic is sound and strong complete w.r.t. residuated lattices. In other words: for all $\Gamma \subseteq \mathscr{F}_{FL}$ and $\phi \in \mathscr{F}_{FL}$, we have $\Gamma \vdash_{ML} \phi$ iff $\Gamma \models_{\mathscr{L}} \phi$ for all residuated lattices $\mathscr{L}$.*

---

9 In [26] there are two axioms instead of the last one. In these two axioms the negation $\neg$ appears (as a unary connective), but not the constant $\overline{0}$. In this work we have chosen another way, namely to define the negation based on the constant $\overline{0}$ (instead of the other way around).

Moreover, also $(\phi\&(\psi\&\chi)) \rightarrow ((\phi\&\psi)\&\chi)$ was listed as an axiom. But FL-formulae of this form can be proven from the other axioms, so it can be left out.

10 An instance of an axiom of ML is any FL-formula obtained by replacing $\phi$, $\psi$ and $\chi$ with FL-formulae. For example, $((p_1 \rightarrow (p_2 \vee \overline{0}))\&(p_2 \wedge q)) \rightarrow (p_1 \rightarrow (p_2 \vee \overline{0}))$ is an instance of $(\phi\&\psi) \rightarrow \phi$.

Each identity or inequality that is valid in residuated lattices can easily be transformed into a scheme[11] of FL-formulae that are provable in ML. Therefore we just need to change such an identity or inequality to an equivalent 'equal to 1 identity'.

- An example with an identity: $x \Rightarrow (y \sqcap z) = (x \Rightarrow y) \sqcap (x \Rightarrow z)$ holds in all residuated lattices. This is equivalent with $(x \Rightarrow (y \sqcap z)) \Leftrightarrow ((x \Rightarrow y) \sqcap (x \Rightarrow z)) = 1$, which can be immediately transformed into the scheme $(\phi \to (\psi \wedge \chi)) \leftrightarrow ((\phi \to \psi) \wedge (\phi \to \chi))$ of FL-formulae that are provable in ML.
- An example with an inequality: $x * y \leq x \sqcap y$ holds in all residuated lattices. This is equivalent with $(x * y) \Rightarrow (x \sqcap y) = 1$, which can be transformed into the scheme $(\phi \& \psi) \to (\phi \wedge \psi)$ of FL-formulae that are provable in ML.

So there is a close connection between identities (and inequalities) in residuated lattices and (schemes of) FL-formulae that are provable in ML.

ML enjoys a so-called local deduction theorem:

**Theorem 3.** *[24] Let $\Gamma \cup \{\phi, \psi\}$ be a set of FL-formulae. Then the following are equivalent:*

- $\Gamma \cup \{\phi\} \vdash_{ML} \psi$,
- *there is an integer $n$ such that $\Gamma \vdash_{ML} \phi^n \to \psi$.*

This local deduction theorem, as well as the soundness and completeness of ML, remain valid in axiomatic extensions of ML. An axiomatic extension of ML is a logic having the same axioms and deduction rule as ML, plus one or more other axioms. Axiomatic extensions of ML are sound and complete w.r.t. residuated lattices that satisfy the identities corresponding to the extra axioms.

### 4.1.2 Monoidal t-Norm Based Logic

Monoidal t-norm based logic (MTL) [16] is an axiomatic extension of ML. The extra axiom is $(\phi \to \psi) \vee (\psi \to \phi)$. This axiom corresponds with the identity $(x \Rightarrow y) \sqcup (y \Rightarrow x) = 1$ (prelinearity) in residuated lattices. Therefore we have the following soundness and completeness theorem.

**Theorem 4.** *[16] Monoidal t-norm based logic is sound and strong complete w.r.t. prelinear residuated lattices (MTL-algebras). In other words: for all $\Gamma \subseteq \mathscr{F}_{FL}$ and $\phi \in \mathscr{F}_{FL}$, we have $\Gamma \vdash_{MTL} \phi$ iff $\Gamma \models_{\mathscr{L}} \phi$ for all MTL-algebras $\mathscr{L}$.*

The provability relation $\vdash_{MTL}$ is defined in the same way as $\vdash_{ML}$, only now also instances of the axiom $(\phi \to \psi) \vee (\psi \to \phi)$ may appear in proofs of FL-formulae.

Because MTL-algebras are isomorphic to subdirect products of MTL-chains, the strong completeness can be strengthened to so-called (strong) chain completeness, i.e., (strong) completeness w.r.t. MTL-chains.

[11] A scheme of FL-formulae consists of all FL-formulae of a particular form (the instances of the scheme). Note that the axioms of ML are schemes.

**Theorem 5.** *[16] Monoidal t-norm based logic is sound and strong chain complete. In other words: for all $\Gamma \subseteq \mathscr{F}_{FL}$ and $\phi \in \mathscr{F}_{FL}$, we have $\Gamma \vdash_{MTL} \phi$ iff $\Gamma \models_{\mathscr{L}} \phi$ for all MTL-chains $\mathscr{L}$.*

Chain completeness and strong chain completeness are properties that remain valid for axiomatic extensions of MTL. Important theorems about MTL that do not necessarily remain valid for axiomatic extensions, are standard completeness and strong standard completeness. Recall that standard MTL-algebras are MTL-algebras on the unit interval.

**Theorem 6.** *[29] Monoidal t-norm based logic is sound and strong standard complete. In other words: for all $\Gamma \subseteq \mathscr{F}_{FL}$ and $\phi \in \mathscr{F}_{FL}$, we have $\Gamma \vdash_{MTL} \phi$ iff $\Gamma \models_{\mathscr{L}} \phi$ for all standard MTL-chains $\mathscr{L}$.*

Because of this theorem, and because standard MTL-chains are induced by left-continuous t-norms, we can say that MTL is 'the logic of left-continuous t-norms'.

MTL enjoys the same local deduction theorem as ML:

**Theorem 7.** *[24] Let $\Gamma \cup \{\phi, \psi\}$ be a set of FL-formulae. Then the following are equivalent:*

- $\Gamma \cup \{\phi\} \vdash_{MTL} \psi$,
- *there is an integer $n$ such that $\Gamma \vdash_{MTL} \phi^n \to \psi$.*

### 4.1.3 Basic Logic

Basic logic (BL) [24] is an axiomatic extension of MTL (and therefore also of ML). The extra axiom is $(\phi \wedge \psi) \to (\phi \& (\phi \to \psi))$. This axiom corresponds with the identity $(x \sqcap y) \Rightarrow (x * (x \Rightarrow y)) = 1$ in residuated lattices, in other words $x \sqcap y \leq x * (x \Rightarrow y)$. This is equivalent with $x \sqcap y = x * (x \Rightarrow y)$ (divisibility) because $x \sqcap y \geq x * (x \Rightarrow y)$ is valid in all residuated lattices. Because BL is an axiomatic extension of MTL and because divisible MTL-algebras are BL-algebras, we immediately have the following soundness and completeness result.

**Theorem 8.** *[24] Basic logic is sound and strong chain complete. In other words: for all $\Gamma \subseteq \mathscr{F}_{FL}$ and $\phi \in \mathscr{F}_{FL}$, we have $\Gamma \vdash_{BL} \phi$ iff $\Gamma \models_{\mathscr{L}} \phi$ for all BL-chains $\mathscr{L}$.*

Note that BL is also (strong) complete w.r.t. all BL-algebras, which is a weaker property than (strong) chain completeness. BL also satisfies standard completeness (which is stronger than chain completeness), but *not* strong standard completeness. This implies that the following theorem in general does not hold for infinite theories $\Gamma$.

**Theorem 9.** *[5] For all finite $\Gamma \subseteq \mathscr{F}_{FL}$ and $\phi \in \mathscr{F}_{FL}$, we have $\Gamma \vdash_{BL} \phi$ iff $\Gamma \models_{\mathscr{L}} \phi$ for all standard BL-chains $\mathscr{L}$.*

Because standard BL-chains are induced by continuous t-norms, we can say BL is 'the logic of continuous t-norms'.

The local deduction theorem is also valid for BL.

**Theorem 10.** *[24] Let $\Gamma \cup \{\phi, \psi\}$ be a set of FL-formulae. Then the following are equivalent:*

- $\Gamma \cup \{\phi\} \vdash_{BL} \psi$,
- *there is an integer n such that* $\Gamma \vdash_{BL} \phi^n \to \psi$.

### 4.1.4 Łukasiewicz Logic

Łukasiewicz logic (Ł) is an axiomatic extension of BL (and therefore also of MTL and ML). The extra axiom is $\neg\neg\phi \to \phi$, in which $\neg\phi$ is a short notation for $\phi \to \overline{0}$. This axiom corresponds with the identity $\neg\neg x \Rightarrow x = 1$ in residuated lattices, in other words $\neg\neg x \leq x$. This is equivalent with $\neg\neg x = x$ (involutive negation) because $x \leq \neg\neg x$ is valid in all residuated lattices. Because Ł is an axiomatic extension of MTL, we immediately have the following soundness and completeness result.

**Theorem 11.** *[24] Łukasiewicz logic is sound and strong chain complete. In other words: for all $\Gamma \subseteq \mathscr{F}_{FL}$ and $\phi \in \mathscr{F}_{FL}$, we have $\Gamma \vdash_{Ł} \phi$ iff $\Gamma \models_{Ł} \phi$ for all MV-chains $\mathscr{L}$.*

Note that Ł is also (strong) complete w.r.t. all MV-algebras, which is a weaker property than (strong) chain completeness. Ł also satisfies standard completeness (which is stronger than chain completeness), but *not* strong standard completeness. This implies that the following theorem in general does not hold for infinite theories.

**Theorem 12.** *[3, 24] For all finite $\Gamma \subseteq \mathscr{F}_{FL}$ and $\phi \in \mathscr{F}_{FL}$, we have $\Gamma \vdash_{Ł} \phi$ iff $\Gamma \models_{\mathscr{L}} \phi$ for all standard MV-chains $\mathscr{L}$.*

Because standard MV-chains are induced by t-norms that are conjugated to the Łukasiewicz t-norm, we can say that Ł is 'the logic of the Łukasiewicz t-norm'.

The local deduction theorem is also valid for Ł.

**Theorem 13.** *[24] Let $\Gamma \cup \{\phi, \psi\}$ be a set of FL-formulae. Then the following are equivalent:*

- $\Gamma \cup \{\phi\} \vdash_{Ł} \psi$,
- *there is an integer n such that* $\Gamma \vdash_{Ł} \phi^n \to \psi$.

### 4.1.5 Classical Logic

Classical logic (CPC[12]) is an axiomatic extension of Ł (and therefore also of BL, MTL and ML). The extra axiom is $\phi \to (\phi \& \phi)$. This axiom corresponds with the identity $x \Rightarrow (x * x) = 1$ in residuated lattices, in other words $x \leq x * x$. This is equivalent with $x = x * x$ (contraction) because $x * x \leq x$ is valid in all residuated lattices. Because CPC is an axiomatic extension of MTL, we immediately have the following soundness and completeness result.

**Theorem 14.** *[34] Classical logic is sound and strong chain complete. In other words: for all $\Gamma \subseteq \mathscr{F}_{FL}$ and $\phi \in \mathscr{F}_{FL}$, we have $\Gamma \vdash_{CPC} \phi$ iff $\Gamma \models_{\mathscr{L}} \phi$ for all linear Boolean algebras $\mathscr{L}$.*

---

[12] The abbreviation, taken from [6], stands for 'classical propositional calculus'.

Note that there is only one linear Boolean algebra (apart from the trivial one with one element), namely the Boolean algebra with two elements, 0 and 1. Therefore CPC cannot satisfy standard completeness, in the sense that CPC is not complete w.r.t. Boolean algebras on the unit interval (because there are no such Boolean algebras).

The local deduction theorem is also valid for CPC. But because in CPC the FL-formulae $\phi$ and $\phi^n$ (with $n$ a strictly positive integer) are equivalent (meaning $\vdash_{CPC} \phi \leftrightarrow \phi^n$ holds[13]), the theorem can be strengthened.

**Theorem 15.** *[34] Let $\Gamma \cup \{\phi, \psi\}$ be a set of FL-formulae. Then the following are equivalent:*

- $\Gamma \cup \{\phi\} \vdash_{CPC} \psi$,
- $\Gamma \vdash_{CPC} \phi \rightarrow \psi$.

### 4.1.6 Other Fuzzy Logics

Apart from the examples in the previous sections, many other fuzzy logics can be defined by adding axioms to ML, MTL,... Examples can be found in e.g. [6]). In [6] it was proven that an axiomatic extension of MTL is strong standard complete iff it has the real-chain embedding property, i.e., iff each countable MTL-chain in its semantics is embeddable in a standard MTL-chain in its semantics.

## 4.2 *Interval-Valued Monoidal Logic: Definition*

As semantics of interval-valued fuzzy logics, we choose triangle algebras. Because triangle algebras have more operators than residuated lattices, IVFL-formulae can contain more connectives than FL-formulae.

**Definition 10.** [41] IVFL-formulae are built up from a countable set of propositional variables (denoted by $p$, $q$, $r$, $p_1$, $p_2$, ...) and the constants $\overline{0}$ and $\overline{u}$. These symbols are IVFL-formulae by definition. The other IVFL-formulae are defined recursively: if $\phi$ and $\psi$ are IVFL-formulae, then so are $(\phi \wedge \psi)$, $(\phi \vee \psi)$, $(\phi \& \psi)$, $(\phi \rightarrow \psi)$, $\Box\phi$ and $\Diamond\phi$. The set of IVFL-formulae is denoted by $\mathscr{F}_{IVFL}$. Note that $\mathscr{F}_{FL} \subseteq \mathscr{F}_{IVFL}$.

In order to avoid unnecessary brackets, we agree on the following priority rules:

- unary operators always take precedence over binary ones, while
- among the connectives, & has the highest priority; furthermore $\wedge$ and $\vee$ take precedence over $\rightarrow$,
- the outermost brackets are not written.

The same notations ($\overline{1}$ is $\overline{0} \rightarrow \overline{0}$, ...) as for FL-formulae are used. Now we are ready to introduce interval-valued monoidal logic (IVML) [41]. Its axioms are those of

---

[13] In the logics we are concerned with in this work, we have the following property. If a subformula of a formula is replaced by an equivalent subformula, the resulting formula is equivalent with the original one. This can be proven using soundness and completeness. For example, in ML and its axiomatic extensions $(p_1 \& p_2) \rightarrow q$ is equivalent with $(p_2 \& p_1) \rightarrow q$ because the subformulae $p_1 \& p_2$ and $p_2 \& p_1$ are equivalent.

ML, i.e., (ML.1)–(ML.12), complemented with axioms corresponding to 13 properties of triangle algebras:

| | |
|---|---|
| | $(IVML.1')\ \phi \to \Diamond\phi$, |
| $(IVML.2)\ \Box\phi \to \Box\Box\phi$, | |
| $(IVML.3)\ (\Box\phi\wedge\Box\psi) \to \Box(\phi\wedge\psi)$, | $(IVML.3')\ (\Diamond\phi\wedge\Diamond\psi) \to \Diamond(\phi\wedge\psi)$, |
| $(IVML.4)\ \Box(\phi\vee\psi) \to (\Box\phi\vee\Box\psi)$, | $(IVML.4')\ \Diamond(\phi\vee\psi) \to (\Diamond\phi\vee\Diamond\psi)$, |
| $(IVML.5)\ \neg\Box\overline{u}$, | $(IVML.5')\ \Diamond\overline{u}$, |
| $(IVML.6)\ \Diamond\phi \to \Box\Diamond\phi$, | $(IVML.6')\ \Diamond\Box\phi \to \Box\phi$, |
| $(IVML.7)\ \Box(\phi \to \psi) \to (\Box\phi \to \Box\psi)$, | |
| $(IVML.9)\ (\Box\phi \to \Box\psi) \to \Box(\Box\phi \to \Box\psi)$, | |
| $(IVML.10)\ \phi \leftrightarrow (\Box\phi \vee (\Diamond\phi \wedge \overline{u}))$. | |

All instances of these axioms are by definition provable in IVML. To determine which other IVFL-formulae are provable, there are three deduction rules: modus ponens (MP, if $\phi$ and $\phi \to \psi$ are provable in IVML, then so is $\psi$), generalization (G, if $\phi$ is provable in IVML, then so is $\Box\phi$) and monotonicity of $\Diamond$ (M$\Diamond$, if $\phi \to \psi$ is provable, then so is $\Diamond\phi \to \Diamond\psi$). Proofs in IVML and the provability relation $\vdash_{IVML}$ are defined in the usual way, similarly as for ML (and the other fuzzy logics from Section 4.1). If $\Gamma$ is a theory, i.e., a set of IVFL-formulae, then a (formal) proof of an IVFL-formula $\phi$ in $\Gamma$ is a finite sequence of IVFL-formulae with $\phi$ at its end, such that every IVFL-formula in the sequence is either an instance of an axiom of IVML, an IVFL-formula of $\Gamma$, or the result of an application of a deduction rule to previous IVFL-formulae in the sequence. If a proof for $\phi$ exists in $\Gamma$, we denote this by $\Gamma \vdash_{IVML} \phi$.

**Definition 11.** [41] Let $\mathscr{A} = (A,\sqcap,\sqcup,*,\Rightarrow,\nu,\mu,0,u,1)$ be a triangle algebra, $\Gamma$ a theory (i.e., a set of IVFL-formulae). An $\mathscr{A}$-evaluation is a mapping $e$ from the set of IVFL-formulae to $A$ that satisfies, for each two IVFL-formulae $\phi$ and $\psi$:

- $e(\phi \wedge \psi) = e(\phi) \sqcap e(\psi)$,
- $e(\phi \vee \psi) = e(\phi) \sqcup e(\psi)$,
- $e(\phi \& \psi) = e(\phi) * e(\psi)$,
- $e(\phi \to \psi) = e(\phi) \Rightarrow e(\psi)$,
- $e(\Box\phi) = \nu e(\phi)$,
- $e(\Diamond\phi) = \mu e(\phi)$,
- $e(\overline{0}) = 0$ and
- $e(\overline{u}) = u$.

If an $\mathscr{A}$-evaluation $e$ satisfies $e(\chi) = 1$ for every $\chi$ in $\Gamma$, it is called an $\mathscr{A}$-model for $\Gamma$. We write $\Gamma \models_{\mathscr{A}} \phi$ if $e(\phi) = 1$ for all $\mathscr{A}$-models $e$ for $\Gamma$.

Soundness, completeness and strong completeness are defined similarly as for formal fuzzy logics. We just have to replace 'residuated lattice' by 'triangle algebra' and 'FL-formula' by 'IVFL-formula'.

Now we introduce some axiomatic extensions of IVML, by adding well-known axioms. Note that these axioms are applied to IVFL-formulae of the form $\Box\phi$

instead of to all IVFL-formulae. As the image of a triangle algebra $(A,\sqcap,\sqcup,*,\Rightarrow,\nu,\mu,0,u,1)$ under $\nu$ is the set $E(\mathscr{A})$ of exact elements[14], this means that the axioms schemes do not hold for all truth values, but only for exact truth values. This is not a drawback. On the contrary, it is precisely what we want because the exact truth values are easier to interpret and handle. Moreover, using Proposition 3, for all axioms equivalent axioms can be found that only involve IVFL-formulae of the form $\Box\phi$ and $\Diamond\phi$, and $\overline{u}$.

**Definition 12.** [38]

- Interval-valued monoidal t-norm based logic (IVMTL) is IVML extended with the axiom scheme pseudo-prelinearity

$$(\Box\phi \to \Box\psi) \vee (\Box\psi \to \Box\phi).$$

- Interval-valued basic logic (IVBL) is IVMTL extended with the axiom scheme pseudo-divisibility

$$(\Box\phi \wedge \Box\psi) \to (\Box\phi \& (\Box\phi \to \Box\psi)).$$

- Interval-valued Łukasiewicz logic (IVŁ) is IVBL extended with the axiom scheme pseudo-involution

$$\neg\neg\Box\phi \to \Box\phi.$$

- Interval-valued classical propositional calculus (IVCPC) is IVŁ extended with the axiom scheme pseudo-contraction

$$\Box\phi \to (\Box\phi \& \Box\phi).$$

## *4.3 Soundness and Completeness*

It is easy to check that IVML is sound w.r.t. the variety of triangle algebras, i.e., that if an IVFL-formula $\phi$ can be proven from a theory $\Gamma$ in IVML ($\Gamma \vdash_{IVML} \phi$), then for every triangle algebra $\mathscr{A}$ and for every $\mathscr{A}$-model $e$ of $\Gamma$, $e(\phi) = 1$ (in other words: for every triangle algebra $\mathscr{A}$, $\Gamma \models_{\mathscr{A}} \phi$). To show that IVML is also strong complete (w.r.t. triangle algebras), i.e., that the converse of soundness also holds, a general result from abstract algebraic logic (shortly AAL, see e.g. [19] for a survey) can be applied. It proceeds by showing that IVML is an implicative logic (in the sense of Rasiowa [33]). From this we can deduce (according to e.g. [18]) that IVML is strong complete w.r.t. the variety of triangle algebras if it is sound w.r.t. it and if in triangle algebras $x = y$ if $x \Rightarrow y = 1$ and $y \Rightarrow x = 1$. Triangle algebras indeed satisfy these conditions, so we can conclude that IVML is sound and strong complete w.r.t. triangle algebras.

---

[14] Note that the image under $\mu$ is also $E(\mathscr{A})$. All axiom schemes in Definition 12 can also be given in an equivalent way by changing $\Box\phi$ to $\Diamond\phi$ and/or $\Box\psi$ to $\Diamond\psi$.

**Theorem 16. (Soundness and strong completeness of IVML)** [41] An IVFL-formula $\phi$ can be deduced from a theory $\Gamma$ in IVML iff for every triangle algebra $\mathscr{A}$ and for every $\mathscr{A}$-model $e$ of $\Gamma$, $e(\phi) = 1$.

A more basic proof for the completeness of IVML, using the Lindenbaum algebra of IVML and inspired by a similar, commonly used procedure for formal fuzzy logics, was given in [36].

Theorem 16 implies similar results for axiomatic extensions (e.g. the interval-valued fuzzy logics in Definition 12), in the same way as the soundness and completeness of ML remains valid for axiomatic extensions. This can be seen by taking the set of all instances of the extra axioms as $\Gamma$ in Theorem 16. In particular, all extensions of IVML introduced in Section 4.2 are sound and (strong) complete w.r.t. their corresponding subvariety of the variety of triangle algebras. For example, IVBL is sound and complete w.r.t. the variety of triangle algebras satisfying $(\nu x \Rightarrow \nu y) \sqcup (\nu y \Rightarrow \nu x) = 1$ and $\nu x \sqcap \nu y \leq \nu x * (\nu x \Rightarrow \nu y)$.

For IVMTL and its axiomatic extensions we can prove a stronger version of completeness, namely strong pseudo-chain completeness. This is similar to axiomatic extensions of MTL being strong chain-complete. Together with Theorem 16, Theorem 1 implies the following result:

**Theorem 17.** *[42] For each set of IVFL-formulae $\Gamma \cup \phi$, the following three statements are equivalent:*

- *$\phi$ can be deduced from a theory $\Gamma$ in IVMTL ($\Gamma \vdash_{IVMTL} \phi$),*
- *for every pseudo-prelinear triangle algebra $\mathscr{A}$ and for every $\mathscr{A}$-model e of $\Gamma$, $e(\phi) = 1$,*
- *for every pseudo-linear triangle algebra $\mathscr{A}$ and for every $\mathscr{A}$-model e of $\Gamma$, $e(\phi) = 1$.*

This completeness result remains valid for axiomatic extensions of IVMTL. The reason is that Theorem 1 also holds for subvarieties of pseudo-prelinear triangle algebras.

Using Theorem 6 and the real-chain embedding property, also the strong standard completeness of IVMTL can be proven.

**Theorem 18.** *[38] (Strong standard completeness) For each set of IVFL-formulae $\Gamma \cup \{\phi\}$, the following four statements are equivalent:*

1. *$\phi$ can be deduced from $\Gamma$ in IVMTL ($\Gamma \vdash_{IVMTL} \phi$),*
2. *for every pseudo-prelinear triangle algebra $\mathscr{A}$, $\Gamma \models_{\mathscr{A}} \phi$ (i.e., for every $\mathscr{A}$-model e of $\Gamma$, $e(\phi) = 1$),*
3. *for every pseudo-linear triangle algebra $\mathscr{A}$, $\Gamma \models_{\mathscr{A}} \phi$,*
4. *for every standard triangle algebra $\mathscr{A}$, $\Gamma \models_{\mathscr{A}} \phi$.*

Because of Proposition 2, every standard triangle algebra is isomorphic to a standard extended IVRL, and every standard extended IVRL is a standard triangle algebra. This result leads to the following corollary of Theorem 18.

**Corollary 1.** *For each set of IVFL-formulae $\Gamma \cup \{\phi\}$, the following statements are equivalent:*

*1.* $\phi$ *can be deduced from* $\Gamma$ *in IVMTL* ($\Gamma \vdash_{IVMTL} \phi$),
*2. for every standard extended IVRL* $\mathscr{A}$, $\Gamma \models_{\mathscr{A}} \phi$.

So we can truly state that IVMTL is an interval-valued fuzzy logic. It is the logic of the t-norms $\mathscr{T}_{T,t}$ in (2), with $T$ a left-continuous t-norm on the unit interval.

Finally, it is also possible to prove the local deduction theorem for IVML (and its axiomatic extensions), which gives a connection between $\vdash_{\mathbf{L}}$ and $\rightarrow$.

**Theorem 19.** *[38] Let* $\Gamma \cup \{\phi, \psi\}$ *be a set of IVFL-formulae, and* **L** *be an axiomatic extension of IVML. Then the following are equivalent:*

- $\Gamma \cup \{\phi\} \vdash_L \psi$,
- *There is an integer n such that* $\Gamma \vdash_L (\Box\phi)^n \rightarrow \psi$.

## 4.4 Interpretation

Interval-valued fuzzy logics as we introduced them, are truth-functional logics: the truth degree of a compound proposition is determined by the truth degree of its constituent parts. This causes some counterintuitive results, if we want to interpret the element $[0,1]$ of an IVRL as uncertainty regarding the actual truth value of a proposition. For example: suppose we don't know anything about the truth value of propositions $p$ and $q$, i.e., $e(p) = e(q) = [0,1]$. Then yet the implication $p \rightarrow q$ is definitely valid: $e(p \rightarrow q) = e(p) \Rightarrow e(q) = [1,1]$. However, if $\neg[0,1] = [0,1]$ [15] (which is intuitively preferable, since the negation of an uncertain proposition is still uncertain), then we can take $q = \neg p$, and obtain that $p \rightarrow \neg p$ is true. Or, equivalently (using the residuation principle), that $p\&p$ is false. This does not seem intuitive, as one would rather expect $p\&p$ to be uncertain if $p$ is uncertain.

Another consequence of $[0,1] \Rightarrow [0,1] = [1,1]$ is that it is impossible to interpret the intervals as a set in which the 'real' (unknown) truth value is contained, and $X \Rightarrow Y$ as the smallest closed interval containing every $x \Rightarrow y$, with $x$ in $X$ and $y$ in $Y$ (as in [15]). Indeed: $1 \in [0,1]$ and $0 \in [0,1]$, but $1 \Rightarrow 0 = 0 \notin [1,1]$.

On the other hand, for t-norms it is possible that $X * Y$ is the smallest closed interval containing every $x * y$, with $x$ in $X$ and $y$ in $Y$, but only if they are t-representable (described by the axiom $\mu(x * y) = \mu x * \mu y$). However, in this case $\neg[0,1] = [0,0]$, which does not seem intuitive ('the negation of an uncertain proposition is absolutely false').

These considerations seem to suggest that IVML and its axiomatic extensions are not suitable to reason with uncertainty. This does not mean that intervals are not a good way for representing degrees of uncertainty, only that they are not suitable as truth values in a truth-functional logical calculus when we interpret them as expressing uncertainty. It might even be impossible to model uncertainty as a truth value in any truth-functional logic. This question is discussed in [12, 13]. However, nothing prevents the intervals in interval-valued fuzzy logics from having more adequate interpretations.

[15] This is for example the case if $\neg$ is involutive.

**Acknowledgements.** Bart Van Gasse and Chris Cornelis would like to thank the Research Foundation–Flanders for funding their research.

## References

1. Alcalde, C., Burusco, A., Fuentes-González, R.: A constructive method for the definition of interval-valued fuzzy implication operators. Fuzzy Sets and Systems 153, 211–227 (2005)
2. Bustince, H.: Indicator of inclusion grade for interval-valued fuzzy sets. application to approximate reasoning based on interval-valued fuzzy sets. International Journal of Approximate Reasoning 23, 137–209 (2000)
3. Chang, C.: Algebraic analysis of many valued logics. Transactions of the American Mathematical Society 88(2), 467–490 (1958)
4. Chang, C.: A new proof of the completeness of the lukasiewicz axioms. Transactions of the American Mathematical Society 93(1), 74–80 (1959)
5. Cignoli, R., Esteva, F., Godo, L., Torrens, A.: Basic fuzzy logic is the logic of continuous t-norms and their residua. Soft Computing 4, 106–112 (2000)
6. Cintula, P., Esteva, F., Gispert, J., Godo, L., Montagna, F., Noguera, C.: Distinguished algebraic semantics for t-norm based fuzzy logics: methods and algebraic equivalencies. Annals of Pure and Applied Logic 160(1), 53–81 (2009)
7. Cornelis, C., Deschrijver, G., Kerre, E.: Implication in intuitionistic fuzzy and interval-valued fuzzy set theory: construction, classification, application. International Journal of Approximate Reasoning 35, 55–95 (2004)
8. Cornelis, C., Deschrijver, G., Kerre, E.: Advances and challenges in interval-valued fuzzy logic. Fuzzy Sets and Systems 157(5), 622–627 (2006)
9. Deschrijver, G.: The Łukasiewicz t-norm in interval-valued fuzzy and intuitionistic fuzzy set theory. In: Atanassov, K.T., Kacprzyk, J., Krawczak, M., Szmidt, E. (eds.) Issues in the Representation and Processing of Uncertain and Imprecise Information. Fuzzy Sets, Intuitionistic Fuzzy Sets, Generalized Nets, and Related Topics, Akademicka Oficyna Wydawnicza EXIT, pp. 83–101 (2005)
10. Deschrijver, G., Cornelis, C., Kerre, E.: On the representation of intuitionistic fuzzy t-norms and t-conorms. IEEE Transactions on Fuzzy Systems 12, 45–61 (2004)
11. Deschrijver, G., Kerre, E.: Classes of intuitionistic fuzzy t-norms satisfying the residuation principle. International Journal of Uncertainty, Fuzziness and Knowledge-Based Systems 11, 691–709 (2003)
12. Dubois, D.: On ignorance and contradiction considered as truth values. Personal communication (2006)
13. Dubois, D., Prade, H.: Can we enforce full compositionality in uncertainty calculi? In: Proceedings of the 11th National Conference on Artificial Intelligence (AAAI 1994), Seattle, Washington, pp. 149–154 (1994)
14. Dummett, M.: A propositional calculus with denumerable matrix. The Journal of Symbolic Logic 24(2), 97–106 (1959)
15. Esteva, F., Garcia-Calvés, P., Godo, L.: Enriched interval bilattices and partial many-valued logics: an approach to deal with graded truth and imprecision. International Journal of Uncertainty, Fuzziness and Knowledge-Based Systems 2(1), 37–54 (1994)
16. Esteva, F., Godo, L.: Monoidal t-norm based logic: towards a logic for left-continuous t-norms. Fuzzy Sets and Systems 124, 271–288 (2001)

17. Esteva, F., Godo, L., Garcia-Cerdaña, A.: On the hierarchy of t-norm based residuated fuzzy logics. In: Fitting, M., Orlowska, E. (eds.) Beyond Two: Theory and Applications of Multiple Valued Logic, pp. 251–272. Physica-Verlag, Heidelberg (2003)
18. Font, J.: Beyond rasiowa's algebraic approach to non-classical logics. Studia Logica. 82(2), 172–209 (2006)
19. Font, J., Jansana, R., Pigozzi, D.: A survey of abstract algebraic logic. Studia Logica. 74, 13–79 (2003)
20. Font, J., Rodriguez, A., Torrens, A.: Wajsberg algebras. Stochastica 8, 5–31 (1984)
21. Gehrke, M., Walker, C., Walker, E.: Some comments on interval-valued fuzzy sets. International Journal of Intelligent Systems 11, 751–759 (1996)
22. Gödel, K.: Zum intuitionistischen aussagenkalkül. In: Anzeiger der Akademie der Wissenschaften in Wien, pp. 65–66 (1932)
23. Gottwald, S.: Mathematical fuzzy logic as a tool for the treatment of vague information. Information Sciences 172, 41–71 (2005)
24. Hájek, P.: Metamathematics of Fuzzy Logic. In: Trends in Logic—Studia Logica Library. Kluwer Academic Publishers, Dordrecht (1998)
25. Heyting, A.: Die formalen Regeln der intuitionistischen Logik. Sitzungsberichte der preuszischen Akademie der Wissenschaften, physikalisch-mathematische Klasse, 42–56, 57–71, 158–169 (1930)
26. Höhle, U.: Commutative, residuated l-monoids. In: Höhle, U., Klement, E. (eds.) Nonclassical Logics and their Applications to Fuzzy Subsets: a Handbook of the Mathematical Foundations of Fuzzy Set Theory, pp. 53–106. Kluwer Academic Publishers, Dordrecht (1995)
27. Huntington, E.: Sets of independent postulates for the algebra of logic. Transactions of the American Mathematical Society 5, 288–309 (1904)
28. Jenei, S.: A more efficient method for defining fuzzy connectives. Fuzzy Sets and Systems 90, 25–35 (1997)
29. Jenei, S., Montagna, F.: A proof of standard completeness for Esteva and Godo's logic MTL. Studia Logica. 70, 1–10 (2002)
30. Łukasiewicz, J., Tarski, A.: Untersuchungen über den Aussagenkalkül. Comptes Rendus de la Société des Sciences et des Lettres de Varsovie, 1–21 (1930)
31. Ohnishi, M., Matsumoto, K.: Gentzen method in modal calculi, part I. Osaka Mathematical Journal 9, 113–130 (1957)
32. Ohnishi, M., Matsumoto, K.: Gentzen method in modal calculi, part II. Osaka Mathematical Journal 11, 115–120 (1959)
33. Rasiowa, H.: An algebraic approach to non-classical logics. Studies in Logic and the Foundations of Mathematics 78 (1974)
34. Shoenfield, J.: Mathematical Logic. Addison-Wesley, Reading (1967)
35. Turunen, E.: Mathematics behind Fuzzy Logic. Physica-Verlag, Heidelberg (1999)
36. Van Gasse, B.: Interval-valued algebras and fuzzy logics. Ph. D. Thesis, Ghent University, Belgium (2010)
37. Van Gasse, B., Cornelis, C., Deschrijver, G., Kerre, E.: Filters of residuated lattices and triangle algebras. Information Sciences 180(16), 3006–3020 (2010)
38. Van Gasse, B., Cornelis, C., Deschrijver, G., Kerre, E.: The standard completeness of interval-valued monoidal t-norm based logic (submitted)
39. Van Gasse, B., Cornelis, C., Deschrijver, G., Kerre, E.: Triangle algebras: towards an axiomatization of interval-valued residuated lattices. In: Greco, S., Hata, Y., Hirano, S., Inuiguchi, M., Miyamoto, S., Nguyen, H.S., Słowiński, R. (eds.) RSCTC 2006. LNCS (LNAI), vol. 4259, pp. 117–126. Springer, Heidelberg (2006)

40. Van Gasse, B., Cornelis, C., Deschrijver, G., Kerre, E.: A characterization of interval-valued residuated lattices. International Journal of Approximate Reasoning 49, 478–487 (2008)
41. Van Gasse, B., Cornelis, C., Deschrijver, G., Kerre, E.: Triangle algebras: A formal logic approach to interval-valued residuated lattices. Fuzzy Sets and Systems 159, 1042–1060 (2008)
42. Van Gasse, B., Cornelis, C., Deschrijver, G., Kerre, E.: The pseudo-linear semantics of interval-valued fuzzy logics. Information Sciences 179, 717–728 (2009)
43. Zadeh, L.: The concept of a linguistic variable and its application to approximate reasoning - i. Information Sciences 8, 199–249 (1975)
44. Zalta, E.: Basic concepts in modal logic (1995), `http://mally.stanford.edu/notes.pdf`

# Implications in Fuzzy Logic: Properties and a New Class

Yun Shi, Bart Van Gasse, and Da Ruan

**Abstract.** An implication in fuzzy logic, commonly defined as a two-place operation on the unit interval, is an extension of the classical binary implication. It plays important roles in both mathematical and applied sides of fuzzy set theory. Besides the basic properties, there are many potential properties for implications, among which eight are widely used in the literature. Different implications satisfying different subgroups of these eight properties can be found. However, certain interrelationships exist between these eight properties. This chapter aims to lay bare the interrelationships between these eight properties. When searching counterexamples to prove the independencies we discover a new class of implications determined only by a negation. We then examine under which conditions the eight properties are satisfied. Finally, we obtain the intersection of the new class of implications with the S- and R- implications.

## 1 Introduction to Implication Operators in Fuzzy Logic

Since Lotfi A. Zadeh introduced the concept of *fuzzy sets* in his pioneering work in 1965 [28], a huge amount of research work about fuzzy sets and *fuzzy logic* has appeared in the literature of mathematics and computer science. Fuzzy logic is an extension of fuzzy sets. Fuzzy logic in the narrow sense refers to a kind of many-valued logic, i.e., logic that maintains more than two truth values. Fuzzy logic in the broad sense refers to the theory of *approximate reasoning* and the theory of *linguistic logic* whose truth values are linguistic terms represented by fuzzy sets (i.e., membership functions) [18, 29].

One of the most important research areas in fuzzy logic is to extend the connective operators NOT($\neg$), AND($\wedge$), OR($\vee$) and IMPLY($\rightarrow$) from binary logic to fuzzy logic. Our research work focuses mainly on implications in fuzzy logic.

### *1.1 Negations, Conjunctions and Disjunctions*

A *negation* in fuzzy logic is an extension of the negation operator $\neg$ in binary logic: A $[0,1] \rightarrow [0,1]$ mapping $N$ is a negation if it satisfies $N(0) = 1$, $N(1) = 0$,

C. Cornelis et al. (Eds.): 35 Years of Fuzzy Set Theory, STUDFUZZ 261, pp. 83–103.
springerlink.com 

and $x \le y \Rightarrow N(x) \ge N(y)$, for all $x, y \in [0,1]$. Moreover, A *strong* negation $N$ is a negation that satisfies: $N(N(x)) = x$, for all $x \in [0,1]$.

Strong negations are always continuous. But the converse is not true. One of the famous classes of strong negations are the Sugeno negations $N_a$: there exists an $a \in ]-1,+\infty[$ such that for all $x \in [0,1]$, $N_a(x) = \frac{1-x}{1+ax}$. Notice that if $a = 0$, then $N_a$ is the standard negation $N_0$, $N_0(x) = 1 - x$. An example of a class of non-continuous negations is:

$$N_A(x) = \begin{cases} 1, & \text{if } x \in A, \\ 0, & \text{if } x \notin A \end{cases}, \quad \text{for all } x \in [0,1], \tag{1}$$

where $A = [0,\alpha[$, with $\alpha \in ]0,1]$, or $A = [0,\alpha]$, with $\alpha \in [0,1[$ . Notice that $N_A$ is the class of negations that take values only in $\{0,1\}$. Another class of negations that will be useful later is:

$$N_{A,\beta}(x) = \begin{cases} 1, & \text{if } x \in A, \\ \frac{1-x}{1+\beta x}, & \text{if } x \notin A \end{cases}, \quad \text{for all } x \in [0,1], \tag{2}$$

where $A = [0,\alpha[$ , with $\alpha \in ]0,1]$, or $A = [0,\alpha]$, with $\alpha \in [0,1[$, and $\beta \in ]-1,+\infty[$ . Notice that $N_{\{0\},\beta}$ is the class of Sugeno negations.

A conjunction in fuzzy logic is an extension of the conjunction operator $\wedge$ in binary logic. Widely used are *triangular norms* (*t-norms* for short). A $[0,1]^2 \rightharpoonup [0,1]$ mapping $T$ is a t-norm if for all $x,y,z \in [0,1]$ it satisfies: $T(x,1) = x$, $y \le z \Rightarrow T(x,y) \le T(x,z)$, $T(x,y) = T(y,x)$ and $T(x,T(y,z)) = T(T(x,y),z)$.

A disjunction in fuzzy logic is an extension of the disjunction operator $\vee$ in binary logic. Widely used are *triangular conorms* (*t-conorms* for short). A $[0,1]^2 \rightharpoonup [0,1]$ mapping $S$ is a t-conorm if for all $x,y,z \in [0,1]$ it satisfies: $S(x,0) = x$, $y \le z \Rightarrow S(x,y) \le S(x,z)$, $S(x,y) = S(y,x)$ and $S(x,S(y,z)) = S(S(x,y),z)$.

## 1.2 Implications

Define a *statement* as a sentence that can be attached with a truth value in $[0,1]$. In binary logic, a statement is either attached with 0, which means that it is false, or attached with 1, which means that it is true. The implication operator in binary logic, denoted by $\rightarrow$, has the following truth table:

| $p$ | $q$ | $p \rightarrow q$ |
|---|---|---|
| 0 | 0 | 1 |
| 0 | 1 | 1 |
| 1 | 0 | 0 |
| 1 | 1 | 1 |

The implication operator $\rightarrow$ represents the meaning of 'if...then...' in a *conditional rule*. An implication $I$ is extended from the implication operator $\rightarrow$ in binary logic. In order to coincide with $\rightarrow$ on the set $\{0,1\}$, it should be a $[0,1]^2 \rightarrow [0,1]$ mapping that at least satisfies: $I(0,0) = I(0,1) = I(1,1) = 1$ and $I(1,0) = 0$. However, there is no standard definition for an implication as for a conjunction, a disjunction and a negation in fuzzy logic. As Kerre [11] states 'One of the main difficulties I have met during the preparation of lecture notes on some basic material concerning fuzzy set theory, consisted of a lack of standard definitions for basic elementary notions'. By taking into account the extensive literature about implications [3, 4, 5, 8, 10, 26], we have proposed in our work the following definition:

An implication $I$ is a $[0,1]^2 \to [0,1]$ mapping that satisfies:

FI1. the first place antitonicity FA:
$(\forall(x_1,x_2,y) \in [0,1]^3)(x_1 < x_2 \Rightarrow I(x_1,y) \geq I(x_2,y))$;
FI2. the second place isotonicity SI:
$(\forall(x,y_1,y_2) \in [0,1]^3)(y_1 < y_2 \Rightarrow I(x,y_1) \leq I(x,y_2))$;
FI3. dominance of falsity of antecedent DF: $(\forall x \in [0,1])(I(0,x) = 1)$;
FI4. dominance of truth of consequent DT: $(\forall x \in [0,1])(I(x,1) = 1)$;
FI5. boundary condition BC: $I(1,0) = 0$.

There are three important ways to construct implications from the other connective operators in fuzzy logic:

1. An strong implication (S-implication for short) $I_{S,N}$:

$$(\forall(x,y) \in [0,1]^2)(I_{S,N}(x,y) = S(N(x),y)).$$

2. An residuated implication (R-implication for short) $I_T$:

$$(\forall(x,y) \in [0,1]^2)(I_T(x,y) = \sup\{t | t \in [0,1] \wedge T(x,t) \leq y\}).$$

3. A quantum logic implication operator (QL-implication operator for short) $I_{S,N,T}$:

$$(\forall(x,y) \in [0,1]^2)(I_{S,N,T}(x,y) = S(N(x),T(x,y))).$$

Notice that a QL-implication operator is not always an implication because it does not always satisfy FI1. For under which conditions does a QL-implication satisfy FI1, i.e., it is an implication, we refer to the articles [7, 15, 23], and Section 2.6 of book [3].

## 2 Dependencies and Independencies between Potential Properties FI6-FI13 of Implications

Besides FI1-FI5, sometimes an implication is required to have some additional properties to fulfill different requirements [3, 4, 5, 9, 8, 17, 20, 27, 30], among which the most important ones are, for all $x$, $y$ and $z \in [0,1]$:

FI6. $I(1,x) = x$ (neutrality of truth, NT for short);
FI7. $I(x,I(y,z)) = I(y,I(x,z))$ (exchange principle, EP for short);
FI8. $I(x,y) = 1 \Leftrightarrow x \leq y$ (ordering principle, OP for short);
FI9. the mapping $N_I$ defined by

$$(\forall x \in [0,1])(N_I(x) = I(x,0)),$$

is a strong negation (strong negation principle, SN for short);

FI10. $I(x,y) \geq y$ (consequent boundary, CB for short);

FI11. $I(x,x) = 1$ (identity, ID for short);

FI12. $I(x,y) = I(N(y),N(x))$, where $N$ is a strong negation (contrapositive principle, CP for short);

FI13. $I$ is a continuous mapping (continuity, CO for short).

Certain interrelationships exist between these eight properties. This section aims to lay bare the interrelationships between these eight properties. The result is instrumental to propose a classification of implications.

## 2.1 Getting Neutrality of Truth (NT) from the Other Properties

**Theorem 1.** *([3], Lemma 1.54(v), Corollary 1.57 (iii)) an implication $I$ satisfying SN and CP w.r.t. a strong negation $N$ satisfies NT iff $N_I = N$.*

In the rest of this section we consider the condition that $N_I \neq N$.

**Proposition 1.** *([1], Lemma 6) An implication $I$ satisfying EP and OP satisfies NT.*

**Proposition 2.** *([3], Lemma 1.56(ii)) an implication $I$ satisfying EP and SN satisfies NT.*

**Proposition 3.** *An implication $I$ satisfying EP and CO satisfies NT.*

*Proof.* Because $I$ satisfies EP, we have for all $x \in [0,1]$,

$$I(1,N_I(x)) = I(1,I(x,0)) = I(x,I(1,0)) = I(x,0) = N_I(x). \tag{3}$$

Because $I$ is a continuous mapping, $N_I$ is a continuous mapping. Thus expression (3) is equivalent to $I(1,a) = a$, for all $a \in [0,1]$. Hence $I$ satisfies NT. □

**Proposition 4.** *There exists an implication $I$ satisfying EP, CB, ID, CP and not NT.*

Counterexample: The implication $I_1$ stated in [6] is defined by

$$I_1(x,y) = \begin{cases} 0 & \text{if } x = 1 \text{ and } y = 0 \\ 1 & \text{else} \end{cases}, \quad \text{for all } x,y \in [0,1].$$

$I_1$ satisfies CB, ID and CP w.r.t. any strong negation $N$. However, in case that $x \neq 1$, $I_1(1,x) = 1 \neq x$. Therefore $I_1$ does not satisfy NT.

**Proposition 5.** *There exists an implication $I$ satisfying OP, SN, CB, ID, CP, CO and not NT.*

Counterexample: Let an implication $I_2$ be defined by

$$I_2(x,y) = \begin{cases} 1 & \text{if } x \leq y \\ \sqrt{1-(x-y)^2} & \text{if } x > y \end{cases}, \quad \text{for all } x,y \in [0,1].$$

$I_2$ satisfies OP, SN, CB, ID, CP w.r.t. the standard strong negation $N_0$, and CO. However, in case that $x \neq 1$ and $x \neq 0$, $I_2(1,x) = \sqrt{2x - x^2} \neq x$. Therefore $I_2$ does not satisfy NT.

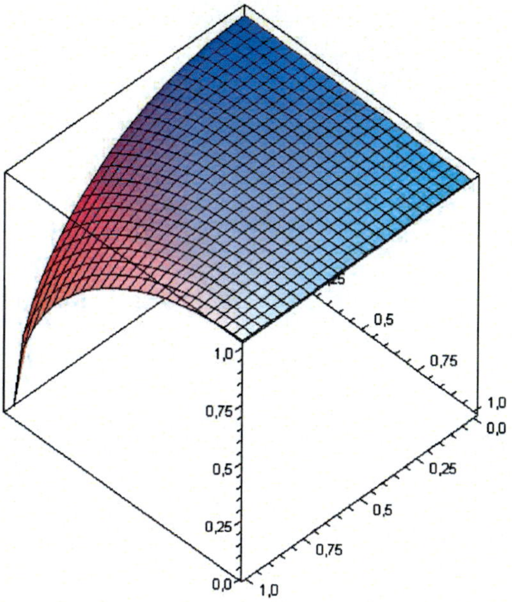


**Fig. 1** The implication $I_2$

So we considered all the possibilities that NT can be implied from the other seven properties. Moreover we stated for each independent case a counterexample.

## 2.2 Getting Exchange Principle (EP) from the Other Properties

**Proposition 6.** *There exists an implication I satisfying NT, OP, SN, CB, ID, CP, CO and not EP.*

Counterexample: Let an implication $I_3$ be defined by

$$I_3(x,y) = \begin{cases} 1 & \text{if } x \leq y \\ 1-(1-y+xy)(x-y) & \text{if } x > y \end{cases}, \text{for all } x,y \in [0,1].$$

$I_3$ satisfies NT, OP, SN, CB, ID, CP w.r.t. the standard strong negation $N_0$, and CO. However, take $x_0 = 0.3$, $y_0 = 0.9$ and $z_0 = 0.1$, we obtain $I(x_0, I(y_0, z_0)) \approx 0.9214$ and $I(y_0, I(x_0, z_0)) \approx 0.9210$. Therefore $I_3$ does not satisfy EP.

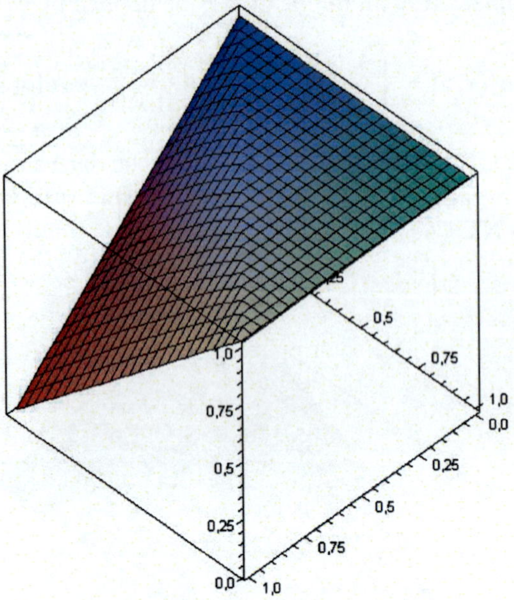


**Fig. 2** The implication $I_3$

EP is thus independent of any of the other seven properties.

## 2.3 *Getting FI8(OP) from the Other Properties*

**Proposition 7.** *There exists an implication I satisfying NT, EP, SN, CB, ID, CP, CO and not OP.*

Counterexample: Given the strong negation $N(x) = \sqrt{1-x^2}$, for all $x \in [0,1]$. The S-implication $I_4$ generated by the t-conorm $S_{\mathbf{L}}(x,y) = \min(x+y,1)$ and the strong negation $N$ is defined by

$$I_4(x,y) = S_{\mathbf{L}}(N(x),y) = \min(\sqrt{1-x^2}+y,1), \text{ for all } x,y \in [0,1].$$

Because $I_4$ is an S-implication generated from a continuous t-conorm and a strong negation, it satisfies NT, EP, SN, CB, CP w.r.t. the strong negation $N$ and CO [8]. Moreover, for all $x \in [0,1]$, $I_4(x,x) = 1$. Therefore $I_4$ also satisfies ID. However, take $x_0 = 0.5$ and $y_0 = 0.4$, we obtain $I(x_0,y_0) = 1$ while $x_0 > y_0$. Therefore $I_4$ does not satisfy OP.

Therefore OP is thus independent of any of the other seven properties.

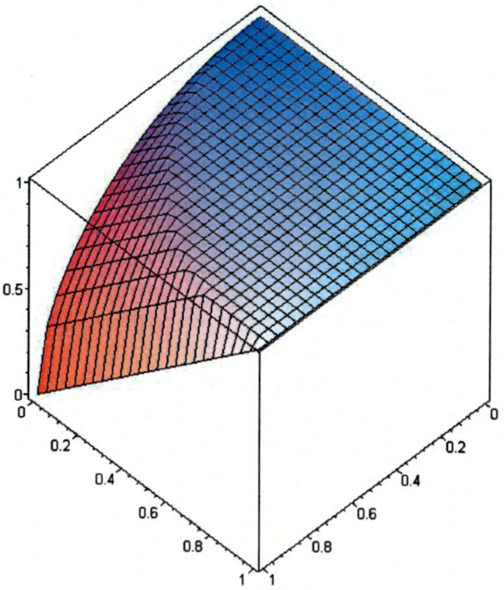


**Fig. 3** The implication $I_4$

## 2.4 Getting Strong Negation Principle (SN) from the Other Properties

**Proposition 8.** *([3], Lemma 1.5.4(v)) An implication I satisfying NT and CP w.r.t. a strong negation N satisfies SN. Moreover,* $N_I = N$.

**Corollary 1.** *An implication I satisfying EP, OP and CP w.r.t. a strong negation N satisfies SN. Moreover,* $N_I = N$.

*Proof.* Straightforward from Propositions 1 and 8. □

**Corollary 2.** *An implication I satisfying EP, CP w.r.t. a strong negation N and CO satisfies SN. Moreover,* $N_I = N$.

*Proof.* Straightforward from Propositions 3 and 8. □

**Proposition 9.** *([1], Lemma 14)([8], Corollary 1.1) An implication I satisfying EP, OP and CO satisfies SN.*

**Proposition 10.** *([8], Table 1.1) There exists an implication satisfying NT, EP, OP, CB, ID and not SN.*

*Proof.* The Gödel implication

$$I_{GD}(x,y) = \begin{cases} 1, & \text{if } x \leq y \\ y, & \text{if } x > y \end{cases}, \quad \text{for all } x,y \in [0,1]. \tag{4}$$

is an R-implication generated by the continuous t-norm $T_{\mathbf{M}}(x,y) = \min(x,y)$. Therefore $I_{GD}$ satisfies NT, EP, OP, CB and ID [8]. However we have for all $x \in [0,1]$:

$$N_{I_{GD}}(x) = I_{GD}(x,0) = \begin{cases} 1, & \text{if } x = 0 \\ 0, & \text{if } x > 0 \end{cases}.$$

Therefore $I_{GD}$ does not satisfy SN. □

**Proposition 11.** *There exists an implication I satisfying NT, EP, CB, ID, CO and not SN.*

Counterexample: Given the negation $N(x) = 1 - x^2$, for all $x \in [0,1]$. The S-implication generated from the t-conorm $S_{\mathbf{L}}$ and the negation $N$ is defined by

$$I_5(x,y) = \min(1 - x^2 + y, 1), \text{ for all } x,y \in [0,1].$$

$I_5$ satisfies NT, CB, ID and CO. Moreover, because $I_5$ is an $(S,N)$-implication generated from the Łukasiewicz t-conorm and the strict negation $N(x) = 1 - x^2$, it then also satisfies EP([3], Proposition 2.4.3(i)). However, we have for all $x \in [0,1]$

$$N_{I_5}(x) = I_5(x,0) = 1 - x^2$$

which is not a strong negation. Therefore $I_5$ does not satisfy SN.

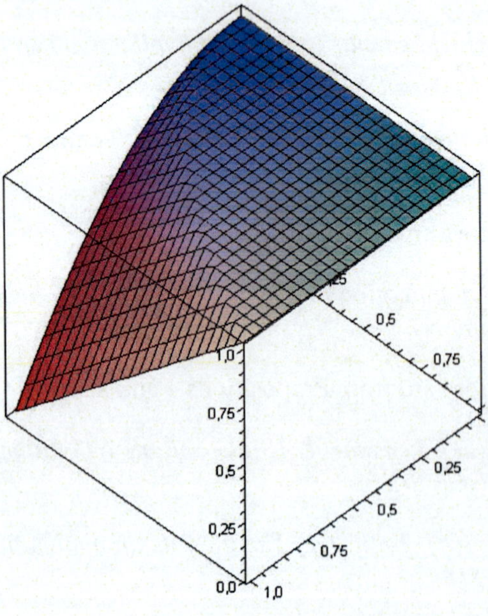


**Fig. 4** The implication $I_5$

**Proposition 12.** *There exists an implication I satisfying NT, OP, CB, ID, CO and not SN.*

Counterexample: Let an implication $I_6$ be defined by

$$I_6(x,y) = \begin{cases} 1, & \text{if } x \le y \\ \frac{y}{1+\sqrt{1-x}} + \sqrt{1-x}, & \text{if } x > y \end{cases}, \quad \text{for all } x, y \in [0,1].$$

$I_6$ satisfies NT, OP, CB, ID and CO. However, we have for all $x \in [0,1]$

$$N_{I_6}(x) = I_6(x,0) = \sqrt{1-x}$$

which is not a strong negation. Therefore $I_6$ does not satisfy SN

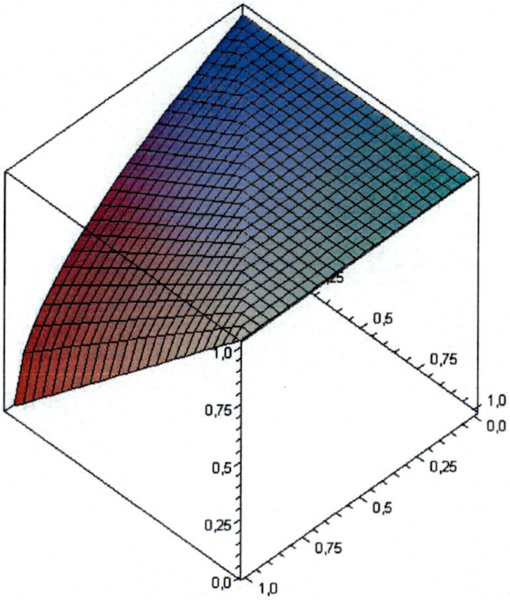


**Fig. 5** The implication $I_6$

**Proposition 13.** *There exists an implication I satisfying EP, CB, ID, CP and not SN.*

The implication $I_1$ stated in the proof of Proposition 4 satisfies EP, CB, ID and CP w.r.t. any strong negation $N$. However, we have

$$N_{I_1}(x) = I_1(x,0) = \begin{cases} 1, & \text{if } x < 1 \\ 0, & \text{if } x = 1 \end{cases}, \quad \text{for all } x \in [0,1],$$

which is not a strong negation. Therefore $I_1$ does not satisfy SN.

**Proposition 14.** *There exists an implication I satisfying OP, CB, ID, CP, CO and not SN.*

Counterexample: Let an implication $I_7$ be defined by

$$I_7(x,y) = \begin{cases} 1, & \text{if } x \le y \\ \sqrt{1-(x-y)}, & \text{if } x > y \end{cases}, \quad \text{for all } x,y \in [0,1].$$

$I_7$ satisfies OP, CB, ID, CP w.r.t. the standard strong negation $N_0$, and CO. However, we have for all $x \in [0,1]$

$$N_{I_7}(x) = I_7(x,0) = \sqrt{1-x},$$

which is not a strong negation. Therefore $I_7$ does not satisfy SN.

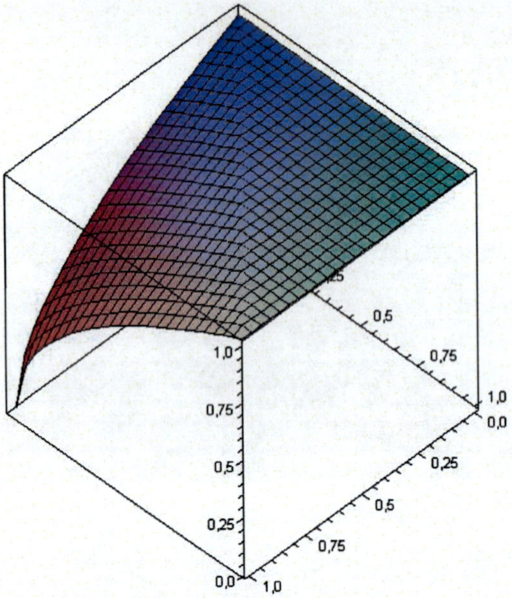


**Fig. 6** The implication $I_7$

So we considered all the possibilities that SN can be implied from the other seven properties. Moreover we stated for each independent case a counterexample.

## 2.5 *Getting Consequent Boundary (CB) from the Other Properties*

**Proposition 15.** *([4],Lemma 1 (viii)) An implication I satisfying NT satisfies CB.*

**Corollary 3.** *An implication I satisfying EP and SN satisfies CB.*

*Proof.* Straightforward from Propositions 2 and 15. □

**Corollary 4.** *An implication I satisfying EP and CO satisfies CB.*

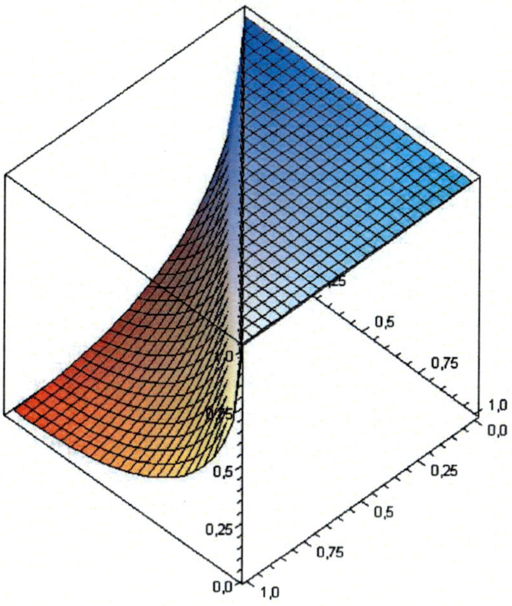


**Fig. 7** The implication $I_9$

*Proof.* Straightforward from Propositions 3 and 15. □

**Proposition 16.** *([1], Lemma 6) An implication I satisfying EP and OP satisfies CB.*

**Proposition 17.** *There exists an implication I satisfying EP, ID, CP and not CB.*

Counterexample: Let an implication $I_8$ be defined by

$$I_8(x,y) = \begin{cases} 1, & \text{if } x \leq 0.5 \text{ or } y \geq 0.5 \\ 0, & \text{else} \end{cases}, \quad \text{for all } x,y \in [0,1].$$

$I_8$ satisfies ID and CP w.r.t. the standard strong negation $N_0$. However, take $x_0 = 1$ and $y_0 = 0.1$, we obtain $I_8(x_0,y_0) = 0 < y_0$. Therefore $I_8$ does not satisfy CB.

**Proposition 18.** *There exists an implication I satisfying OP, SN, ID, CP, CO and not CB.*

Counterexample: Let an implication $I_9$ be defined by

$$I_9(x,y) = \begin{cases} 1, & \text{if } x \leq y \\ (1-\sqrt{x-y})^2, & \text{if } x > y \end{cases}, \quad \text{for all } x,y \in [0,1].$$

$I_9$ satisfies OP, SN, ID, CP w.r.t. the standard strong negation $N_0$, and CO. However, take $x_0 = 1$ and $y_0 = 0.64$, we obtain $I_9(x_0,y_0) = 0.16 < y_0$. Therefore $I_9$ does not satisfy CB.

So we considered all the possibilities CB can be implied from the other seven properties. Moreover we stated for each independent case a counterexample.

## 2.6 Getting Identity (ID) from the Other Properties

**Proposition 19.** *An implication I satisfying OP satisfies ID.*

*Proof.* Straightforward. □

**Proposition 20.** *There exists an implication I satisfying NT, EP, SN, CB, CP, CO and not ID.*

*Proof.* The Kleene-Dienes implication $I_{KD}(x,y) = \max(1-x,y)$, for all $(x,y) \in [0,1]^2$ is an S-implication generated from the t-conorm $S_{\mathbf{M}}(x,y) = \max(x,y)$ and the standard strong negation $N_0$. Therefore $I_{KD}$ satisfies NT, EP, SN, CB, CP w.r.t. the standard strong negation $N_0$, and CO. However, for $x_0 = 0.1$, we obtain $I_{KD}(x_0,x_0) = 0.9 \neq 1$. Therefore $I_{KD}$ does not satisfy ID. □

So we considered all the possibilities that ID can be implied from the other seven properties, and stated for the independent case a counterexample.

## 2.7 Getting Contrapositive Principle (CP) from the Other Properties

**Proposition 21.** *([4],Lemma 1(ix)) An implication I satisfying EP and SN satisfies CP w.r.t. the strong negation $N_I$.*

**Proposition 22.** *([1]) An implication I satisfying EP, OP and CO satisfies CP w.r.t. the strong negation $N_I$.*

**Proposition 23.** *There exists an implication I satisfying NT, EP, OP, CB, ID and not CP.*

According to the proof of Proposition 10, the Gödel implication $I_{GD}$ satisfies NT, EP, OP, CB and ID. However, for any strong negation $N$ we obtain

$$I_{GD}(N(y),N(x)) = \begin{cases} 1, & \text{if } x \leq y \\ N(x), & \text{if } x > y \end{cases} \quad \text{for all } x,y \in [0,1].$$

In case that $x > y$ and $N(x) \neq y$, $I_{GD}(N(y),N(x)) \neq I_{GD}(x,y)$. Therefore $I_{GD}$ does not satisfy CP w.r.t. any strong negation.

**Proposition 24.** *There exists an implication I satisfying NT, EP, CB, ID, CO and not CP.*

The implication $I_5$ stated in the proof of Proposition 11 satisfies NT, EP, CB, ID and CO. However, because for all $x \in [0,1]$, $N_{I_5}(x) = 1-x^2$, which is not a strong negation, according to Corollary 1.5.5 in [3], $I_5$ does not satisfy CP w.r.t. any strong negation.

**Proposition 25.** *There exists an implication I satisfying NT, OP, SN, CB, ID, CO and not CP.*

Counterexample: Let an implication $I_{10}$ be defined by

$$I_{10}(x,y) = \begin{cases} 1, & \text{if } x \le y \\ \frac{y+(x-y)\sqrt{1-x^2}}{x}, & \text{if } x > y \end{cases}, \quad \text{for all } x,y \in [0,1].$$

$I_{10}$ satisfies NT, OP, SN, ID and CO. If $I_{10}$ satisfies CP w.r.t. a strong negation $N$, then for all $x \in [0,1]$, we obtain

$$N(x) = I_{10}(1, N(x)) = I_{10}(x,0) = N_{I_{10}}(x) = \sqrt{1-x^2}.$$

However, take $x_0 = 0.8$ and $y_0 = 0.1$, we obtain $I_{10}(x_0,y_0) = 0.65$ and $I_{10}(N(y_0), N(x_0)) \approx 0.643$. Therefore $I_{10}$ does not satisfy CP w.r.t. any strong negation $N$.

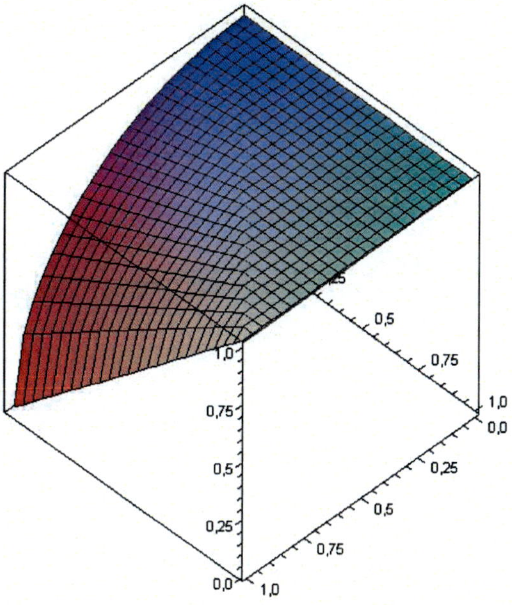


**Fig. 8** The implication $I_{10}$

So we considered all the possibilities that CP can be implied from the other seven properties. Moreover we stated for each independent case a counterexample.

## 2.8 Getting Continuity (CO) from the Other Properties

**Proposition 26.** *There exists an implication I satisfying NT, EP, OP, SN, CB, ID, CP and not CO.*

Counterexample: Let $N$ be a strong negation. Recall the $R_0$-implication stated in [19] which is defined by

$$(I_{\min_0})_N(x,y) = \begin{cases} 1, & \text{if } x \le y \\ \max(N(x),y), & \text{if } x > y \end{cases}, \quad \text{for all } x,y \in [0,1].$$

$(I_{\min_0})_N$ is the R-implication generated by the left-continuous t-norm, nilpotent minimum [9]:

$$(T_{\min_0})_N(x,y) = \begin{cases} \min(x,y), & \text{if } y > N(x) \\ 0, & \text{if } y \le N(x) \end{cases}, \quad \text{for all } x,y \in [0,1].$$

$(I_{\min_0})_N$ satisfies NT, EP, OP, SN, CB, ID and CP w.r.t. $N$, and is right-continuous in the second place [19] but it is not continuous.

Therefore CO is independent of any of the other seven properties.

## 3 A New Class of Implications

In the previous section we have studied dependencies and independencies of eight potential properties for implications, and found different implications satisfying different subgroups of these eight properties, while in these implications $I_6$ and $I_{10}$ actually have the same form. Indeed they can be represented by $I^N$ where $N$ is a negation:

$$I^N(x,y) = \begin{cases} 1, & \text{if } x \le y \\ \frac{(1-N(x))y}{x} + N(x), & \text{if } x > y \end{cases}, \quad \text{for all } x,y \in [0,1]. \tag{5}$$

If $I^N$ is always an implication, then (5) is an interesting new class of implications because it is only determined by a negation. In this section we check that $I^N$ is always an implication, and then study this new class of implications.

### 3.1 Is $I^N$ Defined by (5) Always an Implication?

We examine whether the mapping $I^N$ defined by (5) takes its values in $[0,1]$ and it satisfies properties FI1-FI5. First we rewrite $I^N$ as

$$I^N(x,y) = S_{\mathbf{P}}(N(x), I_{GG}(x,y)), \tag{6}$$

where $I_{GG}$ is the Goguen implication: $I_{GG}(x,y) = \begin{cases} 1, & \text{if } x \le y \\ y/x, & \text{if } x > y \end{cases}$, for all $x,y \in [0,1]$, and $S_{\mathbf{P}}$ is the probabilistic sum: $S_{\mathbf{P}}(x,y) = x + y - xy$, for all $x,y \in [0,1]$. It is then straightforward that $I^N(x,y) \in [0,1]$, and $I^N$ satisfies F1-F5. Therefore, $I^N$ is an implication.

## 3.2 The Potential Properties of the New Class of Implications

Now we work out whether $I^N$ defined by (5) satisfies the potential properties FI6-FI13. If not always, then under which conditions $I^N$ satisfies these properties.
(1)NT: We see immediately from (6) that $I^N$ always satisfies NT.
(2)EP: We obtain the following theorem:

**Theorem 2.** *The implication $I^N$ defined by (5) satisfies EP iff N belongs to one of the following two classes of negations:*

*(1)$N_A$ defined by (1),*
*(2)$N_{A,\beta}$ defined by (2).*

*Proof.* Necessity: Suppose $I^N$ satisfies EP. We will show that if $N$ is not of the form $N_A$, $N$ must be of the form $N_{A,\beta}$. We will do this in three steps: first we will show that we can find a $y_0$ such that $0 < N(y_0) < y_0 < 1$. Second we prove that for $x \geq y_0$, $N(x) = \frac{1-x}{1+\beta x}$ for some fixed $\beta$. And finally we use this second step to prove that for $x < y_0$, $N(x) = 1$ or $N(x) = \frac{1-x}{1+\beta x}$.

Indeed, if $I^N$ satisfies EP, then for all $x, y, z \in [0,1]$, $I^N(x, I^N(y,z)) = I^N(y, I^N(x,z))$. Take $z = 0$, we obtain

$$(\forall (x,y) \in [0,1]^2)(I^N(x,N(y)) = I^N(y,N(x))). \tag{7}$$

Suppose $N \neq N_A$. Then in particular $N \neq N_{[0,1[}$ . So there exists a $y_1 \in [0,1[$ such that $N(y_1) < 1$. Now take $y_0 \in \,]\max(y_1, N(y_1)), 1[$ , then $N(y_0) \leq N(y_1) < y_0 < 1$. We first show that $N(y_0) > 0$. Indeed, if $N(y_0) = 0$, then for all $x \in [0,1]$, we obtain:

$$N(x) = I^N(x, N(y_0)) = I^N(y_0, N(x)) = \begin{cases} 1, & \text{if } y_0 \leq N(x) \\ \frac{N(x)}{y_0}, & \text{if } y_0 > N(x) \end{cases}$$

$$\Rightarrow N(x) \in \{0,1\}, \text{ for all } x \in [0,1]$$
$$\Rightarrow N = N_A, \quad \text{for a certain } A,$$

which we have already excluded. Therefore $N(y_0) > 0$. For all $x \in [y_0, 1[$ , $x > N(y_0)$ and $N(x) < y_0$. We obtain:

$$(7) \Rightarrow \frac{1-N(x)}{x} N(y_0) + N(x) = \frac{1-N(y_0)}{y_0} N(x) + N(y_0)$$
$$\Rightarrow \frac{1-N(x)-x}{x} = \frac{1-N(y_0)-y_0}{y_0 N(y_0)} N(x)$$

If $N(x) = 0$, then $\frac{1-N(x)-x}{x} = 0 \Rightarrow x = 1$, which we have already excluded. Therefore we obtain:

$$\frac{1-N(x)-x}{xN(x)}=\frac{1-N(y_0)-y_0}{y_0N(y_0)}$$
$$\Rightarrow N(x)=\frac{1-x}{1+\beta x}\ (\text{with } \beta=\frac{1-N(y_0)-y_0}{y_0N(y_0)}, \beta\in]-1,+\infty[).$$

Now we prove, for any $x \in ]0,y_0[$ ,that if $N(x) \neq 1$, then $N(x)=\frac{1-x}{1+\beta x}$. In other words that, because $N$ is decreasing, $N=N_{A,\beta}$ defined by (2). Indeed, if $N(x)\neq 1$, then we can take $y$ in $]\max(N(x),y_0),1[$ such that $N(y)\leq x$ (this is possible because we have just proved that for $y\in[y_0,1[$ , $N(y)=\frac{1-y}{1+\beta y}$). We obtain:

$$(7)\Rightarrow \frac{1-N(x)-x}{xN(x)}=\frac{1-N(y)-y}{yN(y)}=\beta.$$

Thus $N(x)=\frac{1-x}{1+\beta x}$.

Sufficiency of $N_A$: We obtain: $I^{N_A}(x,y)=\begin{cases}1, & \text{if } x\in A\\ I_{GG}(x,y), & \text{if } x\notin A\end{cases}$ for all $x,y\in[0,1]$. Thus

$$I^{N_A}(x,I^{N_A}(y,z))=\begin{cases}1, & \text{if } x\in A \text{ or } y\in A\\ I_{GG}(x,I_{GG}(y,z)), & \text{if } x\notin A \text{ and } y\notin A\end{cases}\quad \text{for all } x,y,z\in[0,1].$$

According to [1], $I_{GG}$ satisfies EP. Therefore $I^{N_A}$ satisfies EP.
Sufficiency of $N_{A,\beta}$, $A=[0,\alpha[$ , with $\alpha\in]0,1]$, or $A=[0,\alpha]$, with $\alpha\in[0,1[$ : We obtain: $I^{N_{A,\beta}}(x,y)=\begin{cases}1, & \text{if } x\leq y \text{ or } x\in A\\ \frac{1-x+(1+\beta)y}{1+\beta x}, & \text{if } x>y \text{ and } x\notin A\end{cases}$ for all $x,y\in[0,1]$. Thus for all $x,y,z\in[0,1]$:

$$I^{N_{A,\beta}}(x,I^{N_{A,\beta}}(y,z))=\begin{cases}1, & \text{if } x\in A \text{ or } y\in A \\ & \text{or } x+y+\beta xy\leq 1+z+\beta z\\ \frac{2+\beta-x-y-\beta xy+(1+\beta)^2z}{(1+\beta x)(1+\beta y)}, & \text{else}\end{cases}$$
$$=I^{N_{A,\beta}}(y,I^{N_{A,\beta}}(x,z)).$$ □

(3) OP: We obtain the following theorem:

**Theorem 3.** *The implication $I^N$ defined by (5) satisfies OP iff $x>0\Rightarrow N(x)<1$.*

*Proof.* This follows from, for all $0\leq y<x\leq 1$,

$$I^N(x,y)<1\Leftrightarrow \frac{1-N(x)}{x}y+N(x)<1\Leftrightarrow 1-N(x)>0\Leftrightarrow N(x)<1.$$ □

(4) SN: It is straightforward that $N_{I^N}(x)=I^N(x,0)$ is a strong negation iff $N$ is a strong negation, because $N_{I^N}=N$.
(5) CB: Because $I_{GG}$ satisfies CB, $I^N$ satisfies CB according to (6).

(6) ID: We see immediately through the definition that $I^N(x,x) = 1$ for all $x \in [0,1]$.
(7) CP: We obtain the following theorem:

**Theorem 4.** *The implication $I^N$ defined by (5) satisfies CP w.r.t. a strong negation $N'$ iff $N$ is a Sugeno negation $N_a$, $a \in ]-1,+\infty[$, and $N' = N_a$.*

*Proof.* Necessity: Recall that $I^N$ always satisfies NT. If $I^N$ satisfies CP w.r.t. $N'$, then according to Proposition 8, $I^N$ also satisfies SN, and for all $x \in [0,1]$, $N'(x) = I^N(x,0) = N(x)$. Therefore, $N$ is strong and $I^N$ satisfies CP w.r.t. $N$. We obtain

$$
\begin{aligned}
& I^N(N(y),N(x)) = I^N(x,y) \\
\Rightarrow & (\forall x \in ]0,1[)(\forall y \in ]0,x[)\left(\frac{1-y-N(y)}{N(y)}N(x) = \frac{1-N(x)-x}{x}y\right) \\
\Rightarrow & (\forall x \in ]0,1[)(\forall y \in ]0,x[)\left(\frac{1-y-N(y)}{yN(y)} = \frac{1-x-N(x)}{xN(x)}\right) \\
\Rightarrow & (\exists a \in [-1,+\infty])(\forall x \in ]0,1[)\left(\frac{1-x-N(x)}{xN(x)} = a\right).
\end{aligned}
$$

If $a = -1$ or $a = +\infty$, then $N = N_A$ defined in (1) with $A = [0,1[$ or $A = \{0\}$, which is not a strong negation. Thus $N = N_a$, which is a Sugeno implication.

Sufficiency: If $N = N_a$, then for all $x, y \in [0,1]$: $I^N(x,y) = \begin{cases} 1, & \text{if } x \le y \\ \frac{(1+a)y+1-x}{1+ax}, & \text{if } x > y \end{cases}$, and

$$
\begin{aligned}
I^N(N(y),N(x)) &= \begin{cases} 1, & \text{if } x \le y \\ \frac{1-y}{N(y)}N(x) + y, & \text{if } x > y \end{cases} \\
&= \begin{cases} 1, & \text{if } x \le y \\ \frac{(1+a)y+1-x}{1+ax}, & \text{if } x > y \end{cases}.
\end{aligned}
$$

Hence $I^N(x,y) = I^N(N(y),N(x))$. □

(8) CO: We obtain the following theorem:

**Theorem 5.** *The implication $I^N$ defined by (5) satisfies CO iff $N$ is continuous.*

*Proof.* It is easily verified that if $N$ is continuous, $I^N$ is continuous in each variable. Therefore by Corollary 1.2.2 in [3], $I^N$ is continuous. The converse follows immediately from $I^N(x,0) = N(x)$. □

Combining the four theorems in this section and ([1], Theorem 1), we obtain the following two corollaries:

**Corollary 5.** *For the implication $I^N$ defined in (5), the following four conditions are equivalent:*

*(1) $N$ is a Sugeno negation $N_a$, $a \in ]-1,+\infty[$,*
*(2) $I^N$ satisfies EP and $N$ is a continuous negation,*

*(3) $I^N$ satisfies CP (w.r.t. N),*
*(4) $I^N$ is conjugate with the Łukasiewicz implication $I_L$:*

$$I_L(x,y) = \min(1-x+y,1).$$

Notice that if $a=0$, then $N=N_0$. Then $I^N = I_L$.

**Corollary 6.** *An implication $I^N$ defined by (5) satisfying EP and CO also satisfies OP.*

The converse of Corollary 6 is not true. For example, the implication $I^{N_{\{0\}}} = I_{GG}$: $I_{GG}$ satisfies OP but it is not continuous at the point $(0,0)$.

## 3.3 Intersection of the New Class of Implications with the S- and R- Implications

### 3.3.1 Intersection of the New Class of Implications and S-Implications

In this section we find the intersection of the new class of implications defined in (5) and the class of all S-implications as well as the class of all S-implications generated by a t-conorm and a strong negation.

**Theorem 6.** *The implication $I^N$ defined in (5) is an S-implication $S(N'(x),y)$ iff $N=N'$ and N belongs to one of the following two classes of negations:*

*(1) $N_A$ defined by (1) with $A = [0,1[$,*
*(2) $N_{A,\beta}$ defined by (2).*

*Proof.* Necessity: Because for all $x \in [0,1]$,
$N(x) = I^N(x,0) = S(N'(x),0) = N'(x)$, $N = N'$.

According to ([3], Proposition 2.4.6), any S-implication satisfies EP. Then according to Theorem 2, if $I^N$ is an S-implication, then $N = N_A$, $A = [0,\alpha[$, with $\alpha \in ]0,1]$, or $A = [0,\alpha]$, with $\alpha \in [0,1[$, or $N = N_{A,\beta}$. Nevertheless,

$$I^{N_A}(x,y) = \begin{cases} 1, & \text{if } x \le y \text{ or } x \in A \\ \frac{y}{x}, & \text{if } x > y \text{ and } x \notin A \end{cases}, \quad \text{for all } x,y \in [0,1], \tag{8}$$

while

$$S(N_A(x),y) = \begin{cases} 1, & \text{if } x \in A \\ y, & \text{if } x \notin A \end{cases}, \quad \text{for all } x,y \in [0,1]. \tag{9}$$

If $A \neq [0,1[$, then we can take $x$ and $y$ such that $0 < y < x < 1$ and $x \notin A$. Then $S(N_A(x),y) = y \neq \frac{y}{x} = I^{N_A}(x,y)$. Thus (8)$\neq$(9) provided $A \neq [0,1[$. Therefore $I^{N_{[0,\alpha[}}$ $(\alpha < 1)$ and $I^{N_{[0,\alpha]}}$ are not S-implications.

Sufficiency of $N = N_{[0,1[}$: $I^{N_{[0,1[}}(x,y) = S(N_{[0,1[}(x),y)$ for any t-conorm $S$.

Sufficiency of $N = N_{A,\beta}$: Take $S(x,y) = \min(1, x+y+\beta xy)$. We can verify that $S$ is a t-conorm (for the associativity, for all $x,y,z \in [0,1]$:

$$S(x,S(y,z)) = \min(1, x+y+z+\beta xy+\beta yz+\beta xz+\beta^2 xyz) = S(S(x,y),z),)$$

and that

$$S(N_{A,\beta}(x),y) = \begin{cases} 1, & \text{if } x \in A \text{ or } x \leq y \\ \frac{1-x+y+\beta y}{1+\beta x}, & \text{if } x \notin A \text{ and } x > y \end{cases}$$
$$= I^{N_{A,\beta}}(x,y).$$

Consequently, $I^{N_{A,\beta}}$ is an S-implication. □

Combining Corollary 5 and Theorem 6 we obtain the following corollary.

**Corollary 7.** *For the implication $I^N$ defined by (5), the following three conditions are equivalent:*

*(1) $I^N$ is an S-implication generated by a t-conorm and a strong negation,*
*(2) $N$ is a Sugeno negation $N_a$, $a \in ]-1,+\infty[$,*
*(3) $I^N$ is conjugate with the Łukasiewicz implication $I_L$.*

## 3.4 Intersection of the New Class of Implications and R-Implications

In this section we find the intersection of the new class of implications defined in (5) and the class of the R-implications generated by left-continuous t-norms.

**Theorem 7.** *The implication $I^N$ defined in (5) is an R-implication generated by a left-continuous t-norm iff $N$ belongs to one of the following two classes of negations:*

*(1) a Sugeno negation $N_a$, $a \in ]-1,+\infty[$,*
*(2) $N_A$ defined by (1) with $A = \{0\}$.*

*Proof.* Necessity: If $I^N$ is an R-implication generated by a left-continuous t-norm, then according to ([8], Theorem 1.14), $I^N$ satisfies EP and OP. According to Theorem 2, $N = N_A$ defined by (1), or $N = N_{A,\beta}$ defined by (2). According to Theorem 3, $N(x) < 1$ provided $x > 0$. Therefore $N = N_a$, or $N = N_{\{0\}}$.

Sufficiency of $N = N_a$: According to Corollary 5, if $N = N_a$, then $I^N$ is conjugate with $I_{\mathbf{L}}(x,y) = \max(x+y-1,0)$. According to ([1], Theorem 1), $I^N$ is an R-implication.

Sufficiency of $N = N_{\{0\}}$: $I^{N_{\{0\}}} = I_{GG}$, the R-implication generated by the continuous t-norm $T_{\mathbf{P}}(x,y) = xy$. □

Notice that although $I^{N_{[0,1[}}$ is not an R-implication generated by a left-continuous t-norm, it is the R-implication $I_{LR}$ generated by the non-left-continuous t-norm
$$T_{\mathbf{D}}(x,y) = \begin{cases} \min(x,y), & \text{if } x = 1 \text{ or } y = 1 \\ 0, & \text{otherwise} \end{cases}, \quad \text{for all } x,y \in [0,1].$$

## 3.5 Conclusion Remarks

In this chapter we first studied the complete dependencies and independencies between the eight potential properties FI6-FI13 of implications. Two of the counterexamples we define to show the independencies lead us to a new class of implications determined by a negation $N$, i.e.,

$$I^N(x,y) = \begin{cases} 1, & \text{if } x \le y \\ \frac{(1-N(x))y}{x} + N(x), & \text{if } x > y \end{cases}, \quad \text{for all } x,y \in [0,1].$$

We first checked that $I^N$ is always an implication. For each of the properties FI6-FI13 we found sufficient and necessary conditions in terms of $N$. We also obtained the intersection of $I^N$ with any S-implications and R-implications generated by a left-continuous t-norm. An example of $I^N$ being an R-implication generated by a non-left-continuous t-norm was also given.

It is worth mentioning that if we take $N$ as a Sugeno negation, then $I^N$ is an implication that is conjugate with the Łukasiewicz implication $I_{\mathbf{L}}$.

At the end of this paper we mention a possible generalization of the new class of implications $I^N$. If we replace $S_{\mathbf{P}}$ and $I_{GG}$ in (6) with any t-conorm $S$ and any implication $I$, then we obtain a class of implications defined by $N$, $S$ and $I$:

$$I^{N,S,I}(x,y) = S(N(x), I(x,y)), \text{ for all } x,y \in [0,1].$$

Using the same proof for $I^N$ we obtain that $I^{N,S,I}$ is always an implication. This class of implications helps us to generate new implications from existing ones, which will be the topic of further research.

## References

1. Baczyński, M.: Residual implications revisited. Notes on the Smets-Magrez Theorem. Fuzzy Sets and Systems 145, 267–277 (2004)
2. Baczyński, M., Drewniak, J.: Monotonic fuzzy implications. In: Szczepaniak, P.S., Lisboa, P.J.G., Kacprzyk, J. (eds.) Fuzzy Systems in Medicine, pp. 90–111. Physica, Heidelberg (2000)
3. Baczyński, M., Jayaram, B.: Fuzzy Implications. Springer, Heidelberg (2008)
4. Bustince, H., Burillo, P., Soria, F.: Automorphisms, negations and implication operators. Fuzzy Sets and Systems 134, 209–229 (2003)
5. Bustince, H., Pagola, M., Barrenechea, E.: Construction of fuzzy indices from fuzzy DI-subsethood measures: Application to the global comparison of images. Information Sciences 177, 906–929 (2007)
6. Drewniak, J.: Invariant fuzzy implications. Soft Computing 10, 506–513 (2006)
7. Fodor, J.C.: On fuzzy implication operators. Fuzzy Sets and Systems 42, 293–300 (1991)
8. Fodor, J.C., Roubens, M.: Fuzzy Preference Modelling and Multicriteria Decision Support. Kluwer Academic Publishers, Dordrecht (1994)
9. Fodor, J.C.: Contrapositive symmetry of fuzzy implications. Fuzzy Sets and Systems 69, 141–156 (1995)
10. Jayaram, B.: Rule reduction for efficient inferencing in similarity based reasoning. International Journal of Approximate Reasoning 48, 156–173 (2008)
11. Kerre, E.E.: A call for crispness in fuzzy set theory. Fuzzy Sets and Systems 29, 57–65 (1989)
12. Kerre, E.E., Nachtegael, M.: Fuzzy Techniques in Image Processing. Physica-Verlag, New York (2000)
13. Klement, E.P., Mesiar, R., Pap, E.: Triangular Norms. Kluwer Academic Publishers, Netherlands (2000)

14. Klir, J., Yuan, B.: Fuzzy Sets and Fuzzy Logic, Theory and Applications. Prentice Hall, New Jersey (1995)
15. Mas, M., Monserrat, M., Torrens, J.: QL-implications versus $D$-implications. Kybernetika 42, 956–966 (2006)
16. Mas, M., Monserrat, M., Torrens, J., Trillas, E.: A survey on fuzzy implication functions. IEEE Transactions on Fuzzy Systems 15(6), 1107–1121 (2007)
17. Nachtegael, M., Heijmans, H., Van der Weken, D., Kerre, E.: Fuzzy Adjunctions in Mathematical Morphology. In: Proc. of JCIS 2003, North Carolina, USA, pp. 202–205 (September 2003)
18. Novák, V., Perfilieva, I., Mockor, J.: Mathematical Principles of Fuzzy Logic. Kluwer Academic Publishers, Boston (1999)
19. Pei, D.: $R_0$ implication: characteristics and applications. Fuzzy Sets and Systems 131, 297–302 (2002)
20. Ruan, D., Kerre, E.E.: Fuzzy implication operators and generalized fuzzy method of cases. Fuzzy Sets and Systems 54, 23–37 (1993)
21. Ruan, D., Kerre, E.E.: Fuzzy IF-THEN Rules in Computational Intelligence: Theory and Applications. Kluwer Academic Publishers, Boston (1995)
22. Ruan, D., Kerre, E.: Fuzzy IF-THEN Rules in Computational Intelligence: Theory and Applications. Kluwer Academic Publishers, Boston (2000)
23. Shi, Y., Van Gasse, B., Ruan, D., Kerre, E.E.: Interrelationships among Fuzzy Implication Axioms: Dependence versus Independence. Fuzzy Sets and Systems 161, 1388–1405 (2010)
24. Trillas, E.: Sobre funciones de negación en la teoría de conjuntos difusos. Stochastica 3(1), 47–60 (1979)
25. Trillas, E., Alsina, C., Renedo, E., Pradera, A.: On contra-symmetry and MPT conditionality in fuzzy logic. International Journal of Intelligent Systems 20, 313–326 (2005)
26. Yager, R.R.: On some new classes of implication operators and their role in approximate reasoning. Information Sciences 167, 193–216 (2004)
27. Yan, P., Chen, G.: Discovering a cover set of ARsi with hierarchy from quantitative databases. Information Sciences 173, 319–336 (2005)
28. Zadeh, L.A.: Fuzzy sets. Information and Control 8, 338–353 (1965)
29. Zadeh, L.A.: Fuzzy logic and approximate reasoning. Synthese 30, 407–428 (1975)
30. Zhang, H.Y., Zhang, W.X.: Hybrid monotonic inclusion measure and its use in measuring similarity and distance between fuzzy sets. Fuzzy Sets and Systems 160, 107–118 (2009)

# Part II
# Data Analysis

# On Lattice–Based Fuzzy Rough Sets

Anna Maria Radzikowska

*Dedicated to Etienne E. Kerre*
*on the occasion of his retirement*

**Abstract.** Rough sets were originally proposed by Pawlak as a formal tool for analyzing and processing incomplete information represented in data tables. Later on, fuzzy generalizations of rough sets were introduced and investigated to be able to deal with imprecision. In this paper we present $L$–fuzzy rough sets as a further generalization of rough sets. As an underlying algebraic structure we take an *extended residuated lattice*, that is a residuated lattice endowed with a De Morgan negation. The signature of these structures gives algebraic counterparts of main fuzzy logical connectives. Properties of $L$–fuzzy rough sets are presented. We show that under some conditions families of all lower (resp. upper) $L$–fuzzy rough sets are complete (distributive) lattices. It is also pointed out that in some specific cases lower and upper fuzzy rough approximation operators are $L$–fuzzy topological operators of interior and closure, respectively.

## 1 Preface

I have enjoyed Etienne Kerre's acquaintance since the late nineties. In June 1998 I visited Gent for a short–term scientific mission in the framework of the project COST Action 15 "*Many–Value Logics for Computer Science Applications.* I was impressed and moved by Etienne's scientific open-mindness and his abilities to combine, in a mastery way, pure mathematics with modern approaches in mathematics of fuzziness and their wide applications in diverse areas. With great enthusiasm he

Anna Maria Radzikowska
Faculty of Mathematics and Information Science,
Warsaw University of Technology, Plac Politechniki 1, 00-661 Warsaw, Poland
e-mail: Anna.Radzikowska@mini.pw.edu.pl

C. Cornelis et al. (Eds.): 35 Years of Fuzzy Set Theory, STUDFUZZ 261, pp. 107–126.
springerlink.com 

introduced me into various aspects of fuzzy set theory. We discussed about links between fuzzy set theory and rough set theory, which started our long–lasting research cooperation on fuzzy rough sets, fuzzy modal logics, and fuzzy information relations (see, for example, [4],[5],[22],[26]–[36]) Since then I came to Gent every year in June for 2–3 weeks. Each visit was a very fruitful research experience and, due to Etienne's openness and kindness, it was also a fantastic, unforgettable stay.

I would like to take the opportunity to thank Etienne and Andrea for their great hospitality during my stays. I fondly remember our lunches, when I enjoyed their guidance on delicious Flemish cuisine. Thank you, Etienne, for being a reference, for your stimulating, challenging ideas, for your real, warm friendship – in summary, for being as you are.

## 2 Introduction

In real–life problems the available information is usually incomplete and/or imprecise. On one hand, not all relevant information is known or accessible, on the other hand, user's data may be imprecise or vague (e.g., when expressed by means of linguistic terms like "*quite good*" or "*rather cold*"). Rough set theory was developed by Pawlak ([24],[25]) as a formal tool for analyzing and processing information in data tables. As a consequence, it acts on partial information. This formalism proved to be a natural instrument to inquire into many theoretical and practical problems related to data analysis and knowledge discovery (see, e.g., [21],[39],[40],[45],[47]).

Fuzzy set theory ([46]), on the other hand, offers a wide variety of techniques for analyzing imprecise data. Basically, both theories address the problem of information granulation: the theory of fuzzy sets is centered upon fuzzy information granulation, whereas rough set theory is focused on crisp information granulation. Originally, the basic notion in rough set theory was indistinguishability (i.e., indistinguishability between objects in information systems induced by different values of attributes characterizing these objects), yet in recent extensions ([21]) the focus moved to the notion of similarity, which is in fact a fuzzy concept. It is therefore apparent that these two theories have become much closer to each other, so it seems natural to combine methods developed within both theories in order to construct hybrid structures capable to deal with both incompleteness and imprecision. Such structures, called *fuzzy rough sets* and *rough fuzzy sets*, were proposed in the literature ([8],[9],[23], [41]). In [27] Radzikowska and Kerre investigated fuzzy rough sets taking the unit interval $[0,1]$ as the basic structure.

In [15] Gougen pointed out that in many situations the linearly ordered set may be insufficient to adequately represent degrees of membership. From this reason, *L–fuzzy sets* were introduced, where degrees of membership are elements of a lattice $L$.

As mentioned above, the concept of rough set is recently focused on the notion of *similarity*. In many situations it is hard to specify whether similarity between objects $x$ and $y$ is stronger (resp. weaker) than similarity between $x$ and $z$. Hence it is naturally justified to consider $L$–fuzzy rough sets ([29],[30],[35]). In the present paper these structures are presented and investigated. Residuated lattices endowed with a De Morgan negation, referred to as *extended residuated lattices*, are taken as the underlying algebraic structures. These structures allow us to obtain algebraic counterparts of main fuzzy logical connectives: triangular norms and triangular conorms, $R$–implications and $S$–implications, and two types of fuzzy negations (including a De Morgan negation). Using two classes of fuzzy implications the respective two classes of $L$–fuzzy rough sets are investigated. We show that the families of lower and upper $L$–fuzzy rough sets constitute a lattice structure and point out conditions under which these lattices are distributive. Finally, we point out that under some conditions $L$–fuzzy rough approximation operators are fuzzy interior and fuzzy closure topological operators.

The present paper contains some important results on fuzzy rough sets that were obtained during my long cooperation with Etienne, but there are also some studies that were only a topic of our discussions, but not yet been published. This tribute book is a nice occasion to supplement our joint research results.

## 3 Rough Sets

Let $X$ be a non–empty universe and let $R$ be a non–empty binary relation on $X$. The set $X$ is viewed as a set of *objects* and $R$ represents relationships between these objects determined by their features (attributes). The question addressed in rough set theory is: *how to approximate a concept $A \subseteq X$ using classes of the relation $R$?*

The following notation will be used. Given a subset $A \subseteq X$, $-A$ will stand for the set complement of $A$, i.e., $-A = X \setminus A$. By $2^X$ we will denote the powerset of $X$. For a binary relation $R$ on $X$ and for every $x \in X$, we will write $xR$ to denote the class of $R$ with the representant $x$, that is the set $xR = \{y \in X : (x,y) \in R\}$.

Let $X$ be viewed as a set of objects (examples), and let $R$ be a binary relation on $X$ representing relationships between these objects (indistinguishability or similarity). The pair $\text{AS} = (X,R)$ is called an *approximation space*.

Given an approximation space $\text{AS} = (X,R)$, the following two mappings $\underline{\text{AS}}, \overline{\text{AS}} : 2^X \to 2^X$ are defined as: for every $A \subseteq X$,

$$\underline{\text{AS}}(A) = \{x \in X : xR \subseteq A\} \tag{1}$$

$$\overline{\text{AS}}(A) = \{x \in X : xR \cap A \neq \emptyset\}. \tag{2}$$

It is easily noted that the operations (1) and (2) coincide with modal operators of *necessity* and *possibility*, respectively (see, e.g., [3],[6]). For this reason, if $x \in \underline{\text{AS}}(A)$, then we say that *$x$ certainly belongs to $A$*, whereas if $x \in \overline{\text{AS}}(A)$, then it is said that *$x$ possibly belongs to $A$*.

For any $A \subseteq X$, $\underline{\text{AS}}(A)$, and $\overline{\text{AS}}(A)$ are respectively called a *lower* and an *upper rough approximation of $A$ in* AS. A pair $(A_1, A_2) \in 2^X \times 2^X$ is a *rough set in* AS iff $A_1 = \underline{\text{AS}}(A))$ and $A_2 = \overline{\text{AS}}(A)$ for some $A \subseteq X$. A subset $A \subseteq X$ is called *exact* iff $\underline{\text{AS}}(A) = A = \overline{\text{AS}}(A)$.

Originally, rough sets were defined with respect to approximation spaces with equivalence relations. Later generalizations of this notion lead to so–called *generalized rough sets*. In this paper we simply refer to these structures as *rough sets*.

Let us recall basic properties of rough sets (see, e.g., [6], [21],[24],[25],[45]).

**Theorem 1.** *For every approximation space* $As=(X,R)$,

**(i)** $\underline{\mathrm{AS}}(X)=X,\ \overline{\mathrm{AS}}(\emptyset)=\emptyset$

**(ii)** *for all* $A,B\subseteq X$, $A\subseteq B$ *implies* $\underline{\mathrm{AS}}(A)\subseteq\underline{\mathrm{AS}}(B)$ *and* $\overline{\mathrm{AS}}(A)\subseteq\overline{\mathrm{AS}}(B)$

**(iii)** *for every* $A\subseteq X$, $\underline{\mathrm{AS}}(A)=-\overline{\mathrm{AS}}(-A)$

**(iv)** *for every indexed family* $(A_i)_{i\in I}$ *of subsets of* $X$,

$$\begin{aligned}
\underline{\mathrm{AS}}(\textstyle\bigcap_{i\in I}A_i) &= \textstyle\bigcap_{i\in I}\underline{\mathrm{AS}}(A_i)\\
\overline{\mathrm{AS}}(\textstyle\bigcup_{i\in I}A_i) &= \textstyle\bigcup_{i\in I}\overline{\mathrm{AS}}(A_i)\\
\underline{\mathrm{AS}}(\textstyle\bigcup_{i\in I}A_i) &\supseteq \textstyle\bigcup_{i\in I}\underline{\mathrm{AS}}(A_i)\\
\overline{\mathrm{AS}}(\textstyle\bigcap_{i\in I}A_i) &\subseteq \textstyle\bigcap_{i\in I}\overline{\mathrm{AS}}(A_i).
\end{aligned}$$

It is well–known ([6]) that modal operators, and in consequence lower and upper rough approximation operators, are useful for characterizing properties of binary relations. The following theorem presents these results.

**Theorem 2.** ([6]) *For every approximation space* $\mathrm{AS}=(X,R)$ *and for every* $A\subseteq X$,

**(i)** *R is serial* *iff* $\underline{\mathrm{AS}}(A)\subseteq\overline{\mathrm{AS}}(A)$

**(ii)** *R is reflexive* *iff* $\underline{\mathrm{AS}}(A)\subseteq A$
*iff* $A\subseteq\overline{\mathrm{AS}}(A)$

**(iii)** *R is symmetric* *iff* $A\subseteq\underline{\mathrm{AS}}(\overline{As}(A))$
*iff* $\overline{\mathrm{AS}}(\underline{\mathrm{AS}}(A))\subseteq A$

**(iv)** *R is Euclidean* *iff* $\overline{\mathrm{AS}}(A)\subseteq\underline{\mathrm{AS}}(\overline{\mathrm{AS}}(A))$
*iff* $\overline{\mathrm{AS}}(\underline{\mathrm{AS}}(A))\subseteq\underline{\mathrm{AS}}(A)$

**(v)** *R is transitive* *iff* $\underline{\mathrm{AS}}(A)\subseteq\underline{\mathrm{AS}}(\underline{\mathrm{AS}}(A))$
*iff* $\overline{\mathrm{AS}}(\overline{\mathrm{AS}}(A))\subseteq\overline{\mathrm{AS}}(A)$.

Recall that a relation $R\subseteq X\times X$ is called *serial* iff for every $x\in X$ there is $y\in X$ such that $(x,y)\in R$; it is called *Euclidean* iff for all $x,y,z\in X$, $(x,y)\in R$ and $(x,z)\in R$ imply $(y,z)\in R$.

## 4 Algebraic Foundations

Let $L$ be a non–empty domain, let $1\in L$ be its distinguished element, and let $\otimes$ be a binary operation in $L$. A structure $(L,\otimes,1)$ is called a *monoid* iff $\otimes$ is associative operation satisfying $1\otimes a=a\otimes 1=1$ for every $a\in L$. A monoid is called *commutative* iff $\otimes$ is commutative.

Typical examples of monoid operations are triangular norms and triangular conorms (see [38]). A *triangular norm t* (resp. a *triangular conorm s*) is a commutative, associative and non–decreasing in both arguments $[0,1]^2 - [0,1]$ mapping satisfying the following border condition $t(1,a) = a$ (resp. $s(a,0) = a$) for every $a \in [0,1]$. Three most popular triangular norms are:

◇ the min operation (the Zadeh's triangular norm) $t_Z(a,b) = \min(a,b)$
◇ the algebraic product $t_P(a,b) = a \cdot b$
◇ the Łukasiewicz triangular norm $t_L(a,b) = \max(0, a+b-1)$.

A triangular norm is called *left–continuous* iff its partial mappings $t(a,\cdot)$ and $t(\cdot,a)$ are left–continuous for every $a \in [0,1]$.

The well–known triangular conorms are:

◇ the max operation (the Zadeh's triangular conorm) $s_Z(a,b) = \max(a,b)$
◇ the bounded sum $s_P(a,b) = a+b-a \cdot b$
◇ the Łukasiewicz triangular conorm $s_L(a,b) = \min(1, a+b)$.

Triangular norms and triangular conorms are the basis for constructing main classes of *fuzzy implications* ([19],[20],[44]), that is mappings $i : [0,1]^2 \to [0,1]$, non–increasing in the first argument, non–decreasing in the second argument, which satisfy the boundary conditions: $i(1,1) = i(0,0) = i(0,1) = 1$ and $i(1,0) = 0$. There are two main classes of these connectives:

- *R–implications*, also called a residual implications based on a left–continuous triangular norm $t$, or the residuum of $t$, defined as:
$$i_t(x,y) = \sup\{z \in [0,1] : t(x,z) \leq y\}$$
- *S–implications* based on a triangular conorm $s$ and a fuzzy negation[1] $n$ defined by: $i_{s,n}(x,y) = s(n(x),y)$.

The most popular R–implications, being the residua of $t_Z$, $t_P$ and $t_L$, respectively, are:

◇ the Gödel implication $i_Z(x,y) = 1$ iff $x \leq y$ and $i_Z(x,y) = y$ elsewhere
◇ the Gaines implication $i_P(x,y) = 1$ iff $x \leq y$ and $i_P(x,y) = \frac{y}{x}$ elsewhere
◇ the Łukasiewicz implication $i_L(x,y) = \min(1, 1-x+y)$.

The well–known S–implications are:

◇ the Kleene–Dienes implication $i_{s_Z,\eta} = \max(1-x, y)$ (based on $s_Z$ and $\eta$)
◇ the Reichenbach implication $i_{s_P,\eta}(x,y) = 1-x+x \cdot y$ (based on $s_P$ and $\eta$)
◇ the Łukasiewicz implication (based on $s_L$ and $\eta$)

Let $(L, \leqslant)$ be a poset[2] and let $\circ$ be a unary operation in $L$. We say that $\circ$ is *isotone* (order–preserving) iff for all $a, b \in L$, $a \leq b$ implies $\circ a \leq \circ b$; $\circ$ is called *antitone*

[1] A *fuzzy negation* is a non–increasing mapping $n : [0,1] \to [0,1]$ such that $n(1) = 0$ and $n(0) = 1$. The standard fuzzy negation $\eta$ is the mapping $\eta(x) = 1 - x$, $x \in [0,1]$.

[2] Recall that a *poset* is a structure $(X, \leqslant)$ such that $X \neq \emptyset$ and $\leqslant$ is a partial order, that is $\leqslant$ is reflexive ($x \leqslant x$), transitive ($x \leqslant y \,\&\, y \leqslant z \Longrightarrow x \leqslant z$), and antisymmetric ($x \leqslant y \,\&\, y \leqslant x \Longrightarrow x = y$).

(order–reversing) iff for all $a,b\in L$, $a\leq b$ implies $\circ b\leq\circ a$. We say that $\circ$ is *involution* iff $\circ\circ a=a$ for every $a\in L$.

The following algebra is of main importance, in particular in fuzzy logics (e.g., [7],[10],[11],[14],[16],[43]).

**Definition 1.** *A* residuated lattice *is an algebra* $(L,\wedge,\vee,\otimes,\rightarrow,0,1)$ *such that*

- **(i)** $(L,\wedge,\vee,0,1)$ *is a bounded lattice with the top element* 1 *and the bottom element* 0
- **(ii)** $(L,\otimes,1)$ *is a commutative monoid*
- **(iii)** $\rightarrow$ *is a binary operation in L defined as:*

$$a\rightarrow b=\sup\{c\in L: a\otimes c\leq b\}. \tag{3}$$

*The operation* $\otimes$ *is called a* product *and* $\rightarrow$ *is its* residuum.

Given a residuated lattice $(L,\wedge,\vee,\otimes,\rightarrow,0,1)$, the following *precomplement* operation is defined as: for every $a\in L$,

$$\neg a=a\rightarrow 0.$$

The precomplement operation is a generalization of the pseudo–complement in a lattice ([37]). If $\wedge=\otimes$, then $\rightarrow$ is the relative pseudo–complement, $\neg$ is the pseudo–complement, and $(L,\wedge,\vee,\rightarrow,\neg,0,1)$ is a Heyting algebra.

For recent results of residuated lattices we refer, for example, to [1], [17], and [18].

*Example 1.* Let $\otimes$ be a left–continuous t–norm and let $\rightarrow$ be the residual implication based on $t$. The structure $([0,1],\min,\max,\otimes,\rightarrow,0,1)$ is a residuated lattice.

It is easily noted that the product operation of a residuated lattice is the algebraic counterpart of a triangular norm, $\rightarrow$ corresponds to the residual implication based on $\otimes$, and $\neg$ corresponds to a fuzzy negation (in general, it is not involutive). However, residuated lattices do not give adequate counterparts of triangular conorms, fuzzy $S$–implications and fuzzy involutive negations. For these reasons, so–called *extended residuated lattices* were proposed ([29],[32],[35]).

**Definition 2.** *An* extended residuated lattice *(ER–lattice, for short) is a system* $(L,\wedge,\vee,\otimes,\rightarrow,\sim,0,1)$ *such that*

- **(i)** $(L,\wedge,\vee,\otimes,\rightarrow,0,1)$ *is a residuated lattice*
- **(ii)** $\sim\sim a=a$ *for every* $a\in L$
- **(iii)** $\sim(a\vee b)=\sim a\wedge\sim b$ *for all* $a,b\in L$.

From the above definition it immediately follows that an ER–lattice is a residuated latticed endowed with a De Morgan negation.

An ER–lattice $(L,\wedge,\vee,\otimes,\rightarrow,\sim,0,1)$ is *complete* iff the underlying lattice $(L,\wedge,\vee,0,1)$ is complete.

Given an ER–lattice $(L,\wedge,\vee,\otimes,\rightarrow,\sim,0,1)$, let us define the following binary operations in $L$: for all $a,b\in L$,

$$a\oplus b = \sim(\sim a\otimes\sim b) \tag{4}$$

$$a\Rightarrow b = \sim a\oplus b \tag{5}$$

One can easily note that the operation $\sim$ of an ER–lattice corresponds to a fuzzy De Morgan negation, $\oplus$ and $\Rightarrow$ are algebraic counterparts of a triangular conorm, and a fuzzy $S$–implication, respectively. Therefore, the signature of an ER–lattice gives algebraic counterparts of main fuzzy logical connectives: a triangular norm and a triangular conorm, an $R$–implication and an $S$–implications, and two fuzzy negations (including a De Morgan negation). Properties of ER–lattices can be found, for example, in [35].

Important properties of ER–lattices are listed in the following two propositions.

**Proposition 1.** *Let* $(L,\wedge,\vee,\otimes,\rightarrow,\sim,0,1)$ *be an ER–lattice and let* $\leq$ *be the natural lattice ordering. Then for all* $a,b,c\in L$*, the following conditions hold:*

**(i)** $\otimes$ *and* $\oplus$ *are isotone in both arguments*
**(ii)** $\rightarrow$ *and* $\Rightarrow$ *are antitone in the* $1^{st}$ *and isotone in the* $2^{nd}$ *argument*
**(iii)** $\sim$ *and* $\neg$ *are antitone*
**(iv)** $a\otimes b\leq a\wedge b,\ a\vee b\leq a\oplus b$
**(v)** $a\otimes b\leq a,\ a\leq a\oplus b$
**(vi)** $a\otimes 0=0,\ a\oplus 1=1$
**(vii)** $a\rightarrow a = a\rightarrow 1 = 0\rightarrow a = 1, \quad 1\rightarrow a = a$
**(viii)** $a\Rightarrow 1 = 0\Rightarrow a = 1, \quad 1\Rightarrow a = a$
**(ix)** $a\leq b$ *iff* $a\rightarrow b = 1$
**(x)** $a\otimes(a\rightarrow b)\leq b$
**(xi)** $a\otimes(b\rightarrow c)\leq b\rightarrow(a\otimes c)$
**(xii)** $(a\rightarrow b)\otimes(b\rightarrow c)\leq(a\rightarrow c)$
$(a\Rightarrow c)\leq(a\Rightarrow b)\oplus(b\Rightarrow c)$
**(xiii)** $(a\rightarrow b)\leq(c\rightarrow a)\rightarrow(c\rightarrow b)$
**(xiv)** $a\rightarrow b\leq(a\otimes c)\rightarrow(b\otimes c)$
**(xv)** $b\leq a\rightarrow(a\otimes b)$
**(xvi)** $a\rightarrow(b\rightarrow c) = (a\otimes b)\rightarrow c$
$a\Rightarrow(b\Rightarrow c) = (a\otimes b)\Rightarrow c$
**(xvii)** $a\rightarrow\neg b = \neg(a\otimes b)$
$a\Rightarrow\sim b = \sim(a\otimes b)$
**(xviii)** $a\rightarrow b\leq\neg b\rightarrow\neg a$
$a\Rightarrow b = \sim b\Rightarrow\sim a$
**(xix)** $a\Rightarrow b = \sim(a\otimes\sim b)$
**(xx)** $a\leq\neg\neg a$.

**Proposition 2.** *Let* $(L,\wedge,\vee,\otimes,\rightarrow,\sim,0,1)$ *be an ER–lattice. Then for every* $a\in L$ *and for every family* $(b_i)_{i\in I}$ *of elements of* $L$*, if the respective infima and suprema exist, then the following conditions hold:*

**(i)** $a \otimes \sup_{i\in I} b_i = \sup_{i\in I}(a \otimes b_i)$ $\quad a \oplus \inf_{i\in I} b_i = \inf_{i\in I}(a \oplus b_i)$
**(ii)** $a \rightarrow \inf_{i\in I} b_i = \inf_{i\in I}(a \rightarrow b_i)$ $\quad a \Rightarrow \inf_{i\in I} b_i = \inf_{i\in I}(a \Rightarrow b_i)$
**(iii)** $(\sup_{i\in I} b_i) \rightarrow a = \inf_{i\in I}(b_i \rightarrow a)$ $\quad (\sup_{i\in I} b_i) \Rightarrow a = \inf_{i\in I}(b_i \Rightarrow a)$
**(iv)** $\sup_{i\in I} b_i \leq \neg \inf_{i\in I} \neg b_i$ $\quad \sup_{i\in I} b_i = \sim \inf_{i\in I} \sim b_i$
**(v)** $\inf_{i\in I} \neg b_i = \neg \sup_{i\in I} b_i$.

Throughout the paper we will use the same symbol $L$ to denote the ER–lattice as well as its underlying domain.

Let $L$ be an ER–lattice and let $X \neq \emptyset$. An *L–fuzzy set in* $X$ is a mapping $F : X \rightarrow L$. For any $x \in X$, $F(x)$ is the degree to which $x$ belongs to $F$. The family of all $L$–fuzzy sets in $X$ will be denoted by $\mathscr{F}_L(X)$.

Basic operations on $L$–fuzzy sets in $X$ are defined as follows: for all $A, B \in \mathscr{F}_L(X)$ and for every $x \in X$,

$$
\begin{aligned}
(A \cap_L B)(x) &= A(x) \wedge B(x) \\
(A \sqcap_L B)(x) &= A(x) \otimes B(x) \\
(A \cup_L B)(x) &= A(x) \vee B(x) \\
(A \sqcup_L B)(x) &= A(x) \oplus B(x) \\
(\ominus_L A)(x) &= \ominus A(x), \text{ where } \ominus \in \{\neg, \sim\},
\end{aligned}
$$

Given a complete ER–lattice $L$ and an indexed family $(A_i)_{i\in I}$ of $L$–fuzzy sets in $X$, we write $\bigcup_{i\in I} A_i$ and $\bigcap_{i\in I} A_i$ to denote $L$–fuzzy sets in $X$ defined respectively as: for every $x \in X$,

$$
\begin{aligned}
(\textstyle\bigcup_{i\in I} A_i)(x) &= \sup_{i\in I} A_i(x) \\
(\textstyle\bigcap_{i\in I} A_i)(x) &= \inf_{i\in I} A_i(x).
\end{aligned}
$$

For two $L$–fuzzy sets $A$ and $B$ in $X$, we will write $A \subseteq_L B$ iff $A(x) \leq B(x)$ for every $x \in X$.

A binary *L–fuzzy relation in* $X$ is a mapping $X \times X \rightarrow L$, i.e., an $L$–fuzzy set in $X \times X$. The family of all binary $L$–fuzzy relations in $X$ will be denoted by $\mathscr{R}_L(X)$.

For a binary $L$–fuzzy relation $R$ in $X$ and for $x \in X$, we will write $xR$ to denote the $L$–fuzzy set in $X$ defined as: $(xR)(y) = R(x, y)$ for every $y \in X$.

A binary $L$–fuzzy relation $R$ is called

- *serial* iff $\sup_{y\in X} R(x,y) = 1$ for every $x \in X$
- *reflexive* iff $R(x,x) = 1$ for every $x \in X$
- *symmetric* iff $R(x,y) = R(y,x)$ for all $x, y \in X$
- *L–transitive* iff $R(x,y) \otimes R(y,z) \leq R(x,z)$ for all $x, y, z \in X$
- *L–Euclidean* iff $R(x,y) \otimes R(x,z) \leq R(y,z)$ for all $x, y, z \in X$.

A binary $L$–fuzzy relation $R$ in $X$ is a *tolerance* relation iff it is reflexive and symmetric, it is called an *L–quasi ordering* iff it is reflexive and $L$–transitive, and it is an *L–equivalence* relation iff it is reflexive, symmetric and $L$–transitive.

## 5 L–Fuzzy Rough Sets

Let $L$ be a complete ER–lattice, $X$ be a non–empty universe, and let $R \in \mathscr{R}_L(X)$. A structure $\text{FAS} = (L, X, R)$ is called an *L–fuzzy approximation space.*

Given an $L$–fuzzy approximation space $(L, X, R)$, define the following operations: for every $A \in \mathscr{F}_L(X)$ and for every $x \in X$,

$$\underline{\text{FAS}_R}(A)(x) = \inf_{y \in X}(R(x,y) \rightarrow A(y)), \tag{6}$$

$$\underline{\text{FAS}_S}(A)(x) = \inf_{y \in X}(R(x,y) \Rightarrow A(y)), \tag{7}$$

$$\overline{\text{FAS}}(A)(x) = \sup_{y \in X}(R(x,y) \otimes A(y)). \tag{8}$$

The operation (6) (resp. (7)) is called a *lower $L_R$–fuzzy* (resp. lower *$L_S$–fuzzy) rough approximation of A in* FAS, and the operation (8) is an *upper L–fuzzy rough approximation of A in* FAS. In (6) and (7) the subscripts $R$ and $S$ indicate that the underlying arrow operations correspond to an $R$–implication and an $S$–implication, respectively.

Note that a lower $L_R$–fuzzy (resp. lower $L_S$–fuzzy) rough approximation operation corresponds to a fuzzy box modal operator and an upper $L$–fuzzy rough approximation operation corresponds to a fuzzy diamond modal operator (see, e.g., [12],[13]). In consequence, they have similar intuitive interpretation. Namely, $\underline{\text{FAS}_R}(A)(x)$ (resp. $\underline{\text{FAS}_S}(A)(x)$) is the degree to which $x$ *certainly* belongs to $A$, while $\overline{\text{FAS}}(A)(x)$ is the degree to which $x$ *possibly* belongs to $A$.

**Definition 3.** *Let an L–fuzzy approximation space $(L, X, R)$ be given. We say that $F \in \mathscr{F}_L(X)$ is a* lower $L_R$–fuzzy *(resp. $L_S$–*fuzzy*)* rough set in FAS *iff $F = \underline{\text{FAS}_R}(A)$ (resp. $F = \underline{\text{FAS}_S}(A)$) for some $A \in \mathscr{F}_L(X)$, F is called an* upper $L$–fuzzy rough set in FAS *iff $F = \overline{\text{FAS}}(A)$ for some $A \in \mathscr{F}_L(X)$. By an* $L_R$–fuzzy rough set in FAS *(resp.* $L_S$–fuzzy rough set in FAS*) we mean a pair $(F_1, F_2) \in \mathscr{F}_L(X) \times \mathscr{F}_L(X)$ such that $F_1 = \underline{\text{FAS}_R}(A)$ and $F_2 = \overline{\text{FAS}}(A)$ (resp. $F_1 = \underline{\text{FAS}_S}(A)$ and $F_2 = \overline{\text{FAS}}(A)$) for some $A \in \mathscr{F}_L(X)$.*

The family of all lower $L_R$–fuzzy (resp. $L_S$–fuzzy) rough sets in FAS will be respectively written $\mathfrak{L}_R(\text{FAS})$ and $\mathfrak{L}_S(\text{FAS})$ and the family of all upper $L$–fuzzy rough sets in FAS will be denoted by $\mathfrak{U}(\text{FAS})$.

Basic properties of $L$–fuzzy rough sets are listed below.

**Theorem 3.** ([30],[34],[35]) *For every L–fuzzy approximation space $(L, X, R)$, for all $A, B \in \mathscr{F}_L(X)$, and for every indexed family $(A_i)_{i \in I}$ of L–fuzzy sets in X, the following conditions hold:*

**(i)** $\underline{\text{FAS}_R}(X) = \underline{\text{FAS}_S}(X) = X, \ \overline{\text{FAS}}(\emptyset) = \emptyset$

**(ii)** *if $A \subseteq_L B$, then*

- $\underline{\text{FAS}_R}(A) \subseteq_L \underline{\text{FAS}_R}(B), \ \underline{\text{FAS}_S}(A) \subseteq_L \underline{\text{FAS}_S}(B)$
- $\overline{\text{FAS}}(A) \subseteq_L \overline{\text{FAS}}(B)$

**(iii)** $\underline{\text{FAS}_R}(A) \subseteq_L \neg_L \overline{\text{FAS}}(\neg_L A)$
$\overline{\text{FAS}}(A) \subseteq_L \neg_L \underline{\text{FAS}_R}(\neg_L A)$

$$\underline{\text{FAS}}_S(A) = \sim_L \overline{\text{FAS}}(\sim_L A)$$
$$\overline{\text{FAS}}(A) = \sim_L \underline{\text{FAS}}_S(\sim_L A).$$

**(iv)** $\underline{\text{FAS}}_R(\bigcap_{i\in I} A_i) = \bigcap_{i\in I} \underline{\text{FAS}}_R(A_i), \quad \underline{\text{FAS}}_S(\bigcap_{i\in I} A_i) = \bigcap_{i\in I} \underline{\text{FAS}}_S(A_i)$

$\underline{\text{FAS}}_R(\bigcup_{i\in I} A_i) \; {}_L\!\supseteq \bigcup_{i\in I} \underline{\text{FAS}}_R(A_i), \; \underline{\text{FAS}}_R(\bigcup_{i\in I} A_i) \; {}_L\!\supseteq \bigcup_{i\in I} \underline{\text{FAS}}_R(A_i)$

$\overline{\text{FAS}}(\bigcap_{i\in I} A) \subseteq_L \bigcap_{i\in I} \overline{\text{FAS}}(A_i)$

$\overline{\text{FAS}}(\bigcup_{i\in I} A_i) = \bigcup_{i\in I} \overline{\text{FAS}}(A_i).$

The property **(ii)** of the above theorem states the monotonicity of $L$–fuzzy approximation operators and **(iii)** states the duality (resp. weak duality) between $\underline{\text{FAS}}_S$ (resp. $\underline{\text{FAS}}_R$) and $\overline{\text{FAS}}$. Observe that for $L_S$–fuzzy rough sets all properties mentioned above correspond to the ones in the crisp case (see Theorem 1) For $L_R$–fuzzy rough sets, however, we have a weaker form of duality only.

Depending on classes of $L$–fuzzy relations $L$–fuzzy rough sets have more specific properties.

**Theorem 4.** *Let* $(L,X,R)$ *be an L–fuzzy approximation space.*

**(i)** *$R$ is serial* *iff* $\underline{\text{FAS}}_R(A) \subseteq_L \overline{\text{FAS}}(A)$
*iff* $\underline{\text{FAS}}_S(A) \subseteq_L \overline{\text{FAS}}(A)$

**(ii)** *$R$ is reflexive* *iff* $\underline{\text{FAS}}_R(A) \subseteq_L A$
*iff* $\underline{\text{FAS}}_S(A) \subseteq_L A$
*iff* $A \subseteq_L \overline{\text{FAS}}(A)$

**(iii)** *$R$ is symmetric* *iff* $\overline{\text{FAS}}(\underline{\text{FAS}}_R(A)) \subseteq_L A$
*iff* $A \subseteq_L \underline{\text{FAS}}_R(\overline{\text{FAS}}(A))$

**(iv)** *$R$ is L–transitive* *iff* $\underline{\text{FAS}}_R(A) \subseteq_L \underline{\text{FAS}}_R(\underline{\text{FAS}}_R(A))$
*iff* $\underline{\text{FAS}}_S(A) \subseteq_L \underline{\text{FAS}}_S(\underline{\text{FAS}}_S(A))$
*iff* $\overline{\text{FAS}}(\overline{\text{FAS}}(A)) \subseteq_L \overline{\text{FAS}}(A)$

**(v)** *$R$ is L–Euclidean* *iff* $\overline{\text{FAS}}(A) \subseteq_L \underline{\text{FAS}}_R(\overline{\text{FAS}}(A))$
*iff* $\overline{\text{FAS}}(\underline{\text{FAS}}_R(A)) \subseteq_L \underline{\text{FAS}}_R(A).$

*Proof.* By way of example we show **(ii)** and **(iv)**.
**(ii)** $(\Rightarrow)$ Assume that $R$ is reflexive. Then for every $A \in \mathscr{F}_L(X)$ and for every $x \in X$,

$$\begin{aligned}\underline{\text{FAS}}_R(A)(x) &= \inf_{y\in X}(R(x,y) \to A(y)) \\ &\le R(x,x) \to A(x) = 1 \to A(x) = A(x).\end{aligned}$$

Hence $\underline{\text{FAS}}_R(A) \subseteq_L A$. Also,

$$\begin{aligned}\underline{\text{FAS}}_S(A)(x) &= \inf_{y\in X}(R(x,y) \Rightarrow A(y)) \\ &\le R(x,x) \Rightarrow A(x) = \sim 1 \oplus A(x) = 0 \oplus A(x) = A(x),\end{aligned}$$

so $\underline{\text{FAS}}_S(A) \subseteq_L A$. Furthermore,

$$\begin{aligned}\overline{\text{FAS}}(A)(x) &= \sup_{y\in X}(R(x,y) \otimes A(y)) \\ &\ge R(x,x) \otimes A(x) = 1 \otimes A(x) = A(x),\end{aligned}$$

where $\ge = \le^{-1}$. Therefore, $A \subseteq_L \overline{\text{FAS}}(A)$.

($\Leftarrow$) Assume that $R$ is not reflexive, i.e., $R(x_0,x_0)\neq 1$ for some $x_0\in X$. For $A=x_0R$, we have:

$$\begin{aligned}\underline{\text{FAS}_R}(A)(x_0) &= \inf_{y\in X}(R(x_0,y)\to A(y))\\ &= \inf_{y\in X}(R(x_0,y)\to R(x_0,y))\\ &= 1 && \text{by Proposition 1(\textbf{vii}).}\end{aligned}$$

However, $A(x_0)=R(x_0,x_0)<1$. Hence $\underline{\text{FAS}_R}(A)\not\subseteq_L A$.

Next, let us take $A=\{x_0\}$. Then we have:

$$\begin{aligned}\underline{\text{FAS}_S}(A)(x_0) &= \inf_{y\in X}(R(x_0,y)\Rightarrow A(y))\\ &= \inf_{y\in X}(\sim R(x_0,y)\oplus A(y)) && \text{by definition (5)}\\ &= \inf_{y\in X}(\sim(\sim\sim R(x_0,y)\otimes\sim A(y))) && \text{by definition (4)}\\ &= \inf_{y\in X}(\sim(R(x_0,y)\otimes\sim A(y))\\ &= \sim\inf_{y\in X}(R(x_0,y)\otimes\sim A(y)) && \text{by Proposition 1(\textbf{iv})}\\ &= \sim R(x_0,x_0)\\ &> 0=A(x_0)\end{aligned}$$

so $\underline{\text{FAS}_S}(A)\not\subseteq_L A$.

Finally, for $A=\{x_0\}$, we have:

$$\begin{aligned}\overline{\text{FAS}}(A)(x_0) &= \sup_{y\in X}(R(x_0,y)\otimes A(y))\\ &= R(x_0,x_0)\otimes A(x_0)=R(x_0,x_0)\otimes 1=R(x_0,x_0)<1=A(x_0),\end{aligned}$$

thus $A\not\subseteq_L\overline{\text{FAS}}(A)$.

**(iv)** ($\Rightarrow$) Assume that $R$ is $L$–transitive. Then for every $A\in\mathscr{F}_L(X)$ and for every $x\in X$,

$$\begin{aligned}&\underline{\text{FAS}_R}(\underline{\text{FAS}_R}(A))(x)\\ &= \inf_{y\in X}(R(x,y)\to(\inf_{z\in X}(R(y,z)\to A(z)))\\ &= \inf_{y\in X}\inf_{z\in Z}(R(x,y)\to(R(y,z)\to A(z))) && \text{by Proposition 2(\textbf{xv})}\\ &= \inf_{y\in X}\inf_{z\in X}(R(x,y)\otimes R(y,z)\to A(z)) && \text{by Proposition 1(\textbf{ii})}\\ &\geq \inf_{z\in X}(R(x,z)\to A(z)) && \text{by } L\text{–transitivity of } R\\ &= \underline{\text{FAS}_R}(A)(x).\end{aligned}$$

Hence $\underline{\text{FAS}_R}(A)\subseteq_L\underline{\text{FAS}_R}(\underline{\text{FAS}_R}(A))$.
In the similar way one can show that $\underline{\text{FAS}_S}(A)\subseteq_L\underline{\text{FAS}_S}(\underline{\text{FAS}_S}(A))$. Also,

$$\begin{aligned}&\overline{\text{FAS}}(\overline{\text{FAS}}(A))(x)\\ &= \sup_{y\in X}(R(x,y)\otimes\sup_{z\in X}(R(y,z)\otimes A(z)))\\ &= \sup_{y\in X}\sup_{z\in X}(R(x,y)\otimes R(y,z)\otimes A(z)) && \text{by Proposition 2(\textbf{i})}\\ &\leq \sup_{z\in X}(R(x,z)\otimes A(z)) && \text{by } L\text{–transitivity of } R\\ &= \overline{\text{FAS}}(A)(x),\end{aligned}$$

thus $\overline{\text{FAS}}(\overline{\text{FAS}}(A))\subseteq_L\overline{\text{FAS}}(A)$.

($\Leftarrow$) Assume that $R$ is not $L$–transitive, that is $R(x_0,y_0)\otimes R(y_0,z_0)\not\leq R(x_0,z_0)$ for some $x_0,y_0,z_0\in X$. By Proposition 1(**ix**), $(R(x_0,y_0)\otimes R(y_0,z_0))\to R(x_0,z_0)\not\leq 1$ for some $x_0, y_0, z_0 \in X$. Let us take $A = x_0R$. By Proposition 1(**vii**) we get $\underline{\text{FAS}}_R(A)(x_0) = \inf_{y\in X}(R(x_0,y) \to R(x_0,y)) = 1$. Furthermore,

$$\begin{aligned}
&\underline{\text{FAS}}_R(A)(x_0) \to \underline{\text{FAS}}_R(\underline{\text{FAS}}_R(A))(x_0)\\
&\quad = 1 \to \underline{\text{FAS}}_R(\underline{\text{FAS}}_R(A))(x_0)\\
&\quad = \underline{\text{FAS}}_R(\underline{\text{FAS}}_R(A))(x_0) && \text{by Proposition 1(\textbf{vii})}\\
&\quad = \inf_{y\in X}(R(x_0,y) \to \inf_{z\in X}(R(y,z) \to R(x_0,z)))\\
&\quad = \inf_{y\in X}\inf_{z\in X}(R(x_0,y) \to (R(y,z) \to R(x_0,z))) && \text{by Proposition 2(\textbf{ii})}\\
&\quad = \inf_{y\in X}\inf_{z\in X}(R(x_0,y) \otimes R(y,z) \to R(x_0,z)) && \text{by Proposition 1(\textbf{xvi})}\\
&\quad \le (R(x_0,y_0) \otimes R(y_0,z_0)) \to R(x_0,z_0))\\
&\quad \not\le 1,
\end{aligned}$$

so again, by Proposition 1(**ix**), $\underline{\text{FAS}}_R(A)(x_0) \not\le \underline{\text{FAS}}_R(\underline{\text{FAS}}_R(A))(x_0)$. Hence we have $\underline{\text{FAS}}_R(A) \not\subseteq_L \underline{\text{FAS}}_R(\underline{\text{FAS}}_R(A))$, as required.

Now we show that for $A = X\setminus\{z_0\}$ it holds $\underline{\text{FAS}}_S(A) \not\subseteq_L \underline{\text{FAS}}_S(\underline{\text{FAS}}_S(A))$. Note that $R(x_0,y_0) \otimes R(y_0,z_0) \not\le R(x_0,z_0)$ implies $\sim R(x_0,z_0) \not\le \sim(R(x_0,y_0) \otimes R(y_0,z_0))$, since $\sim$ is order–reversing. Then, by Proposition 1(**ix**),

$$\sim R(x_0,z_0) \to \sim(R(x_0,y_0) \otimes R(y_0,z_0)) \ne 1. \tag{9}$$

Also, for every $y \in X$, we have

$$\begin{aligned}
\underline{\text{FAS}}_S(A)(y) &= \inf_{z\in X}(R(y,z) \Rightarrow A(z))\\
&= \inf_{z\in X} \sim(R(y,z) \otimes \sim A(z)) && \text{by Proposition 1(\textbf{xix})}\\
&= \sim \sup_{z\in X}(R(y,z) \otimes \sim A(z)) && \text{by Proposition 2(\textbf{iv})}\\
&= \sim R(y,z_0).
\end{aligned}$$

Now we have:

$$\begin{aligned}
&\underline{\text{FAS}}_S(A)(x_0) \to \underline{\text{FAS}}_S(\underline{\text{FAS}}_S(A))(x_0)\\
&\quad = \sim R(x_0,z_0) \to \inf_{y\in X}(R(x_0,y) \Rightarrow \underline{\text{FAS}}_S(A)(y))\\
&\quad = \sim R(x_0,z_0) \to \inf_{y\in X}(R(x_0,y) \Rightarrow \sim R(y,z_0))\\
&\quad = \sim R(x_0,z_0) \to \inf_{y\in X} \sim(R(x_0,y) \otimes \sim\sim R(y,z_0)) && \text{by Proposition 1(\textbf{xix})}\\
&\quad = \sim R(x_0,z_0) \to \sim \sup_{y\in X}(R(x_0,y) \otimes R(y,z_0)) && \text{by Proposition 2(\textbf{vi})}\\
&\quad \le \sim R(x_0,z_0) \to \sim(R(x_0,y_0) \otimes R(y_0,z_0))\\
&\quad \ne 1 && \text{by (9).}
\end{aligned}$$

Hence, by Proposition 1(**ix**), $\underline{\text{FAS}}_S(A)(x_0) \not\le \underline{\text{FAS}}_S(\underline{\text{FAS}}_S(A))(x_0)$, which immediately implies $\underline{\text{FAS}}_S(A) \not\subseteq_L \underline{\text{FAS}}_S(\underline{\text{FAS}}_S(A))$, as expected.

Finally, take $A = \{z_0\}$. Observe that $\overline{\text{FAS}}(A)(y) = \sup_{z\in X}(R(y,z) \otimes A(z)) = R(y,z_0)$ for every $y \in X$. Now,

$$\begin{aligned}
&\overline{\text{FAS}}(\overline{\text{FAS}}(A))(x_0) \to \overline{\text{FAS}}(A)(x_0)\\
&\quad = \sup_{y\in X}(R(x_0,y) \otimes \overline{\text{FAS}}(A)(y)) \to \overline{\text{FAS}}(A)(x_0)\\
&\quad = \sup_{y\in X}(R(x_0,y) \otimes R(y,z_0)) \to R(x_0,z_0)\\
&\quad \le (R(x_0,y_0) \otimes R(y_0,z_0)) \to R(x_0,z_0) && \text{by Proposition 1(\textbf{ii})}\\
&\quad \ne 1,
\end{aligned}$$

so $\overline{\text{FAS}}(\overline{\text{FAS}}(A))(x_0) \not\le \overline{\text{FAS}}(A)(x_0)$ by Proposition 1(**ix**), thus $\overline{\text{FAS}}(\overline{\text{FAS}}(A)) \not\subseteq_L \overline{\text{FAS}}(A)$, as required. □

It is worth emphasizing that all properties mentioned in Theorem 4 hold for lower $L_R$–fuzzy rough approximation operators. As stated in Theorem 2, they characterize rough approximation operators. Then these operations are adequate generalizations of their crisp counterparts. However, in the general case properties **(iii)** and **(v)** of the above theorem do not hold for lower $L_S$–fuzzy rough approximations, as the following example shows.

*Example 2.* Let us take a lattice $(L, \min, \max, t_Z, i_{t_Z}, \eta, 0, 1)$ with $L = [0,1]$. Then $\oplus$ is given by $s_Z(x,y) = \max(x,y)$. Let $X = \{x_1, x_2, x_3\}$ and let $R$ be defined as: $R(x_1,x_1) = R(x_3,x_3) = 1$ and $R(x,y) = \frac{1}{2}$ elsewhere. Clearly, $R$ is symmetric. Consider $\text{FAS} = (L,X,R)$ and $A = \{x_3\}$. By simple calculations we get $\overline{\text{FAS}}(\underline{\text{FAS}}_S(A)) = \underline{\text{FAS}}_S(\overline{\text{FAS}}(A)) = (x_1 : \frac{1}{2}, x_2 : \frac{1}{2}, x_3 : \frac{1}{2})$ Obviously, neither $\overline{\text{FAS}}(\underline{\text{FAS}}_S(A)) \subseteq A$ nor $A \subseteq \underline{\text{FAS}}_S(\overline{\text{FAS}}(A))$ holds. Hence, for lower $L_S$–fuzzy rough approximation operation, the property **(iii)** of Theorem 4 does not hold.

Also, it is easy to verify that $R$ is an $L$–Euclidean relation. Since $\underline{\text{FAS}}_S(A) = (x_1 : 0, x_2 : \frac{1}{2}, x_3 : \frac{1}{2})$ and $\overline{\text{FAS}}(A) = (x_1 : \frac{1}{2}, x_2 : \frac{1}{2}, x_3 : 1)$, neither $\overline{\text{FAS}}(A) \subseteq \underline{\text{FAS}}_S(\overline{\text{FAS}}(A))$ nor $\overline{\text{FAS}}(\underline{\text{FAS}}_S) \subseteq \underline{\text{FAS}}_S(A)$.

As mentioned in Theorem 3, lower $L_S$–fuzzy rough approximation operators are dual to upper $L$–fuzzy rough approximation operators, which in general is not the case when lower $L_R$–fuzzy and upper $L$–fuzzy rough approximations are taken. Recall that the Łukasiewicz implication is both an $R$–implication and an $S$–implication. For such operators all properties mentioned in Theorem 3 and 4 are satisfied. Therefore, lattice–based fuzzy rough sets, where such operators are taken, are straightforward generalizations of Pawlak's rough sets.

From the above theorem the following corollary easily follows.

**Corollary 1.** *Let $(L,X,R)$ be an $L$–fuzzy approximation space. Then for every $A \in \mathscr{F}_L(X)$,*

**(i)** *if $R$ is an $L$–quasi ordering, then*

- $\underline{\text{FAS}}_R(\underline{\text{FAS}}_R(A)) = \underline{\text{FAS}}_R(A)$
- $\underline{\text{FAS}}_S(\underline{\text{FAS}}_S(A)) = \underline{\text{FAS}}_S(A)$
- $\overline{\text{FAS}}(\overline{\text{FAS}}(A)) = \overline{\text{FAS}}(A)$.

**(ii)** *if $R$ is a tolerance relation, then*

- $\underline{\text{FAS}}_R(A) \subseteq_L \overline{\text{FAS}}(\underline{\text{FAS}}_R(A)) \subseteq_L A$
- $A \subseteq_L \underline{\text{FAS}}_R(\overline{\text{FAS}}(A)) \subseteq_L \overline{\text{FAS}}(A)$

**(iii)** *if $R$ is an $L$–equivalence relation, then*

- $\overline{\text{FAS}}(\underline{\text{FAS}}_R(A)) = \underline{\text{FAS}}_R(A)$
- $\underline{\text{FAS}}_R(\overline{\text{FAS}}(A)) = \overline{\text{FAS}}(A)$.

Therefore, in the case of a tolerance relation $\overline{\text{FAS}}\underline{\text{FAS}}_R$ and $\underline{\text{FAS}}_R\overline{\text{FAS}}$ give a tighter approximation of $A$ than $\underline{\text{FAS}}_R$ and $\overline{\text{FAS}}$, respectively.

We conclude this section by the following observation. In classical settings it is well-known that box and diamond modal operators coincide with interior and closure topological operators, respectively, provided that the underlying relation is a quasi–ordering. i.e., it is reflexive and transitive. The analogous result holds in the fuzzy case.

First, let us recall (e.g., [2],[34]) the definitions of $L$–fuzzy topological operators of interior and closure. Let $int_L, cl_L : \mathscr{F}_L(X) \to \mathscr{F}_L(X)$. We say that $int_L$ is an *L–fuzzy interior operator* iff

**(I1)** $int_L(X) = X$
**(I2)** $int_L(A) \subseteq_L A$ for every $A \in \mathscr{F}_L(X)$
**(I3)** $int_L(int_L(A)) = int_L(A)$ for every $A \in \mathscr{F}_L(X)$
**(I4)** $int_L(A \cap_L B) = int_L(A) \cap_L int_L(B)$.

We say that $cl_L$ is an *L–fuzzy closure operator* iff

**(C1)** $cl_L(\emptyset) = \emptyset$
**(C2)** $A \subseteq_L cl_L(A)$ for every $A \in \mathscr{F}_L(X)$
**(C3)** $cl_L(cl_L(A)) = cl_L(A)$ for every $A \in \mathscr{F}_L(X)$
**(C4)** $cl_L(A \cup_L B) = cl_L(A) \cup_L cl_L(B)$.

From Theorem 3 and Corollary 1(**i**) we immediately get the following result.

**Theorem 5.** *Let* $\text{FAS} = (L, X, R)$ *be an L–fuzzy approximation space such that R is an L–quasi ordering. Then* $\underline{\text{FAS}}_R$ *and* $\underline{\text{FAS}}_S(A)$ *are L–fuzzy interior operators and* $\overline{\text{FAS}}$ *is an L–fuzzy closure operator.*

## 6 Lattices of Lower and Upper $L$–Fuzzy Rough Sets

Let an $L$–fuzzy approximation space $(L, X, R)$ be given. In this section we will consider algebraic structures for lower $L_R$–fuzzy rough sets, lower $L_S$–fuzzy rough sets, and upper $L$–fuzzy rough sets.

Let $(F_i)_{i \in I}$ be the family of lower $L_R$–fuzzy rough sets (or lower $L_S$–fuzzy rough sets). By Theorem 3(**iv**) it follows that $\bigcap_{i \in I} F_i \in \mathfrak{L}_R(\text{FAS})$ and $\bigcap_{i \in I} F_i \in \mathfrak{L}_S(\text{FAS})$, yet in general neither $\bigcup_{i \in I} F_i \in \mathfrak{L}_R(\text{FAS})$ nor $\bigcup_{i \in I} F_i \in \mathfrak{L}_S(\text{FAS})$, as the following example shows.

*Example 3.* Let $X = \{x_1, x_2, x_3\}$ and let $R$ be a crisp relation given by: $R = \{(x_1, x_1), (x_1, x_2), (x_2, x_1), (x_2 . x_3), (x_3, x_2), (x_3, x_3)\}$. Let $(L, \min, \max, t, \ i_t, n, 0, 1)$ be an extended residuated lattice such that where $L = [0, 1]$, $t$ is an arbitrary left–continuous triangular norm, $i_t$ is the residual implication based on $t$, and $n$ is the standard fuzzy negation $n(x) = 1 - x$. Let $\text{FAS} = (L, X, R)$. Consider two fuzzy sets $A$ and $B$ defined as $A = (x_1 : 0.1, x_2 : 0.8, x_3 : 0.3)$ and $B = (x_1 : 0.9, x_2 : 0.4, x_3 : 0.2)$, respectively. By simple calculations one can easily verify that $\underline{\text{FAS}}_R(A) = \underline{\text{FAS}}_S(A) = (x_3 : 0.1, x_2 : 0.1, x_3 : 0.3)$ and $\underline{\text{FAS}}(B) = (x_1 : 0.4, x_2 : 0.2, x_3 : 0.2)$. Put $F = \underline{\text{FAS}}_R(A) \cup_L \underline{\text{FAS}}_R(B)$, or equivalently, $F = \underline{\text{FAS}}_R(A) \cup_L \underline{\text{FAS}}_R(B)$. Then $F = (x_1 : 0.4, x_2 : 0.2, x_3 : 0.3)$. Assume that $F \in \mathfrak{L}_R(\text{FAS})$. Then there exists $C \in \mathscr{F}_L(X)$ such that $F = \underline{\text{FAS}}_R(C)$. However,

(a) $F(x_1) = \min\{1 \to C(x_1), 1 \to C(x_2), 0 \to C(x_3)\} = \min\{C(x_1), C(x_2)\} = 0.4$,
(b) $F(x_2) = \min\{1 \to C(x_1), 0 \to C(x_2), 1 \to C(x_3)\} = \min\{C(x_1), C(x_3)\} = 0.2$,
(c) $F(x_3) = \min\{0 \to C(x_1), 1 \to C(x_2), 1 \to C(x_3)\} = \min\{C(x_2), C(x_3)\} = 0.3$.

From (a), $C(x_1) \geq 0.4$, so by (b), $C(x_3) = 0.2$, which contradicts (c). Hence, for any $C \in \mathscr{F}_L(X)$, $F \neq \underline{\mathrm{FAS}}_R(C)$, which means that $F \notin \mathfrak{L}_R(\mathrm{FAS})$. Using similar arguments one can easily check that for every $C \in \mathscr{F}_L(X)$, $F \notin \mathfrak{L}_R(\mathrm{FAS})$.

Similarly, for an arbitrary family $(H_k)_{k\in K}$ of upper $L$–fuzzy rough sets it holds, $\bigcup_{k\in K} H_k \in \mathfrak{U}(\mathrm{FAS})$, but not necessarily $\bigcap_{k\in K} H_k \in \mathfrak{U}(\mathrm{FAS})$.

Let $(F_i)i \in I$ be the family of lower $L_R$–fuzzy rough sets, let $(G_j)_{j\in J}$ be the family of lower $L_S$–fuzzy rough set, and let $(H_k)_{k\in K}$ be the family of upper $L$–fuzzy rough sets. Consider the sets

$$
\begin{aligned}
Ub_R(\mathrm{FAS}, \textstyle\bigcup_{i\in I} F_i) &= \{F \in \mathfrak{L}_R(\mathrm{FAS}) : \textstyle\bigcup_{i\in I} F_i \subseteq_L F\} \\
Ub_S(\mathrm{FAS}, \textstyle\bigcup_{j\in J} G_J) &= \{G \in \mathfrak{L}_S(\mathrm{FAS}) : \textstyle\bigcup_{j\in J} G_j \subseteq_L G\} \\
Lb(\mathrm{FAS}, \textstyle\bigcap_{k\in K} H_k) &= \{H \in \mathfrak{U}(\mathrm{FAS}) : H \subseteq_L \textstyle\bigcap_{k\in K} H_k\}.
\end{aligned}
$$

The set $Ub_R(\mathrm{FAS}, \bigcup_{i\in I} F_i)$ is the family of all upper bounds of $\bigcup_{i\in I} F_i$ in $\mathfrak{L}_R(\mathrm{FAS})$ w.r.t. $\subseteq_L$, the set $Ub_S(\mathrm{FAS}, \bigcup_{j\in J} G_j)$ is the family of all upper bounds of $\bigcup_{j\in J} G_j$ in $\mathfrak{L}_S(\mathrm{FAS})$ w.r.t. $\subseteq_L$, and the set $Lb(\mathrm{FAS}, \bigcap_{k\in K} H_k)$ is the family of all lower bounds of $\bigcap_{k\in K} H_k$ in $\mathfrak{U}(\mathrm{FAS})$ w.r.t. $\subseteq_L$. Put

$$
\begin{aligned}
F^\star(\mathrm{FAS}, \textstyle\bigcup_{i\in I} F_i) &= \textstyle\bigcap Ub_R(\mathrm{FAS}, \bigcup_{i\in I} F_i) \\
G^\star(\mathrm{FAS}, \textstyle\bigcup_{j\in J} G_j) &= \textstyle\bigcap Ub_S(\mathrm{FAS}, \bigcup_{j\in J} G_j) \\
H^\star(\mathrm{FAS}, \textstyle\bigcap_{k\in K} H_k) &= \textstyle\bigcup Lb(\mathrm{FAS}, \bigcap_{k\in K} H_k).
\end{aligned}
$$

By Proposition 3(**iv**), $F^\star(\mathrm{FAS}, \bigcup_{i\in I} F_i) \in \mathfrak{L}_R(\mathrm{FAS})$ and $G^\star(\mathrm{FAS}, \bigcup_{j\in J} G_j) \in \mathfrak{L}_S(\mathrm{FAS})$. Clearly, $\bigcup_{i\in I} F_i \subseteq_L F^\star(\mathrm{FAS}, \bigcup_{i\in I} F_i)$ and $\bigcup_{j\in J} G_j \subseteq_L G^\star(\mathrm{FAS}, \bigcup_{j\in J} G_j)$. Therefore, $F^\star(\mathrm{FAS}, \bigcup_{i\in I} F_i) \in Ub_R(\mathrm{FAS}, \bigcup_{i\in I} F_i)$ and $G^\star(\mathrm{FAS}, \bigcup_{j\in J} G_j) \in Ub_S(\mathrm{FAS}, \bigcup_{j\in J} G_j)$. Also, $F^\star(\mathrm{FAS}, \bigcup_{i\in I} F_i) \subseteq_L F$ for any $F \in Ub_R(\mathrm{FAS}, \bigcup_{i\in I} F_i)$ and $G^\star(\mathrm{FAS}, \bigcup_{j\in J} G_j) \subseteq_L G$ for any $G \in Ub_S(\mathrm{FAS}, \bigcup_{j\in J} G_j)$. Hence, $F^\star(\mathrm{FAS}, \bigcup_{i\in I} F_i)$ is the least upper bound of $\bigcup_{i\in I} F_i$ in $\mathfrak{L}_R(\mathrm{FAS})$ w.r.t. $\subseteq_L$ and $G^\star(\mathrm{FAS}, \bigcup_{j\in J} G_j)$ is the least upper bound of $\bigcup_{j\in J} G_j$ in $\mathfrak{L}_S(\mathrm{FAS})$ w.r.t. $\subseteq_L$. Analogously, $H^\star(\mathrm{FAS}, \bigcap_{k\in K} H_k) \in \mathfrak{U}(\mathrm{FAS})$, $H^\star(\mathrm{FAS}, \bigcap_{k\in K} H_k) \subseteq_L \bigcap_{k\in K} H_k$, and $H \subseteq_L H^\star(\mathrm{FAS}, \bigcap_{k\in K} H_k)$ for every $H \in Lb(\mathrm{FAS}, \bigcap_{k\in K} H_k)$. Therefore, $H^\star(\mathrm{FAS}, \bigcap_{k\in K} H_k)$ is the greatest lower bound of $\bigcap_{k\in K} H_k$ in $\mathfrak{U}(\mathrm{FAS})$ w.r.t. $\subseteq_L$.

For all $F_1, F_2 \in \mathfrak{L}_R(\mathrm{FAS})$, for all $G_1, G_2 \in \mathfrak{L}_S(\mathrm{FAS})$, and for all $H_1, H_2 \in \mathfrak{U}(\mathrm{FAS})$, let us denote

$$
\begin{aligned}
F_1 \,\underline{\wedge}_R\, F_2 &= F_1 \cap_L F_2 \\
F_1 \,\underline{\vee}_R\, F_2 &= \textstyle\bigcap Ub_R(\mathrm{FAS}, \{F_1 \cup_L F_2\}) \\
G_1 \,\underline{\wedge}_S\, G_2 &= G_1 \cap_L G_2 \\
G_1 \,\underline{\vee}_S\, G_2 &= \textstyle\bigcap Ub_S(\mathrm{FAS}, \{G_1 \cup_L G_2\}) \\
H_1 \,\overline{\wedge}\, H_2 &= \textstyle\bigcup Lb(\mathrm{FAS}, \{H_1 \cap_L H_2\}) \\
H_1 \,\overline{\vee}\, H_2 &= H_1 \cup_L H_2.
\end{aligned}
$$

From the above discussion we get the following

**Theorem 6.** *The algebras* $(\mathfrak{L}_R(\text{FAS}), \underline{\triangle}_R, \underline{\vee}_R, \emptyset, X)$, $(\mathfrak{L}_S(\text{FAS}), \underline{\triangle}_S, \underline{\vee}_S, \emptyset, X)$, *and* $(\mathfrak{U}(\text{FAS}), \overline{\wedge}, \overline{\vee}, \emptyset, X)$ *are complete lattices.*

Let an $L$–fuzzy approximation space $\text{FAS} = (L, X, R)$ be such that $R$ is an $L$–quasi ordering. Assume that $F_1, F_2 \in \mathfrak{L}_R(\text{FAS})$. Then there are two $L$–fuzzy sets, say $A_1$ and $A_2$, such that $F_1 = \underline{\text{FAS}}_R(A_1)$ and $F_2 = \underline{\text{FAS}}_R(A_2)$. So we have:

$$\begin{aligned} F_1 \cup_L F_2 &= \underline{\text{FAS}}_R(A_1) \cup_L \underline{\text{FAS}}_R(A_2) \\ &= \underline{\text{FAS}}_R(\underline{\text{FAS}}_R(A_1)) \cup_L \underline{\text{FAS}}_R(\underline{\text{FAS}}_R(A_2)) && \text{by Corollary 1(i)} \\ &\subseteq_L \underline{\text{FAS}}_R(\underline{\text{FAS}}_R(A_1) \cup_L \underline{\text{FAS}}_R(A_2)) && \text{by Proposition 3(iv)} \\ &= \underline{\text{FAS}}_R(F_1 \cup_L F_2) \\ &\subseteq_L F_1 \cup_L F_2 && \text{by Proposition 4(ii).} \end{aligned}$$

Hence $\underline{\text{FAS}}_R(F_1 \cup_L F_2) = F_1 \cup_L F_2$. Also, $F_1 \cup_L F_2 \in Ub_R(\text{FAS}, F_1 \cup_L F_2)$. Clearly, $F_1 \cup_L F_2 = \bigcap Ub_R(\text{FAS}, F_1 \cup_L F_2)$, so $F_1 \underline{\vee}_R F_2 = F_1 \cup_L F_2$. Since $F_1 \underline{\triangle}_R F_2 = F_1 \cap_L F_2$, we get

$$F_1 \underline{\triangle}_R (F_2 \underline{\vee}_R F_3) = (F_1 \underline{\triangle}_R F_2) \underline{\vee}_R (F_1 \underline{\triangle}_R F_3)$$
$$F_1 \underline{\vee}_R (F_2 \underline{\triangle}_R F_3) = (F_1 \underline{\vee}_R F_2) \underline{\triangle}_R (F_1 \underline{\vee}_R F_3)$$

for all $F_1, F_2, F_3 \in \mathfrak{L}_R(\text{FAS})$. Using similar arguments it is easy to show that for all $G_1, G_2, G_3 \in \mathfrak{L}_S(\text{FAS})$,

$$G_1 \underline{\triangle}_S (G_2 \underline{\vee}_S G_3) = (G_1 \underline{\triangle}_S G_2) \underline{\vee}_S (G_1 \underline{\triangle}_S G_3)$$
$$G_1 \underline{\vee}_S (G_2 \underline{\triangle}_S G_3) = (G_1 \underline{\vee}_S G_2) \underline{\triangle}_S (G_1 \underline{\vee}_S G_3).$$

Similarly, for all $H_1, H_2, H_3 \in \mathfrak{U}(\text{FAS})$,

$$H_1 \underline{\triangle} (H_2 \underline{\vee} H_3) = (H_1 \underline{\triangle} H_2) \underline{\vee} (H_1 \underline{\triangle} H_3)$$
$$H_1 \underline{\vee} (H_2 \underline{\triangle} H_3) = (H_1 \underline{\vee} H_2) \underline{\triangle} (H_1 \underline{\vee} H_3).$$

Therefore, we get the following

**Proposition 3.** *Let* $\text{FAS} = (L, X, R)$ *be an L–fuzzy approximation space such that R is an L–quasi ordering. Then the structures* $(\mathfrak{L}_R(\text{FAS}), \subseteq_L)$, $(\mathfrak{L}_S(\text{FAS}), \subseteq_L)$, *and* $(\mathfrak{U}(\text{FAS}), \subseteq_L)$ *are complete, distributive lattices.*

We complete this section by showing some link between the lattices of $\mathfrak{L}_S(\text{FAS})$ and $\mathfrak{U}(\text{FAS})$. To this end, let us define the following notion (see [42]). We say that two lattices $(X, \leqslant_X)$ and $(Y, \leqslant_Y)$ are *dually isomorphic* iff there is a bijection $\phi : X \to Y$ such that for all $x_2, x_2 \in X$, $x_1 \leqslant_X x_2$ iff $\phi(x_2) \leqslant_Y \phi(x_1)$.

**Proposition 4.** *For every L–fuzzy approximation space* $\text{FAS} = (L, X, R)$, *the lattices* $(\mathfrak{L}_S(\text{FAS}), \subseteq_L)$ *and* $(\mathfrak{U}(\text{FAS}), \subseteq_L)$ *are dually isomorphic.*

*Proof.* Let $F \in \mathfrak{L}_S(\text{FAS})$. Put $\Gamma(F) = \{A \in \mathfrak{L}_S(\text{FAS}) : \underline{\text{FAS}}_S(A) = F\}$ and define a mapping $\phi : \mathfrak{L}_S(\text{FAS}) \to \mathfrak{U}(\text{FAS})$ by: for every $F \in \mathfrak{L}_S(\text{FAS})$,

$$\phi(F) = \overline{\text{FAS}}(\sim_L \bigcap \Gamma(F)).$$

Note that for every $F \in \mathfrak{L}_S(\text{FAS})$ we have:

$$\begin{aligned} \phi(F) &= \overline{\text{FAS}}(\sim_L \bigcap \Gamma(F)) & \\ &= \sim_L \underline{\text{FAS}}_S(\bigcap \Gamma(F)) & \text{by Theorem 3(iii)} \\ &= \sim_L F & \text{by Theorem 3(iv).} \end{aligned}$$

We show that $\phi$ is a bijection. Let $F_1, F_1 \in \mathfrak{L}_S(\text{FAS})$ be such that $F_1 \neq F_2$. Since $\sim$ is an involution, $\sim_L F_1 \neq \sim_L F_2$. Hence $\phi(F_1) \neq \phi(F_2)$. Now take an arbitrary $G \in \mathfrak{U}(\text{FAS})$. Then for some $A \in \mathscr{F}_L(X)$, $G = \overline{\text{FAS}}(A)$. Therefore, by Theorem 3(**iii**), $G = \sim_L \underline{\text{FAS}}_S(\sim_L A)$. Whence there is $F \in \mathfrak{L}_S(\text{FAS})$, namely $F = \underline{\text{FAS}}_S(\sim_L A)$, such that $G = \sim_L F = \phi(F)$.

Finally, let us take $F_1, F_2 \in \mathfrak{L}_S(\text{FAS})$. Then we have the following equivalences: $F_1 \subseteq_L F_2$ iff $\sim_L F_2 \subseteq_L \sim_L F_1$ iff $\phi(F_2) \subseteq_L \phi(F_1)$, which completes the proof. □

## 7 Conclusions

In this paper we have presented an algebraic extension of fuzzy rough sets. As an underlying algebraic structure a residuated lattice endowed with a De Morgan negation has been taken. The signature of these structures gives algebraic counterparts of main fuzzy logical connectives: triangular norms and triangular conorms, $R$–implication and $S$–implications, and two fuzzy negations, including a De Morgan negation. Taking either of arrow operations we obtain two classes of lower $L$–fuzzy rough sets, namely lower $L_R$–fuzzy rough sets and lower $L_S$–fuzzy rough sets. Moreover, depending on a choice of a class of binary $L$–fuzzy relations, we get further respective classes of $L$–fuzzy rough sets. Properties of these structures have been studied. As it turned out, for $L_R$–fuzzy rough sets most properties of crisp rough sets hold (except duality). Since the Łukasiewicz implication is both an $R$–implication and an $S$–implication, for such operators all properties of $L$–fuzzy rough sets are straightforward generalizations of Pawlak's rough sets. We have proved that the family of lower (resp. upper) $L$–fuzzy rough sets is a lattice w.r.t. $L$–fuzzy inclusion $\subseteq_L$, and, under specific conditions, they are distributive lattices. It was also shown that the family of lower $L$–fuzzy (resp. $L_S$–fuzzy) rough sets are dually isomorphic to the family of upper $L$–fuzzy rough sets. Finally, it has been pointed out that under some assumptions lower and upper $L$–fuzzy rough approximation operations are $L$–fuzzy interior and $L$–fuzzy closure operators, respectively.

It is well–known that the family of all (crisp) rough sets defined on the basis of an equivalence relation, constitute a regular double Stone algebra. The question is what kind of algebraic structure $L$–fuzzy rough sets are. Since I deeply hope that the research cooperation with Etienne will be going on, it may be a challenging topic for further joint studies.

## References


1. Blount, K., Tsinakis, C.: The structure of residuated lattices. Int. J. of Algebra Comput. 13(4), 437–461 (2003)
2. Boixader, D., Jacas, J., Recasens, J.: A new map closely related to the structure of a T–indistinguishability operator. In: Proceedings of the 8th International Conference on Information Processing and Management of Uncertainty in Knowledge–Based Systems IPMU 2000, Madrid, Spain, pp. 841–848 (2000)
3. Chellas, B.F.: Modal logic: An Introduction. Cambridge University Press, Cambridge (1980)
4. De Cock, M., Radzikowska, A.M., Kerre, E.E.: Modelling Linguistic Modifiers using Fuzzy–Rough Structures. In: Proceedings of the 8th International Conference on Information Processing and Management of Uncertainty in Knowledge–Based Systems IPMU 2000, Madrid, Spain, pp. 1735–1742 (2000)
5. De Cock, M., Radzikowska, A.M., Kerre, E.E.: A fuzzy–rough approach to the representation of linguistic hedges. In: Bouchon–Meunier, B., et al. (eds.) Technologies of Constructing Intelligent Systems, vol. 2, pp. 33–43. Springer, Heidelberg (2002)
6. Demri, S., Orłowska, E.: Incomplete Information: Structure, Inference, Complexity. EATCS Monographs in Theoretical Computer Science. Springer, Heidelberg (2002)
7. Dilworth, R.P., Ward, N.: Residuated lattices. Transactions of the American Mathematical Society 45, 335–354 (1939)
8. Dubois, D., Prade, H.: Rough Fuzzy Sets and Fuzzy Rough Sets. Int. J. of General Systems 17(2-3), 191–209 (1990)
9. Dubois, D., Prade, H.: Putting fuzzy sets and rough sets together. In: Słowiński, R. (ed.) Intelligent Decision Support, pp. 203–232. Kluwer Academic, Dordrecht (1992)
10. Esteva, F., Godo, L.: Monoidal t–norm based logic: towards a logic for left–continuous t–norms. Fuzzy Sets and Systems 124, 271–288 (2001)
11. Esteva, F., Godo, L., Garcia–Cerdana, A.: On the hierarchy of t–norm based residuated fuzzy logics. In: Fitting, M., Orłowska, E. (eds.) Beyond Two: Theory and Applications of Multiple-Valued Logic, Studies in Fuzziness and Soft Computing, vol. 114, pp. 251–272. Physica Verlag, Heidelberg (2003)
12. Fitting, M.: Many–valued modal logics. Fundamenta Informaticae 15, 235–254 (1991)
13. Fitting, M.: Many–valued modal logics (II). Fundamenta Informaticae 17, 55–73 (1992)
14. Godo, L., Rodriguez, R.O.: A Fuzzy Modal Logic for Similarity Reasoning. In: Chen, G., Ying, M., Cai, K.-Y. (eds.) Fuzzy Logic and Soft Computing, pp. 33–48. Kluwer Academic Publishers, Dordrecht (1999)
15. Gougen, J.A.: L–fuzzy sets. Journal of Mathematical Analysis and Applications 18, 145–174 (1967)
16. Hajek, P.: Metamathematics of Fuzzy Logic. Kluwer, Dordrecht (1998)
17. Hart, J.B., Rafter, L., Tsinakis, C.: The structure of commutative residuated lattices. Int. J. Algebra Comput. 12(4), 509–525 (2001)
18. Jipsen, P., Tsinaksis, C.: A Survey of Residuated Lattices. In: Martinez, J. (ed.) Ordered Algebraic Structures, pp. 19–56. Kluwer Academic Publishers, Dordrecht (2003)
19. Kerre, E.E.: Introduction to the basic principles of fuzzy set theory and some of its applications. Communication and Cognition, Gent (1993)
20. Klir, G.J., Yuan, B.: Fuzzy Logic: Theory and Applications. Prentice–Hall, Englewood Cliffs (1995)
21. Orłowska, E. (ed.): Incomplete Information: Rough Set Analysis. Springer, Heidelberg (1998)

22. Nachtegael, M., Radzikowska, A.M., Kerre, E.E.: On Links between Fuzzy Morphology and Fuzzy Rough Sets. In: Proceedings of the 8th International Conference on Information Processing and Management of Uncertainty in Knowledge–Based Systems IPMU 2000, Madrid, Spain, pp. 1381–1388 (2000)
23. Nakamura, A.: Fuzzy Rough Sets. Note on Multiple–Valued Logic in Japan 9(8), 1–8 (1988)
24. Pawlak, Z.: Rough sets. Int. J. of Computer and Information Science 11(5), 341–356 (1982)
25. Pawlak, Z.: Rough Sets – Theoretical Aspects of Reasoning about Data. Kluwer Academic Publishers, Dordrecht (1991)
26. Radzikowska, A.M., Kerre, E.E.: Towards studying of fuzzy information relations. In: Proceedings of International Conference in Fuzzy Logics and Technology EUSFLAT 2001, Leicester, UK, pp. 365–369 (2001)
27. Radzikowska, A.M., Kerre, E.E.: A Comparative Study of Fuzzy Rough Sets. Fuzzy Sets and Systems 126, 137–155 (2001)
28. Radzikowska, A.M., Kerre, E.E.: A fuzzy generalisation of information relations. In: Orłowska, E., Fitting, M. (eds.) Beyond Two: Theory and Applications of Multiple–Valued Logics, pp. 264–290. Springer, Heidelberg (2002)
29. Radzikowska, A.M., Kerre, E.E.: On L–Fuzzy Rough Sets. In: Rutkowski, L., Siekmann, J.H., Tadeusiewicz, R., Zadeh, L.A. (eds.) ICAISC 2004. LNCS (LNAI), vol. 3070, pp. 526–531. Springer, Heidelberg (2004)
30. Radzikowska, A.M., Kerre, E.E.: Fuzzy rough sets based on residuated lattices. In: Peters, J.F., Skowron, A., Dubois, D., Grzymała-Busse, J.W., Inuiguchi, M., Polkowski, L. (eds.) Transactions on Rough Sets II. LNCS, vol. 3135, pp. 278–297. Springer, Heidelberg (2004)
31. Radzikowska, A.M., Kerre, E.E.: Algebraic characterisations of some fuzzy information relations. In: Atanassov, K.T., Hryniewicz, O., Kacprzyk, J. (eds.) Soft Computing: Foundations and Theoretical Aspects, pp. 337–352. Akademicka Oficyna Wydawnicza EXIT, Warsaw (2004)
32. Radzikowska, A.M., Kerre, E.E.: Lattice–based fuzzy information relations and operators. In: De Baets, B., De Caluve, R., De Tre, G., Fodor, J., Kacprzyk, J., Zadrożny, S. (eds.) Current Issues in Data and Knowledge Engineering, pp. 433–444. Akademicka Oficyna Wydawnicza EXIT, Warsaw (2004)
33. Radzikowska, A.M., Kerre, E.E.: An algebraic characterisation of fuzzy rough sets. In: Proceedings of IEEE International Conference on Fuzzy Systems FUZZ–IEEE 2004, vol. 1, pp. 115–120 (2004)
34. Radzikowska, A.M., Kerre, E.E.: Characterisations of main classes of fuzzy relations using fuzzy modal operators. Fuzzy Sets and Systems 152(2), 223–247 (2005)
35. Radzikowska, A.M., Kerre, E.E.: An Algebraic Approach to Fuzzy Modalities. In: Hryniewicz, O., Kacprzyk, J., Kuchta, D. (eds.) Issues in Soft Computing – Decision and Operation Research, pp. 71–86. Akademicka Oficyna Wydawnicza EXIT, Warsaw (2005)
36. Radzikowska, A.M., Kerre, E.E.: Fuzzy Information Relations and Operators: An Algebraic Approach Based on Residuated Lattices. In: de Swart, H., Orłowska, E., Schmidt, G., Roubens, M. (eds.) TARSKI 2006. LNCS (LNAI), vol. 4342, pp. 162–184. Springer, Heidelberg (2006)
37. Rasiowa, H., Sikorski, R.: The Mathematics of Metamathematics. PWN, Warsaw (1970)
38. Schweizer, B., Sklar, A.: Probabilistic Metric Spaces. North Holland, Amsterdam (1983)
39. Skowron, A., Polkowski, L.: Rough Sets and Knowledge Discovery, vol. 1, 2. Springer, Heidelberg (1998)

40. Słowiński, R. (ed.): Intelligent Decision Support – Handbook of Applications and Advances of the Rough Sets Theory. Kluwer Academic Publishers, Dordrecht (1992)
41. Thiele, H.: On the definition of modal operators in fuzzy logics. In: Proceedings of International Symposium of Multiple–Valued Logics ISMVL 1993, Sacramento, California, pp. 62–67 (1993)
42. Thiele, H.: Generalizing the Explicit Concept of Rough Set on the Basis of Modal Logic. In: Reusch, B., Temme, K.-H. (eds.) Computational Intelligence in Theory and Practice, Advances in Soft Computing, pp. 93–120. Physica–Verlag, Heidelberg (2001)
43. Turunen, E.: Mathematics Behind Fuzzy Logics. Physica–Verlag, Heidelberg (1999)
44. Wang, X., Ruan, D., Kerre, E.E.: Mathematics of fuzziness – basic issues. Studies in Fuzziness and Soft Computing, vol. 245. Springer, Heidelberg (2009)
45. Yao, Y.Y., Wang, S.K.M., Lin, T.Y.: A Review of Rough Set Models. In: Lin, T.Y., Cercone, N. (eds.) Rough Sets and Data Mining: Analysis of Imprecise Data, pp. 47–75. Kluwer Academic Publishers, Dordrecht (1997)
46. Zadeh, L.A.: Fuzzy Sets. Information and Control 8, 338–358 (1965)
47. Ziarko, W.P. (ed.): Rough Sets, Fuzzy Sets and Knowledge Discovery Workshop in Computing. Springer, London (1994)

# Graduality, Uncertainty and Typicality in Formal Concept Analysis

Yassine Djouadi, Didier Dubois, and Henri Prade

**Abstract.** There exist several proposals for extending formal concept analysis (FCA) to fuzzy settings. They focus mainly on mathematical aspects and assume generally a residuated algebra in order to maintain the required algebraic properties for the definition of formal concepts. However, less efforts have been devoted for discussing what are the possible reasons for introducing degrees in the relation linking objects and properties (which defines a formal context in the FCA sense), and thus what are the possible meanings of the degrees and how to handle them in agreement with their intended semantics. The paper investigates three different semantics, namely i) the graduality of the link associating properties to objects, pointing out various interpretations of a fuzzy formal context; ii) the uncertainty pervading this link (in case of binary properties) when only imperfect information is available and represented in the framework of possibility theory; and lastly, iii) the typicality of objects and the importance of definitional properties within a class. Remarkably enough, the uncertainty semantics has been hardly considered in the FCA setting, and the third semantics apparently not. Moreover, we provide an algorithm for building the whole fuzzy concept lattice based on Gödel implication for handling gradual properties in a qualitative manner.

## 1 Introduction

Formal concept analysis (FCA) was independently introduced by Wille in the 1980's [28] but its mathematical basis had been pioneered in the setting

Yassine Djouadi
University of Tizi-Ouzou, BP 17, RP, 15000 Tizi-Ouzou, Algeria
e-mail: ydjouadi@mail.ummto.dz

Didier Dubois · Henri Prade
IRIT, Université Paul Sabatier, 118 Route de Narbonne,
31062 Toulouse Cedex 09, France
e-mail: dubois@irit.fr, prade@irit.fr

C. Cornelis et al. (Eds.): 35 Years of Fuzzy Set Theory, STUDFUZZ 261, pp. 127–147.
springerlink.com 

of social sciences by Barbut and Monjardet [1] (see also [21]). Since then, it has become increasingly popular among various methods of conceptual data analysis and knowledge representation. It is formulated by means of the notion of a formal context, which is a triple $(\mathcal{O}, \mathcal{P}, \mathcal{R})$, where $\mathcal{O}$ is a set of objects, $\mathcal{P}$ is a set of properties and $\mathcal{R}$ specifies a binary relation between the sets $\mathcal{O}$ and $\mathcal{P}$ ($\mathcal{R} \subseteq \mathcal{O} \times \mathcal{P}$). In the classical setting, it is always assumed that for any pair $(o, p)$: i) $o\mathcal{R}p$ means that it is *well-known* that the object $o$ satisfies the property $p$, ii) $o\overline{\mathcal{R}}p$ means that it is *well-known* that the object $o$ does not satisfy the property $p$, iii) exclusively one of the two above cases applies.

During the last years, FCA has been applied in many different areas like psychology, sociology, anthropology, medicine, biology, linguistics, etc. In such cases, FCA unavoidably deals with relational information structures (formal contexts) derived from human investigation (judgement, observation, measure, etc.). However, it is widely agreed that such information is often of a gradual nature and possibly tainted with uncertainty, while FCA has been almost exclusively developed in a Boolean setting.

FCA constitutes a particular instance of successful application of the calculus of relations, and can be formally extended to fuzzy relations. As such, it may be paralleled with other applications of fuzzy relational calculus [22], ranging from approximate reasoning [10, 17] to image processing [23].

Several fuzzy extensions of formal concept analysis have nevertheless been addressed in the literature, however to different extents. A central notion in existing approaches is the so-called fuzzy formal context whose entries become now degrees from a totally ordered scale $L$ (generally $L$=[0,1]), whereby property satisfaction becomes a matter of degree. Since the first paper by Burusco and Fuentes-Gonzalez [8], different fuzzy logic settings have been proposed, especially by Bělohlávek (see [2, 3] for a survey, [18] for a fuzzy inference perspective). In a different direction, Burmeister and Holzer [7] have made a proposal that introduces a third value in the relational context that stands for the case where it is not known if a property holds or not for an object. Messai *et al.* [25] handle many-valued attribute domains based on a fuzzy logic encoding.

The different fuzzy extensions of FCA that have been proposed are mainly justified by the requirement of preserving good mathematical (algebraic) properties (i.e. closure and opening properties), without always sufficiently discussing what the extension is good for from a knowledge representation point of view. In particular, one may introduce degrees in a relational context, with many different intended meanings. For instance, Table 1 illustrates different generalizations of a formal context which may naturally appear. Entries in the first column of Table 1 express a satisfaction degree of the gradual property *Young*. Most existing approaches deal with this kind of formal context [2, 3, 8, 9, 20, 24, 26]. Entries in the second column give the proficiency level of a given student in the *English* course. This level may be precisely known (e.g like for *Pierre* and *Mike*) or just approximated using an interval

**Table 1** Different formal context generalizations.

| $\mathcal{R}_1$ | *Young* | *English Proficiency* | *Married* |
|---|---|---|---|
| *Pierre* | 1 | 0.9 | × |
| *Sophie* | 0.7 | (0,1] | ? |
| *Mike* | 0.6 | 0 | |
| *Nahla* | 1 | [0.2,0.4] | (0.7; 0) |
| *Peter* | 0.2 | [0,1] | (0; 0) |

(e.g., like for *Sophie*, *Nahla*, and *Peter*). An interesting feature of the use of intervals is that they provide the ability to encode partial/total ignorance [12, 13]. For instance, *Peter*'s proficiency level in English is totally unknown whereas we just know that it is not true that *Sophie* doesn't at all master English.

The third column of Table 1 illustrates incompleteness and uncertainty. For instance, as it is well known that *Pierre* is married and *Mike* is not. However, nothing may be stated for *Sophie*, which is represented using a question mark. In the presence of total ignorance, Burmeister and Holzer [7] have proposed to consider the "?" symbol as third value. In the possibility theory setting, both partial and total ignorance may be taken into account. For instance, the pair (0.7 ; 0) indicates that the certainty of *Nahla* being married is 0 (no certainty at all), and the certainty that she is not is 0.7, whereas the pair (0; 0) illustrates total ignorance about *Peter* and corresponds to the question mark "?".

In a preliminary version of this paper [11], we have distinguished different possible situations leading to the introduction of degrees in fuzzy formal contexts. We have also provided a preliminary discussion of a possibilistic handling of uncertainty, and introduced the idea of taking into account the typicality of objects and the importance of properties in the setting of formal concept analysis. In this paper, we enlarge the discussion about these three motivations and define different graded extensions of formal concept analysis: accounting for gradual properties, handling uncertainty, and acknowledging typicality. The paper is structured as follows. After a background on fuzzy extensions of formal concept analysis in Section 2 and a discussion of suitable operators, Section 3 both provides an introductory discussion of different situations where graduality is encountered, and an algorithm for building the whole fuzzy concept lattice based on Gödel implication for handling gradual properties in a qualitative manner. Section 4 presents the possibilistic view of uncertain formal concepts, while Section 5 illustrates the potential interest of tolerating missing properties that are not compulsory for non-typical objects in a formal concept.

## 2 Fuzzy Extensions of Formal Concept Analysis

In FCA, a formal context $\mathcal{K} = (\mathcal{O}, \mathcal{P}, \mathcal{R})$ is represented by a Boolean table where rows correspond to objects and columns to properties (or conversely), and a table entry contains "×" or "blank mark", depending on whether the object has or not the corresponding property. Let $R(o) = \{p \in \mathcal{P} \mid o\mathcal{R}p\}$ be the set of properties satisfied by object $o$. Similarly, let $R^{-1}(p) = \{o \in \mathcal{O} \mid o\mathcal{R}p\}$ be the set of objects that satisfy the property $p$. We can define powerset operators $(.)^{\uparrow} : 2^{\mathcal{O}} \longrightarrow 2^{\mathcal{P}}$ and $(.)^{\downarrow} : 2^{\mathcal{P}} \longrightarrow 2^{\mathcal{O}}$ as:

$$O^{\uparrow} = \{p \in \mathcal{P} \mid \forall o \in \mathcal{O} : (o \in O \Rightarrow o\mathcal{R}p)\} = \{p \in \mathcal{P} \mid R^{-1}(p) \supseteq O\} \quad (1)$$

$$P^{\downarrow} = \{o \in \mathcal{O} \mid \forall p \in \mathcal{P} : (p \in P \Rightarrow o\mathcal{R}p)\} = \{o \in \mathcal{O} \mid R(o) \supseteq P\} \quad (2)$$

The formal concept analysis problem is that of extracting formal concepts from object/property relations. A formal concept is a pair $\langle O, P\rangle$, such that $O = \{o \in \mathcal{O} \mid R(o) \supseteq P\}$ and $P = \{p \in \mathcal{P} \mid R^{-1}(p) \supseteq O\}$. Thus, $P$ is the set of properties shared by all the objects in $O$, and $O$ is the set of objects that possess all the properties in $P$. The set $O$ (resp. $P$) is called extent (resp. intent). It is easy to remark that both extents (resp. intents) are fixed points w.r.t. to the composition $(.)^{\downarrow\uparrow}$ (resp. $(.)^{\uparrow\downarrow}$). Following Birkhoff's result [5], the set of all formal concepts (denoted $\mathfrak{L}(\mathcal{K})$) is a complete lattice. A formal concept can be also equivalently defined as a maximal pair $(O, P)$ (in the sense of set inclusion) that satisfies the condition $O \times P \subseteq \mathcal{R}$.

Fuzzy extensions of FCA deal mainly with a fuzzy relation $\mathcal{R} \in L^{\mathcal{O}\times\mathcal{P}}$ which is a mapping: $\mathcal{O} \times \mathcal{P} \longrightarrow L$. Whereas both operators $(.)^{\uparrow} : L^{\mathcal{O}} \longrightarrow L^{\mathcal{P}}$ and $(.)^{\downarrow} : L^{\mathcal{P}} \longrightarrow L^{\mathcal{O}}$ become now based on a fuzzy implication $\mathcal{I}$ as defined below [8]:

$$O^{\uparrow}(p) = \bigwedge_{o\in\mathcal{O}} \Big(\mathcal{I}(O(o), \mathcal{R}(o,p))\Big) \qquad P^{\downarrow}(o) = \bigwedge_{p\in\mathcal{P}} \Big(\mathcal{I}(P(p), \mathcal{R}(o,p))\Big) \quad (3)$$

However, the development of a fuzzy formal concept analysis theory requires an appropriate algebra of fuzzy sets [3] in order to maintain the closure property for the composition of the above defined derivation operators. This property, recalled in the following, turns out to be crucial for the soundness of the theory.

**Definition 1.** *Given a universe $\mathcal{U}$, a mapping $\Phi$: $L^{\mathcal{U}} \longrightarrow L^{\mathcal{U}}$ is a fuzzy closure operator iff $\forall\, U, V \in L^{\mathcal{U}}$ it satisfies:*
(CL1): $U \subseteq V \Longrightarrow \Phi(U) \subseteq \Phi(V)$ *(isotone)*
(CL2): $U \subseteq \Phi(U)$ *(extensive)*
(CL3): $\Phi(\Phi(U)) = \Phi(U)$ *(idempotent)*

In order to satisfy the above property, the existing approaches use residuated algebras. Let us recall that a residuated algebra [27], is an algebra $\mathbf{L} = (L, \wedge, \vee, *, \rightarrow)$ s.t. the pair $(*, \rightarrow)$ satisfies the adjointness property (namely, $p \leqslant q \to r$ iff $p * q \leqslant r$). For maintaining the closure property of the compositions $(.)^{\uparrow\downarrow}$ and $(.)^{\downarrow\uparrow}$ Theorem 1 proposes a sufficient condition (R1), weaker than the residuation principle. Consequently this theorem allows us to use more general fuzzy algebras $\mathbf{L} = (L, \wedge, \vee, \to, *, \sim)$, since requirement (R1) does not refer to the conjunction $*$ nor the negation $\sim$. Let us recall that a fuzzy implication operator ($\to$) is decreasing (in the broad sense) in its first component and increasing (in the broad sense) in its second component and verifies boundaries conditions ($0 \to 0 = 0 \to 1 = 1 \to 1 = 1$ and $1 \to 0 = 0$). Whereas, a fuzzy conjunction ($*$) is a binary increasing operator (in the broad sense) which verifies identity condition ($p * 1 = 1 * p = p$) and boundaries conditions ($0 * 0 = 0 * 1 = 1 * 0 = 0$). Note that $*$ is not necessarily a t-norm.

**Theorem 1.** (Djouadi and Prade [13]). *Given an algebra $\mathbf{L} = (L, \wedge, \vee, \to)$ with a fuzzy implication $\to$, the composition $(.)^{\uparrow\downarrow}$ (symmetrically $(.)^{\downarrow\uparrow}$) is a fuzzy closure operator if the following property is satisfied $\forall\, p, q \in L$:*

$$(R1)\ p \leqslant (p \to q) \to q$$

It is noticeable that Theorem 1 does not restrict the set of admissible implications to the set of residuated implications. The following counter-example shows a tri-valued implication ($\rightarrowtail$) which satisfies the above condition (R1) although it is not residuated.

*Example 1.* Let us consider the tri-valued implication (denoted here $\rightarrowtail$) defined in $\{0, a, 1\}$ as shown in Table 2:

It is easy to check that the implication $\rightarrowtail$ verifies the condition (R1). However, this implication is not residuated, as now checked. Indeed, let us try to prove one half of the residuation principle (namely, $p*q \leqslant r \Rightarrow p \leqslant q \to r$). Since the (fuzzy) conjunction $*$ verifies the boundary condition $p * 1 = p$, by substituting $p$ by 1 and $q$ and $r$ by $a$ we get:

$$\begin{aligned} & 1 * a \leqslant a \Rightarrow 1 \leqslant a \rightarrowtail a \\ \Longleftrightarrow \quad & a \leqslant a \Rightarrow 1 \leqslant a \end{aligned}$$

Thus, it is obvious that the half ($\Rightarrow$) of the residuation principle does not hold. Then, the implication $\rightarrowtail$ is not residuated.

**Table 2** Left: Truth values of $p \rightarrowtail q$. Right: Truth values of $(p \rightarrowtail q) \rightarrowtail q$

| $\rightarrowtail$ | 0 | $a$ | 1 | $(p \rightarrowtail q) \rightarrowtail q$ | 0 | $a$ | 1 |
|---|---|---|---|---|---|---|---|
| 0 | 1 | 1 | 1 | 0 | 0 | 0 | 1 |
| $a$ | $a$ | $a$ | 1 | $a$ | $a$ | $a$ | 1 |
| 1 | 0 | 0 | 1 | 1 | 1 | 1 | 1 |

*Remark 1.* Admissible algebras of the form $\mathbf{L} = (L, \wedge, \vee, \rightarrow, *, \sim)$ are not restricted to algebras with pairs $(\rightarrow, *)$ s.t. $\rightarrow$ is the residuum of $*$. For instance, a fuzzy algebra equipped with Gödel implication ($p \rightarrow q = 1$ if $p \leqslant q$, $q$ elsewhere) may involve any conjunction $*$ rather than the min $t-$norm.

We give now an example of implication that satisfies the condition (R1), and which as such is suitable for the definition of $(.)^{\uparrow\downarrow}$ (symmetrically $(.)^{\downarrow\uparrow}$).

*Example 2.* Let us consider a fuzzy algebra $\mathbf{L} = (L, \wedge, \vee, \twoheadrightarrow, *, \sim)$ where the implication $\twoheadrightarrow$ is defined as:

$$(p \twoheadrightarrow q) = \begin{cases} 1 & \text{if } \ p \leqslant q \\ (1-p) \vee q & \text{elsewhere} \end{cases}$$

This implication is remarkable since it can be obtained both as the $\vee$-based disjunction of Gödel and its contrapositive, and as $\sim (p \otimes \sim q)$, where $\otimes$ is the nilpotent minimum and $\sim$ is an involutive negation ($p \otimes q = 0$ if $p+q \leqslant 1$, $p \wedge q$ if $p+q > 1$) [19]. Note that the pair $(\twoheadrightarrow, *)$ is a residuated pair when taking the conjunction $* = \otimes$, namely the nilpotent minimum which is a non-continous $t-$norm [14]. In any case, this implication is admissible according to Theorem 1 since property $p \leqslant (p \rightarrow q) \rightarrow q$ is verified.

*Remark 2.* Note that condition (R1) is satisfied in a residuated structure, when $*$ is commutative as claimed in the following proposition.

**Proposition 1.** *Condition (R1) is satisfied by any residuated algebra if the conjunction $*$ is commutative.*

*Proof.* We have to prove that: $\langle p \leqslant q \rightarrow r \Leftrightarrow p * q \leqslant r \,,\, * \text{ commutative} \rangle \Rightarrow p \leqslant (p \rightarrow q) \rightarrow q$.
Since $*$ is commutative, we have: $p \leqslant q \rightarrow r \Leftrightarrow p * q \leqslant r \Leftrightarrow q * p \leqslant r \Leftrightarrow q \leqslant p \rightarrow r$.
Since the previous equivalences hold $\forall p, q, r \in L$, assuming $q = p \rightarrow q$ and $r = q$, we get condition (R1):

$$p \leqslant (p \rightarrow q) \rightarrow q \Leftrightarrow p \rightarrow q \leqslant p \rightarrow q$$ □

The following example illustrates the case where a particular implication, residuated w.r.t. a *non commutative* conjunction, does not satisfy the condition (R1).

*Example 3.* The Dienes implication (i.e. $p \hookrightarrow q = (1-p) \vee q$) is a remarkable implication since it forms a residuated pair with the *non commutative* conjunction $*$ (defined as: $p * q = p$ if $p + q > 1$; 0 elsewhere) as we shall prove in the following. Let us first determine the left member of the equivalence $p \leqslant q \hookrightarrow r \Leftrightarrow p * q \leqslant r$:

$$
\begin{aligned}
& p \leqslant q \hookrightarrow r \\
\iff & p \leqslant (1-q) \vee r \\
\iff & \begin{cases} p \leqslant r \text{ if } & p+q>1 \\ \top \quad\ \text{if } & p+q \leqslant 1 \end{cases}
\end{aligned}
\tag{4}
$$

Let us now determine the right member:

$$
\begin{aligned}
& p * q \leqslant r \\
\iff & \begin{cases} p \leqslant r \text{ if } & p+q>1 \\ 0 \leqslant r \text{ if } & p+q \leqslant 1 \end{cases}
\end{aligned}
\tag{5}
$$

Thus, it is easy to state that Expression 4 is equivalent to Expression 5. However, this fuzzy implication does *not* satisfy the condition (R1). Indeed, taking $p = 0.8$ and $q = 0.7$, we have $\big((p \hookrightarrow q) \hookrightarrow q\big) = \big((1 - 0.8) \vee 0.7) \hookrightarrow 0.7\big) = 0.7 \not\geqslant p = 0.8$. However, as shown by Example 2, it can be mixed with the requirement that $p \to q = 1$ when $p \leqslant q$ into a suitable implication.

*Remark 3.* Gödel, Goguen and Lukasiewicz implications are admissible implications since they are residuated with respect to a commutative *continuous* triangular norm.

Assuming a sound algebra w.r.t. the closure property, fuzzy FCA consists then of extracting pairs of $L$-sets $\langle O, P\rangle$ from a fuzzy context $\mathcal{K} = (L, \mathcal{O}, \mathcal{P}, \mathcal{R})$, where $L$-sets $O$ and $P$ determine each other pointwisely, being s.t. $O(o) = P^{\downarrow}(o)$, $\forall o \in \mathcal{O}$ and $P(p) = O^{\uparrow}(p)$, $\forall p \in \mathcal{P}$. The set of all formal $L$-concepts is also a complete lattice.

## 3 Gradual Link between a Property and an Object

This section highlights first different gradual interpretations of a fuzzy formal context. We emphasize the representation aspect and the underlying issues of achieving a well-suited representation for many valued contexts (conceptual scaling vs. fuzzy properties). Lastly, we take advantage of both semantical and computational aspects of Gödel implication and provide an algorithm for building the whole concept lattice.

### *3.1 Unipolar vs. Bipolar Scale Interpretation*

A "fuzzy" or graded extension of binary formal contexts may convey different semantics. In a first interpretation, the values in the table (which are scalars in $L$) may be understood as providing a refinement of the cross marks. Namely, they represent to which extent an object has a property, while in the classical model, this relationship was not a matter of degree. It is important to remark that in this view, we do not refine the absence of a property for an object (the blank mark is always replaced by the bottom element 0 of $L$). This view

**Table 3** Many valued relation

| $\mathcal{R}_2$ | *Pierre* | *Sophie* | *Mike* | *Nahla* |
|---|---|---|---|---|
| *Age* | 22 | 28 | 30 | 22 |
| *Salary* | 1100 | 1300 | 1500 | 1500 |

**Table 4** Context subsumption

| $\mathcal{R}_3$ | *Pierre* | *Sophie* | *Mike* | *Nahla* |
|---|---|---|---|---|
| $age \geq 20$ | × | × | × | × |
| $age \geq 25$ | | × | × | |
| $age \geq 30$ | | | × | |
| $salaire \geq 1000$ | × | × | × | × |
| $salaire \geq 1200$ | | × | × | × |
| $salaire \geq 1400$ | | | × | × |

will be referred to as the *positive unipolar interpretation*. In this interpretation, $R^{-1}(p)$ (resp. $R(o)$) is considered as the support of the fuzzy set of objects (resp. properties) satisfying the property $p$ (resp. the object $o$). One could also consider the opposite convention namely the *negative unipolar interpretation* where degrees would represent to which extent an object does not have a property and equivalently provide a refinement of the blank marks.

The most commonly used interpretations, through existing FCA proposals, are implicitly based on the positive unipolar interpretation that allows to map a formal context with quantitative attributes into a fuzzy formal context. In this spirit, conceptual scale theory [29] may be used to achieve a suited (Boolean) representation by successive subsumptions. For instance, the formal context illustrated in Table 4 is obtained from Table 3 by a conceptual scaling of both many-valued attributes "Age" and "Salary". As can be seen, we have two sets of properties with obvious subsumption relations between them. Pairs $(\{Peter, Sophie, Mike, Joe\}, \{age \geq 20, salary \geq 1000\})$, $(\{Sophie, Mike\}, \{age \geq 20,\ age \geq 25, salary \geq 1000, salary \geq 1200\})$, or $(\{Mike\}, \{age \geq 20, age \geq 25, age \geq 30, salary \geq 1000, salary \geq 1200, salary \geq 1400\})$ are formal concepts.

Knowing the ages and the salaries, the formal context $\mathcal{R}_3$ can be re-encoded in a more compact way, using two fuzzy sets 'young' and 'small' with decreasing membership functions, as illustrated in Table 5.

Observe also that $\mathcal{R}_4$ offers a more precise representation of initial data than Table 4. The context in Table 5, event tough more compact than Table 4 highlights the fact that *Mike*, and to a lesser extent *Sophie* are not very young and have a salary that is not really low. It constitutes in some sense the negative of the picture shown on Table 3. Note that the type of representation on Table 5 can be obtained even without providing interpretable fuzzy sets

**Table 5** Context summarization

| $\mathcal{R}_4$ | *Pierre* | *Sophie* | *Mike* | *Nahla* |
|---|---|---|---|---|
| *age 'young'* | 1 | 0.7 | 0.6 | 1 |
| *salary 'low'* | 1 | 0.8 | 0.6 | 0.6 |

and thus, by normalizing in $L$ the domain of attribute values. This approach is used in [25].

Another interpretation of the degrees, maybe more in the standard spirit of fuzzy logic would be to replace both the *cross* marks and the *blank* marks by values in the scale $L$ ($L = [0, 1]$). Then $L$ possesses a mid-point acting as a pivoting value between the situations where the object possesses the property to some extent and the converse situation where the object possesses the opposite property to some extent. Under this view, a fuzzy formal concept should be learnt together with its negation. This view corresponds to a *bipolar scale interpretation.*

## 3.2 Concept Lattice Based on Gödel Implication

From a semantic point of view, the use of Gödel implication, which is at the basis of the expression of a graduality in fuzzy FCA, can also refine fuzzy contexts into sub-contexts for which a gradual rule [16] is satisfied. Thus, in the above example the pair: ({*Pierre*, *Sophie*, *Mike*, *Nahla*}, {$young_{0.6}$, $low_{0.6}$}) is a formal concept. Note however that *Nahla* violates the rule "the more the person is young the more the salary is low", which is however completely satisfied for *Pierre*, *Sophie* and *Mike*: indeed, $\mathcal{I}_G(young, low) = 1$ (where $\mathcal{I}_G$ denotes Gödel implication) if the extension of this concept is limited to *Pierre*, *Sophie* and *Mike.* The search for fuzzy concepts where a gradual rule is satisfied can be a topic of interest.

Computing the whole concept lattice (i.e. the set of all $L$-closed elements) remains the main objective of FCA. However, it is noticeable that the set of all fixed points is infinite under an arbitrary implication and computing this whole set is a non envisageable task. Some approaches [4], [30] compute only the closure of a given $L$-set $O \in L^{\mathcal{O}}$ (or $P \in L^{\mathcal{P}}$). Another approach [9] assumes a finite scale $L$ with cardinality $k$ in order to compute the whole (finite) set of closed elements of $L^{\mathcal{O}}$. The complexity is thus reduced to $k^{|\mathcal{O}|}$ where $|\mathcal{O}|$ is the cardinality of the set $\mathcal{O}$.

Beyond the "gradual" semantics of the Gödel implication, an important feature is that the set of all fixed points is finite under the use of this implication as given in the following proposition. Due to this important property, the computation of the whole concept lattice then becomes feasible.

**Proposition 2.** *Let $\mathcal{K} = (L, \mathcal{O}, \mathcal{P}, \mathcal{R})$ a $L$-context where $L$ is an arbitrary scale (not necessarily finite). Then, the $L$-concept lattice $\mathfrak{L}(\mathcal{K})$, is a finite set under Gödel implication.*

*Proof.* It is a well known result that the sets of all intents and the set of all extents are isomorphic complete lattices [2]. Thus, it is sufficient to prove that one of them is a finite set, namely the set of all extents. Thus, we have:

$$O^{\uparrow\downarrow}(o) = \bigwedge_{p\in\mathcal{P}} \mathcal{I}_G(O^{\uparrow}(p), \mathcal{R}(o,p))$$
$$= \bigwedge_{p\in\mathcal{P}} \left( \begin{cases} 1 & \text{if } O^{\uparrow}(p) \leqslant \mathcal{R}(o,p) \\ \mathcal{R}(o,p) & \text{elsewhere} \end{cases} \right)$$

Thus, necessarily $O^{\uparrow\downarrow}(o) \in \{\mathcal{R}(o_i, p_j)\} \cup \{1\}$ where $i, j$ are integers bounded by the numbers of lines and columns of $\mathcal{R}$. This implies that the set of all extents is finite. □

It has been shown through the proof of Proposition 2 that $O^{\uparrow\downarrow}(o) \in \{\mathcal{R}(o_i, p_j)\} \cup \{1\}$ (symmetrically $P^{\downarrow\uparrow}(p) \in \{\mathcal{R}(o_i, p_j)\} \cup \{1\}$). This means that each closed element is valued on the set of entries of the relation $\mathcal{R}$. Thus, in order to determine all closed elements, it is then sufficient to enumerate $L$-sets valued in $\{\mathcal{R}(o,p), o \in \mathcal{O}, p \in \mathcal{P}\}$ from the bottom element of the concept lattice to the top element (or conversely). The following proposition characterizes these elements. In the following, we assume that $1_{\mathcal{O}}$ and $0_{\mathcal{O}}$ (resp. $1_{\mathcal{P}}$ and $0_{\mathcal{P}}$) stands for the extreme membership values to the corresponding $L$-set s.t. $1_{\mathcal{O}}(o) = 1$ and $0_{\mathcal{O}}(o) = 0\ \forall o \in \mathcal{O}$ (resp. $1_{\mathcal{P}}(p) = 1$ and $0_{\mathcal{P}}(p) = 0\ \forall p \in \mathcal{P}$). We also denote by $\mathcal{O}^{\wedge}$ (resp. $\mathcal{O}^{\vee}$) the particular $L$-sets s.t. $\mathcal{O}^{\wedge}(o) = \bigwedge_{p\in\mathcal{P}} \mathcal{R}(o,p)$ (resp. $\mathcal{O}^{\vee}(o) = \bigvee_{p\in\mathcal{P}} \mathcal{R}(o,p)$) $\forall o \in \mathcal{O}$, and by $\mathcal{P}^{\wedge}$ (resp. $\mathcal{P}^{\vee}$) the particular $L$-sets s.t. $\mathcal{P}^{\wedge}(p) = \bigwedge_{o\in\mathcal{O}} \mathcal{R}(o,p)$ (resp. $\mathcal{P}^{\vee}(p) = \bigvee_{o\in\mathcal{O}} \mathcal{R}(o,p)\ \ \forall p \in \mathcal{P}$).

**Proposition 3.** *Under Gödel implication, the lower and upper bound of the $L$-concept lattice $\mathfrak{L}(\mathcal{K})$ are given by:*

$$\bigwedge_{\langle O,P\rangle} \mathfrak{L}(\mathcal{K}) = \langle \mathcal{O}^{\wedge}, 1_{\mathcal{P}} \rangle \quad \text{and} \quad \bigvee_{\langle O,P\rangle} \mathfrak{L}(\mathcal{K}) = \langle 1_{\mathcal{O}}, \mathcal{P}^{\wedge} \rangle$$

*Proof.* $(.)^{\uparrow\downarrow}$ is isotone, so $\forall o \in \mathcal{O} : 0 \leqslant O(o) \leqslant 1 \Rightarrow 0_{\mathcal{O}}^{\uparrow\downarrow}(o) \leqslant O^{\uparrow\downarrow}(o) \leqslant 1_{\mathcal{O}}^{\uparrow\downarrow}(o)$.
i) let us determine the $L$-set $0_{\mathcal{O}}^{\uparrow\downarrow}$:

$$0_{\mathcal{O}}^{\uparrow\downarrow}(o) = \bigwedge_{p\in\mathcal{P}} \mathcal{I}_G\Big( \bigwedge_{o\in\mathcal{O}} \mathcal{I}_G(0, \mathcal{R}(o,p)), \mathcal{R}(o,p) \Big)$$
$$= \bigwedge_{p\in\mathcal{P}} \mathcal{I}_G(1, \mathcal{R}(o,p)) \quad \text{since } \mathcal{I}_G(0,v) = 1 \quad \forall v \in L$$
$$= \bigwedge_{p\in\mathcal{P}} \mathcal{R}(o,p) \quad \text{since } \mathcal{I}_G(1,v) = v \quad \forall v \in L$$

ii) let us determine the $L$-set $1_{\mathcal{O}}^{\downarrow\uparrow}$:

$$1_{\mathcal{O}}^{\uparrow\downarrow}(o) = \bigwedge_{p\in\mathcal{P}} \mathcal{I}_G\Big(\bigwedge_{o\in\mathcal{O}} \mathcal{I}_G(1,\mathcal{R}(o,p)),\mathcal{R}(o,p)\Big)$$
$$= \bigwedge_{p\in\mathcal{P}} \mathcal{I}_G\Big(\bigwedge_{o\in\mathcal{O}} \mathcal{R}(o,p),\mathcal{R}(o,p)\Big)$$
$$= \bigwedge_{p\in\mathcal{P}} 1 = 1$$

Top and bottom intents are similarly determined. □

The following proposition will be useful for pruning useless potential fixed points in the algorithm.

**Proposition 4.** (Djouadi and Prade [13]) *Let $O_1, O_2 \in L^{\mathcal{O}}$ (resp. $P_1, P_2 \in L^{\mathcal{P}}$) two $L$-sets such that $O_1 \subseteq O_2$ (resp. $P_1 \subseteq P_2$). If $O_2 \subseteq O_1^{\uparrow\downarrow}$ (resp. $P_2 \subseteq P_1^{\downarrow\uparrow}$) then $O_2^{\downarrow\uparrow} = O_1^{\downarrow\uparrow}$ (resp. $P_2^{\downarrow\uparrow} = P_1^{\downarrow\uparrow}$)*

### 3.2.1 Construction Algorithm

The proposed algorithm is intended to build the lattice $\mathfrak{L}(\mathcal{K})$ of all $L$-formal concepts (i.e. pairs $\langle O,P\rangle$ of fixed points s.t. $O = P^{\downarrow}$ and $P = O^{\uparrow}$). Since the set of extents and the set of intents are isomorphic complete lattices, we may just build one of them, for instance the intent-concept lattice. The lattice construction process is organized through two procedures. The main procedure CLOSED_LATTICE computes top and bottom intent-concept and calls the recursive procedure RECURSIVE_CLOSURE which calls itself in order to generate all potential closed elements. The following notations and functions are used by the procedures.

| | |
|---|---|
| $m$: | The cardinality of the set $\mathcal{P}$ ($m = \lvert\mathcal{P}\rvert$) |
| *INTENT_SET*: | Set containing all intent concepts. |
| *Closed*: | Corresponds to a closed $L$-set. |
| $DIR\_SUCC(Closed, p_j)$: | Returns the direct successor of the $L$-set *Closed* w.r.t the property $p_j$. The result (denoted $V$) is defined s.t. $V(p_k) = Closed(p_k)$ when $k \neq j$, $V(p_j) = Closed(p_j)$ when $Closed(p_j) = \mathcal{P}^{\vee}(p_j)$, $V(p_j) = \beta$ s.t. $\beta > Closed(p_j)$ and $\nexists\, \beta' : \beta > \beta' > Closed(p_j)$, elsewhere. |
| *Successor*: | This $L$-set contains a direct successor of the last computed closed element. |
| *Closure*: | This $L$-set contains the closure of *Successor*. |
| *Next*: | Contains the direct successor of the last computed closure w.r.t the attribute $p_j$. |
| *FLAG*: | This Boolean is used to prune already computed closed elements. |

**Algorithm.** CLOSED_LATTICE($\mathcal{K}$)

**Input**: A fuzzy context $\mathcal{K} = (L, \mathcal{X}, \mathcal{A}, \mathcal{R})$.
**Output**: The set $INTENT\_SET$ of all fuzzy intent concepts.
**Begin**
$Top(p_j) \longleftarrow 1_{\mathcal{P}}(p_j)\ \forall p_j \in \mathcal{P}$;
$Bottom(p_j) \longleftarrow \mathcal{P}^{\wedge}(p_j)\ \forall p_j \in \mathcal{P}$;
$INTENT\_SET \longleftarrow \{Top\} \cup \{Bottom\}$;
$RECURSIVE\_CLOSURE(Bottom)$;
**End**

Proposition 4 is used by the procedure RECURSIVE_CLOSURE for pruning potential closed elements for which the closure is already computed. It allows also to fix the next potential closed element. This avoids to have to enumerate all potential closed elements and significantly reduces complexity, as shown through Example 4. The Boolean $FLAG$ prevents the procedure RECURSIVE_CLOSURE from calling itself for already computed closed elements. Note that the use of a recursive procedure provides the ability to build the whole lattice structure while computing the closed elements.

**Algorithm.** RECURSIVE_CLOSURE(*Closed*)

**Begin**
$j \longleftarrow 1$; /* index $j$ allows to enumerate the attributes */
$FLAG \longleftarrow TRUE$;
**While** $(j \leqslant m)$ and $FLAG$ **Do**
**Begin**
  $Next \longleftarrow DIR\_SUCC(Closed, p_j)$;
  **While** $(Next \neq \mathcal{P}^{\vee}(a_j))$ and $FLAG$ **Do**
  **Begin** /* there exists a direct successor */
    $Successor \longleftarrow Closed$;
    $Successor(p_j) \longleftarrow DIR\_SUCC(Closed, p_j)$;
    $Closure \longleftarrow (Successor)^{\downarrow\uparrow}$;
    **If** $\neg(Closure \in INTENT\_SET)$
      **Then**
        $INTENT\_SET \longleftarrow INTENT\_SET \cup \{Closure\}$;
        $RECURSIVE\_CLOSURE(Closure)$;
        $Next \longleftarrow DIR\_SUCC(Closure, p_j)$;
      **Else**
        $FLAG \longleftarrow FALSE$;
    **End If**
  **End While**
  $j++$;
**End While**
**End**

**Proposition 5.** *The procedure CLOSED_LATTICE is complete i.e. it finds all closed sets $P \in L^{\mathcal{P}}$ s.t. $P = P^{\downarrow\uparrow}$.*

*Proof.* The proof follows from Proposition 2, Proposition 4 and the structure of the procedure RECURSIVE_CLOSURE which enumerates recursively all non useless potential closed elements. □

### 3.2.2 Illustrative Example

An example is given to illustrate the proposed algorithm. It is based on a fuzzy formal context that expresses relationships about meteorological observations. A special notation $P = p_1^{\alpha_1} p_2^{\alpha_2} \ldots p_m^{\alpha_m}$ where $\alpha_j = P(p_j)$, (resp. $O = o_1^{\beta_1} o_2^{\beta_2} \ldots o_n^{\beta_n}$ where $\beta_i = O(o_i)$) will be used to represent $L$-sets $P \in L^{\mathcal{P}}$ (resp. $O \in L^{\mathcal{O}}$).

**Table 6** Fuzzy formal context

| $\mathcal{R}$ | Low Temperature | Low Pressure | High Humidity |
|---|---|---|---|
| Warm Day | 0.0 | 0.2 | 0.3 |
| Cold Day | 1 | 0.9 | 0.6 |
| Rather Dry Day | 0.0 | 0.4 | 0.1 |
| Violent Wind Day | 0.7 | 0.9 | 0.5 |

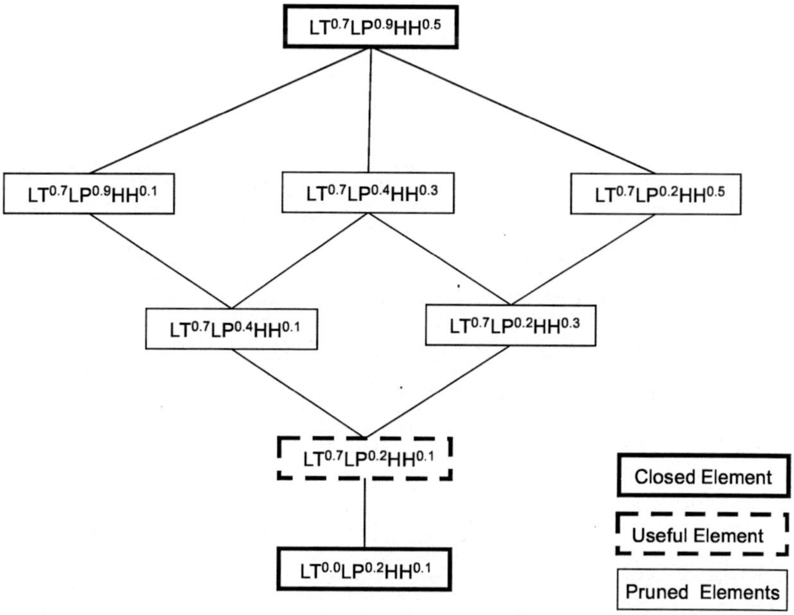


**Fig. 1** Pruning useless $L$-sets.

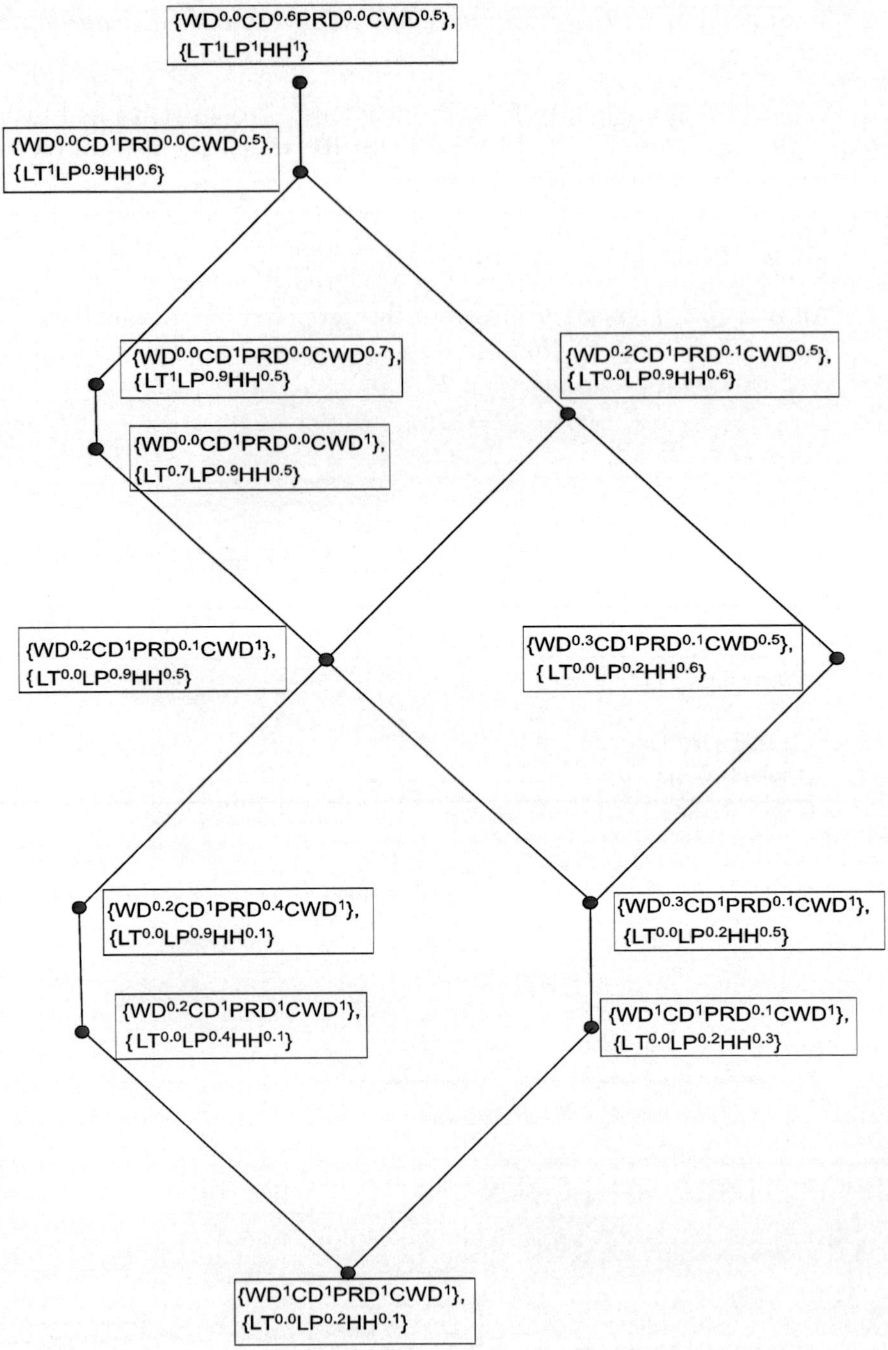


**Fig. 2** Whole lattice of fuzzy formal concepts.

*Example 4.* Let us consider a fuzzy context which consists of:

- A set of properties $\mathcal{P}$={Low Temperature, Low Pressure, High Humidity} respectively abbreviated as LT, LP, HH.
- A set of objects $\mathcal{O}$={Warm Day, Cold Day, Rather Dry Day, Violent Wind Day} respectively abbreviated as WD, CD, RDD, VWD.
- A fuzzy relation $\mathcal{R}$ illustrated in Table 6.
- A scale $L$=[0,1].

The proposed algorithm proceeds as follows (the variable $m$ is equal to 3):

*Step 1*: "*Top*" and "*Bottom*" on the intent lattice are determined. $Top = LT^1LP^1HH^1$. $Bottom = LT^{0.0}LP^{0.2}HH^{0.1}$.

*Step 2*: The procedure RECURSIVE_CLOSURE($LT^{0.0}LP^{0.2}HH^{0.1}$) is called. The variable $Next$ takes the value 0.7 that corresponds to the direct successor of the degree 0.0 w.r.t the attribute LT in the relation $\mathcal{R}$. The variable $Successor$ is first assigned the $L$-set $LT^{0.0}LP^{0.2}HH^{0.1}$ in line 8. Whereas in line 9, $Successor$ is assigned the degree $Next$ and becomes $LT^{0.7}LP^{0.2}HH^{0.1}$. In line 10, the closure of the $L$-set $Successor$ is computed and the variable $Closure$ is assigned the $L$-set $LT^{0.7}LP^{0.9}HH^{0.5}$.

*Step 3*: Since the intent-concept $LT^{0.7}LP^{0.9}HH^{0.5}$ has not been already generated, it will be added in the set INTENT_SET and the Boolean $FLAG$ remains true. At this level the Proposition 4 will be used to prune all useless potential closed elements (i.e. which will not generate new closed elements). In this precise case, Figure 1 illustrates a lattice of useless $L$-sets which will be pruned.

*Step 4*: The procedure RECURSIVE_CLOSURE is again called with ($LT^{0.7}$ $LP^{0.8}HH^{0.5}$) as parameter since the next $L$-set which may generate (or may be) a closed element is inevitably a direct successor of the $L$-set $LT^{0.7}LP^{0.8}HH^{0.5}$.

*Step n*: The procedure ends when all potentially closed elements are generated.

The whole lattice structure is depicted in Figure 2.

## 4 Uncertainty

Neither the standard FCA approach nor its fuzzy extension are equipped for representing situations of partial or complete ignorance. To this end, in the Boolean case, we need to introduce a proper representation of partial uncertainty including ignorance in the relational table of the formal context. One may think of introducing gradations of uncertainty by changing crosses and blanks in the table into probability degrees, or by possibility or necessity degrees. In the probabilistic case, one number shall assess the probability that a considered property holds for a given object (its complement to 1 corresponding to the probability it does not hold). However, this is assuming precise knowledge on the probability values, which is not really appropriate if we have to model the state of complete ignorance. It is why we investigate the use of the possibilistic setting in the following.

In the possibilistic setting, crosses may be replaced by positive degrees of necessity for expressing some certainty that an object satisfies a property. The blanks could be refined by possibility degrees less than 1, expressing that it is little possible that an object satisfies a property. However, this convention using a single number in the unit interval for each entry in the context may be misleading as according to whether the number replaces a blank or a cross the meaning of the number is not the same.

In the possibilistic setting, possibility and necessity functions are related by the duality relation $N(A) = 1 - \Pi(\overline{A})$, that holds for any event $A$, where $\overline{A}$ denotes the opposite event [15]. Then, for entries $(o, p)$ in the table, we use a representation as *a pair* of necessity degrees $(\alpha, 1-\beta)$ where $\alpha = N(o\mathcal{R}p)$ (resp. $1 - \beta = N(o\overline{\mathcal{R}}p)$) corresponds to the necessity (certainty) that object $o$ has (resp. does not have) property $p$. Moreover, we should respect the property $\min(\alpha, 1-\beta) = 0$, since $\min(N(A), N(\overline{A})) = 0$ in agreement with complete ignorance, in which case nothing (i.e., neither $A$ nor $\overline{A}$) is even somewhat certain. Pairs (1,0) and (0,1) correspond to completely informed situations where it is known that object $o$ has, respectively does not have, property $p$. The pair (0,0) reflects total ignorance, whereas pairs $(\alpha, 1-\beta)$ s.t. $1 > \max(\alpha, 1-\beta) > 0$ correspond to partial ignorance.

An uncertain formal context is thus represented by

$$\mathcal{R}^U = \{(\alpha(o,p), 1-\beta(o,p)) \mid o \in \mathcal{O}, p \in \mathcal{P}\}$$

where $\alpha(o,p) \in [0,1]$, $\beta(o,p) \in [0,1]$. A relational database with fuzzily-known attribute values is *theoretically* equivalent to the fuzzy set of all ordinary databases corresponding to the different possible ways of completing the information consistently with the fuzzy restrictions on the attribute values [6]. In the same way, an uncertain formal context may be viewed as a weighted family of all standard formal contexts obtained by changing uncertain entries into sure ones. More precisely, one may consider all the completions of an uncertain formal context. This is done by substituting entries $(o,p)$ that are uncertain, i.e., such that $1 > \max(\alpha(o,p), 1-\beta(o,p))$ by a pair (1,0), or a pair (0,1). Replacing $(\alpha(o,p), 1-\beta(o,p))$ by $(1,0)$ is possible at degree $\beta(o,p)$, the possibility that $o$ has property $p$. Similarly, replacing $(\alpha(o,p), 1-\beta(o,p))$ by $(0,1)$ is possible at degree $1-\alpha(o,p)$, the possibility that $o$ does *not* have the property $p$. In this way, one may determine to what extent a particular completion (a context $\mathcal{C}$) is possible, by aggregating the possibility degrees associated with each completed entry (using min operator). Formally, one can write

$$\pi(\mathcal{C}) = \min(\min_{(o,p):o\mathcal{C}p}\beta(o,p), \min_{(o,p):o\overline{\mathcal{C}}p}1-\alpha(o,p)).$$

Likewise the degree of possibility that $(X, Y)$ is a formal context of $\mathcal{R}^U$ is

$$\pi(X,Y) = \sup\{\pi(\mathcal{C}) : \mathcal{C} \text{ such that } (X,Y) \text{ is a formal context of } \mathcal{C}\}.$$

Useful completions are those where partial certainty becomes full certainty. Indeed, given an uncertain formal context and a threshold pair $(u, v)$, let us replace all entries of the form $(\alpha, 0)$ such that $\alpha \geqslant u$ with (1, 0) and entries of the form $(0, 1-\beta)$ such that $1-\beta \geqslant v$ with (0, 1). All such replacements have possibility 1 according to the above formula. Remaining entries, which are more uncertain, can be systematically substituted either by (1,0), or by (0,1). Considering, the two extreme cases where *all* such entries are changed into (1,0) and the case when where *all* such entries are changed into (0,1) gives birth to upper and lower completions, respectively. In this way, two classical (Boolean) formal contexts, denoted $\mathcal{R}^*_{(u,v)}$ and $\mathcal{R}_{*(u,v)}$ are obtained as respective results of the two completions. They allow to determine, for a given threshold $(u, v)$, maximal extensions (resp. minimal intensions) and minimal extensions (resp. maximal intensions) of uncertain formal concepts. It is clear that $\mathcal{R}_{*(u,v)} \subseteq \mathcal{R}^*_{(u,v)}$. Let us illustrate the idea with an example.

*Example 5.* Table 7 exhibits a formal context where some entries are pervaded with uncertainty. Let us examine the situation regarding formal concepts. Take $u = 0.7, v = 0.5$ for instance. In context $\mathcal{R}_{*(0.7,0.5)}$, examples of formal concepts are pairs $(\{6,7,8\},\{c,d,e\})$, or $(\{5,6,7,8\},\{d,e\})$, or $(\{2,3,4\},\{g,h\})$, although with $u = 0.9$, the last formal concept would reduce to $(\{2,3\},\{g,h\})$, i.e. the extent of the concept is smaller.

Now consider $\mathcal{R}^*_{(0.7,0.5)}$, where the entries with low certainty levels (either in favor or against the existence of the link between $o$ and $p$) are turned into positive links. Then, $(\{2,3,4\},\{g,h\})$ remains unchanged as a formal concept, while a larger concept now emerges, namely $(\{5,6,7,8\},\{c,d,e\})$. However, one may prefer to consider the results obtained from $\mathcal{R}_{*(0.7,0.5)}$, where only the almost certain information is changed into positive links. In the example, if we move down $u$ to 0.5, and use $\mathcal{R}_{*(0.5,0.5)}$ we still validate the larger former concept $(\{5,6,7,8\},\{c,d,e\})$. This illustrates the fact that becoming less and less demanding on the level of certainty, may enable the fusion of close concepts (here $(\{6,7,8\},\{c,d,e\})$, and $(\{5,6,7,8\},\{d,e\})$, providing a more synthetic view of the formal context.

This small example is intended to illustrate several points. First of all, it should be clear that being uncertain about the existence of a link between an object and a property is not the same as being certain about a gradual link. Second, under uncertainty, there are formal concepts whose boundaries

**Table 7** Uncertain formal concepts

| | 1 | 2 | 3 | 4 | 5 | 6 | 7 | 8 |
|---|---|---|---|---|---|---|---|---|
| $a$ | | | | | | | × | |
| $b$ | | | | | × | × | | |
| $c$ | | | | | (0.5,0) | × | × | × |
| $d$ | | | | | × | × | × | × |
| $e$ | | | | | × | × | × | × |
| $f$ | | (0, 0.8) | | × | (0, 0.3) | | | |
| $g$ | | × | × | (0.8, 0) | | | | |
| $h$ | | × | × | (0.8, 0) | | | | |
| $i$ | × | | | | | | | |

are not affected by uncertainty, while others are. Lastly, regarding certain enough pieces of information as certain may help simplifying the analysis of the formal context. Besides, the proposed setting may also handle inconsistent information by relaxing the constraint $\min(\alpha, 1-\beta) = 0$. This would amount to introducing paraconsistent links between objects and properties.

## 5 Typical Objects and Less Important Properties

Beyond the formal concepts which are strongly related to the mathematical notion of closure, one may consider that a formal context conveys an implicit semantics related to the typicality of objects and to the importance of properties. The notions of typicality and importance are dually stated according to the following principles:

A) an object $o$ is all the more typical with respect to a set of properties $P$ as it has all the properties $p \in P$ that are sufficiently important;
B) a property $p$ is all the more important with respect to a set of objects $O$ as all the objects $o \in O$ that are sufficiently typical possess it.

Let us illustrate these notions through an example. We consider the set of animals $\mathcal{O} = \{albatross, parrot, penguin, kiwi, turtle\}$, and the set of properties $\mathcal{P} = \{\text{'}laying\ eggs\text{'}, \text{'}having\ two\ legs\text{'}, \text{'}flying\text{'}, \text{'}having\ feathers\text{'}\}$. Table 8 describes these animals with respect to relevant properties. It can be easily seen from a FCA point of view w. r. t. $\mathcal{R}_5$, there are a set of "regular birds" (here 'albatross', 'parrot'), a set of "more or less regular birds" (here 'albatross', 'parrot', 'penguin'), a set of "less regular birds" including 'kiwis' and, a set of animals which are not at all categorized as "birds". Thus, the idea of a "bird" (as in Table 8) may appear as a *vague* concept if on the one hand one considers that some birds (e.g. 'albatross', 'parrot') are more *typical* than others (e.g. 'penguin', 'kiwi'), and on the other hand that the satisfaction of some properties (e.g. 'laying eggs', 'having two legs') is more *important* than others (e.g. 'flying', or even 'having feathers'). More formally, given a formal context $\mathcal{K} = (\mathcal{O}, \mathcal{P}, \mathcal{R})$, the definitions of typicality and importance of a concept $\mathcal{C}$ are given as:

**Table 8** Birds example

| $\mathcal{R}_5$ | *eggs* | *2 legs* | *feather* | *fly* |
|---|---|---|---|---|
| *albatross* | × | × | × | × |
| *parrot* | × | × | × | × |
| *penguin* | × | × | × | |
| *kiwi* | × | × | | |
| *turtle* | × | | | |

**Definition 2.** *Given a formal context $\mathcal{K}$, the typicality $\tau_\mathcal{C}$ (resp. the importance $\iota_\mathcal{C}$) of a concept $\mathcal{C}$ is a mapping $\tau_\mathcal{C} : \mathcal{O} \to L$ (resp. $\iota_\mathcal{C} : \mathcal{P} \to L$). Typicality and importance define themselves mutually as:*

$$\iota_\mathcal{C}(p) = \bigwedge_{o \in \mathcal{O}} \tau_\mathcal{C}(o) \to \mathcal{R}(o,p) \qquad \text{or} \qquad \tau_\mathcal{C}(o) = \bigwedge_{p \in \mathcal{P}} \iota_\mathcal{C}(p) \to \mathcal{R}(o,p)$$

It is interesting to remark that under some conditions, the pair $(\tau_\mathcal{C}, \iota_\mathcal{C})$ forms a Galois connection. Indeed, let us assume that the typical extension of a concept $\mathcal{C}$ is given by $\tau_\mathcal{C}(albatross) = \tau_\mathcal{C}(parrot) = 1$, $\tau_\mathcal{C}(penguin) = \alpha$, $\tau_\mathcal{C}(kiwi) = \beta$, and $\tau_\mathcal{C}(turtle) = 0$ with $1 > \alpha > \beta > 0$. Let us compute the fuzzy set of important properties according to Definition 2, i.e. $\iota_\mathcal{C}(p) = \bigwedge_{o \in \mathcal{O}} \tau_\mathcal{C}(o)$, with the condition $a \to 1 = 1$ and $a \to 0 = 1 - a$. This choice expresses the idea that a bird not having property $p$ makes a property all the less important in the definition of the concept *bird* as this bird is considered more typical. Then, we obtain $\iota_\mathcal{C}(\text{'}laying\ eggs\text{'}) = \iota_\mathcal{C}(\text{'}having\ two\ legs\text{'}) = 1$; $\iota_\mathcal{C}(\text{'}flying\text{'}) = 1 - \alpha$; $\iota_\mathcal{C}(\text{'}having\ feathers\text{'}) = 1 - \beta$. Thus, 'having feathers' is more important than 'flying', since $1 - \alpha < 1 - \beta$.

Let us now compute the fuzzy set of typical objects w.r.t. the obtained fuzzy concept intension $\iota_\mathcal{C}(p)$. We get $\tau_\mathcal{C}(albatross) = \tau_\mathcal{C}(parrot) = 1$; $\tau_\mathcal{C}(penguin) = \alpha$; $\tau_\mathcal{C}(kiwi) = \beta$; $\tau_\mathcal{C}(turtle) = 0$ (since $(1 - a) \to 0 = a$). We thus recover $\left( \bigwedge_{p \in \mathcal{P}} \left( \bigwedge_{o \in \mathcal{O}} \tau_\mathcal{C}(o) \to \mathcal{R}(o,p) \right) \to \mathcal{R}(o,p) \right) = \tau_\mathcal{C}(o)\ \forall o$, and we recognize a fuzzy Galois connection.

This example illustrates the idea that, viewing a formal concept as a maximal rectangle included in the formal context, one may tolerate some missing object-property links in such a rectangle, provided that they pertain to non-fully typical objects that may miss non-fully compulsory properties. This may facilitate the emergence of larger formal concepts, with wider range of applicability, but having exceptions.

## 6 Conclusion

We claim that the semantics of the grades that may be introduced in FCA is crucial and influences the way these grades should be handled. Starting with the view of a formal concept in a formal context as a maximal subset where the relation holds everywhere, we have suggested how the ideas of gradualness, of uncertainty, of typicality or of importance may be used for refining formal concepts in a graded way.

## References


1. Barbut, M., Monjardet, B.: Ordre et Classification. Algèbre et Combinatoire. Tome 2. Hachette, Paris (1970)
2. Bělohlávek, R.: Fuzzy Galois connections. Math. Logic Quart 45, 497–504 (1999)
3. Bělohlávek, R., Vychodil, V.: What is a fuzzy concept lattice. In: Proc. CLA 2005, Olomounc. Czech Republic, pp. 34–45 (2005)
4. Bělohlávek, R., Vychodil, V.: Graded LinClosure and its role in relational data analysis. In: Yahia, S.B., Nguifo, E.M., Belohlavek, R. (eds.) CLA 2006. LNCS (LNAI), vol. 4923, pp. 139–154. Springer, Heidelberg (2008)
5. Birkhoff, G.: Théorie et applications des treillis. Annales de l'IHP 11(5), 227–240 (1949)
6. Bosc, P., Pivert, O.: About yes/no queries against possibilistic databases. Int. J. Intell. Syst. 22(7), 691–721 (2007)
7. Burmeister, P., Holzer, R.: Treating incomplete knowledge in formal concepts analysis. In: Ganter, B., Stumme, G., Wille, R. (eds.) Formal Concept Analysis. LNCS (LNAI), vol. 3626, pp. 114–126. Springer, Heidelberg (2005)
8. Burusco, A., Fuentes-González, R.: The study of the L-fuzzy concept lattice. Mathware & Soft Computing 3, 209–218 (1994)
9. Burusco, A., Fuentes-González, R.: Construction of the L-fuzzy concept lattice. Fuzzy Sets and Systems 97(1), 109–114 (1998)
10. De Baets, B., Kerre, E.: Fuzzy relational compositions. Fuzzy Sets Syst. 60, 109–120 (1993)
11. Djouadi, Y., Dubois, D., Prade, H.: On the possible meanings of degrees when making formal concept analysis fuzzy. In: Proc. EUROFUSE 2009, pp. 253–258 (2009)
12. Djouadi, Y., Prade, H.: Interval-valued fuzzy formal concept analysis. In: Rauch, J., Raś, Z.W., Berka, P., Elomaa, T. (eds.) Foundations of Intelligent Systems. LNCS(LNAI), vol. 5722, pp. 592–601. Springer, Heidelberg (2009)
13. Djouadi, Y., Prade, H.: Interval-valued fuzzy Galois connections: Algebraic requirements and concept lattice construction. Fundamenta Informaticae 99(2), 169–186 (2010)
14. Dubois, D., Prade, H.: A theorem on implication functions defined from triangular norms. Stochastica 8, 267–279 (1984)
15. Dubois, D., Prade, H.: Possibility Theory. Plenum Press, New York (1988)
16. Dubois, D., Prade, H.: Gradual inference rules in approximate reasoning. Information Sciences 61, 103–122 (1992)

17. Dubois, D., Prade, H.: Upper and lower images of a fuzzy set induced by a fuzzy relation - Applications to fuzzy inference and diagnosis. Information Sciences 64, 203–232 (1992)
18. Fan, S.Q., Zhang, W.X., Ma, J.M.: Fuzzy inference based on fuzzy concept lattice. Fuzzy Sets and Systems 157(24), 3177–3187 (2006)
19. Fodor, J.C.: Nilpotent minimum and related connectives for fuzzy logic. In: Fourth IEEE Int. Conf. Conference on Fuzzy Systems, Yokohama, Japan, pp. 2077–2082 (1995)
20. Georgescu, G., Popescu, A.: Non-dual fuzzy connections. Arch. Math. Log. 43(8), 1009–1039 (2004)
21. Guénoche, A.: Construction du treillis de Galois d'une relation binaire. Mathématiques et Sciences Humaines 109, 41–53 (1990)
22. Kerre, E.E.: An overview of fuzzy relational calculus and its applications. In: Torra, V., Narukawa, Y., Yoshida, Y. (eds.) MDAI 2007. LNCS (LNAI), vol. 4617, pp. 1–13. Springer, Heidelberg (2007)
23. Kerre, E.E., Nachtegael, M.: Fuzzy relational calculus and its application to image processing. In: Gesù, V.D., Pal, S.K., Petrosino, A. (eds.) Proceedings of Fuzzy Logic and Applications, 8th International Workshop, WILF 2009. LNCS, vol. 5571, pp. 179–188. Springer, Heidelberg (2009)
24. Medina, J., Ojeda-Aciego, M., Ruiz-Calviño, J.: Formal concept analysis via multi-adjoint concept lattices. Fuzzy Sets and Systems 160(2), 130–144 (2009)
25. Messai, N., Devignes, M., Napoli, A., Tabbone, M.: Many-valued concept lattices for conceptual clustering and information retrieval. In: Proc. 18th Europ. Conf. on Artif. Intellig., Patras, pp. 722–727 (2008)
26. Pollandt, S.: Fuzzy Begriffe. Springer, Heidelberg (1997)
27. Ward, M., Dilworth, R.P.: Residuated lattices. Trans. AMS 45, 335–354 (1939)
28. Wille, R.: Restructuring lattice theory: an approach based on hierarchies of concepts. In: Rival, I. (ed.) Ordered Sets, pp. 445–470. Reidel, Dordrecht (1982)
29. Wolff, K.E.: Concepts in fuzzy scaling theory: order and granularity. Fuzzy Sets and Systems 132, 63–75 (2002)
30. Xie, C., Yi, L., Du, Y.: An algorithm for fuzzy concept lattices building with application to social navigation. In: ISKE 2007, International Conference on Intelligent Systems and Knowledge Engineering, China (2007)

# Choice Functions in Fuzzy Environment: An Overview*

Xuzhu Wang, Caiping Wu, and Xia Wu

**Abstract.** In this chapter, we present an overview of choice functions. Firstly, we introduce some important research topics on classic choices which serve as a guideline for the fuzzification research of choice functions. Then we begin with the fuzzy choice function defined by Banerjee. After a brief introduction to Banerjee's work, various preferences derived from a choice function are investigated. Main rationality conditions are presented and the relationships between them are presented. Some necessary and sufficient conditions for $T$-transitive rationality are summarized. Afterwards, we turn to a special family of choice functions which are associated with fuzzy preferences. In this case, the final choices can be exact or fuzzy. For the former, various characterizations of the Orlovsky choice function are primary topics. For the latter, Roubens's work is mainly introduced. Finally, the research around Georgescu fuzzy choice functions is surveyed. The investigation includes three parts: rationality characterization, rationality conditions and rationality indicators. In the first part, $G$-rationality, $G$-normality, $M$-rationality and $M$-normality

Wang Xuzhu
Department of Mathematics, Taiyuan University of Technology, Taiyuan, Shanxi, P.R. China
Tel.: 0086-351-6014819
e-mail: wangxuzhu2006@126.com

Wu Caiping
Department of Mathematics, Taiyuan University of Technology, Taiyuan, Shanxi, P.R. China
Tel.: 0086-351-6699148
e-mail: tywcp@sohu.com

Wu Xia
Department of Mathematics, Taiyuan University of Technology, Taiyuan, Shanxi, P.R. China Tel.: 0086-351-6373577
e-mail: zhaojiangxin@sohu.com

* The work is supported by Shanxi Natural Science Foundation 200901106.

C. Cornelis et al. (Eds.): 35 Years of Fuzzy Set Theory, STUDFUZZ 261, pp. 149–169.
springerlink.com 

are discussed. In the second part, some important fuzzy versions of rationality conditions are displayed and their interrelationships are presented. The third part is devoted to the introduction of various rationality and rationality condition indicators.

**Keywords:** Fuzzy choice function, Revealed preference, Rationality condition, Rationalization.

## 1 Classic Choice Functions

In the real world, we are frequently confronted with the problem of how to choose one or more alternatives from a set $S$ of alternatives. This problem is called a choice problem and the set of the chosen alternatives from $S$ is normally called a choice set which is denoted by $C(S)$. A choice function is a mapping (function) that associates a non-empty choice set $C(S)$ to any $S$ in a certain domain $\mathcal{B}$ which is a subset of the set of all alternatives. The concept of a choice function is initiated for the research on Economics, both a consumer [3] and a demand function [1] in Economics are choice functions in essence. In consumer theory, the involved domain $\mathcal{B}$ is a class of convex polyhedra representing "budget sets" which are obtained by taking the income and prices into account. It is generally assumed [1] that $\mathcal{B}$ contains all finite subsets of $X$, where $X$ is the set of all alternatives. For a close examination of the domain $\mathcal{B}$, the reader may refer to [4]. For simplicity, we assume that the set $X$ of all alternatives is finite and $\mathcal{B}$ is the set of all non-empty subsets of $X$, which will be denoted by $P(X)$ in the sequel.

Given a choice function, there are many ways to derive preferences among which we mention the following important ones:

(1) Revealed preference $R$: $xRy \iff \exists S \in P(X), x, y \in S$ and $x \in C(S)$.
(2) Generated preference $\bar{R}$: $x\bar{R}y \iff x \in C(\{x, y\})$.
(3) Strict revealed preference $\tilde{P}$: $x\tilde{P}y \iff \exists S \in P(X), x \in C(S)$ and $y \notin C(S)$.

Those preferences were introduced in different names in the literature. For example, generated preference is called "relation generated by $C$" by Arrow [1] and "base relation" by Herzberger [15]. Since each of $R$, $\bar{R}$ and $\tilde{R}$ reveals the consumer's (or the decision-maker's) preference from a different angle, all of them are commonly called revealed preferences by many authors [3, 10, 11] in different senses. We give them the current names just to distinguish them for the sake of convenience. In the literature, other preferences derived from a choice function can be found, e.g., the transitive closure $t(R)$ and $t(\tilde{P})$ of $R$ and $\tilde{P}$ in [3, 4], $R_*$ in [10], $\hat{P}$ in [17] etc.

It is easily checked that both $R$ and $\bar{R}$ are reflexive and complete since all singletons and all the sets of two elements are included in the domain $P(X)$. In addition, $R$ is acyclic [22]. An important role of $R$ is to describe normality of choice functions. A choice function $C$ is called normal if $\forall S \in P(X), C(S) = \{x \in S | \forall y \in S, xRy\}$. The normality indicates that the choice function can be generated by its revealed preference $R$.

In the study of choice functions, the most important issue is their rationality. To deal with this issue, various rationality conditions have been presented by a number of authors. In the following, we introduce some important ones.

(1) Condition $\alpha$: $\forall S, T \in P(X)$, $x \in C(T)$ and $S \subseteq T$ imply $x \in C(S)$. Its interpretation is: If an alternative is chosen from a set, then it is chosen from any subset of it.
(2) Condition $\beta$: If $S \subseteq T$ and $x, y \in C(S)$, then $x \in C(T)$ iff $y \in C(T)$. Its interpretation is: If both $x$ and $y$ are chosen from $S$, a subset of $T$, then $x$ is chosen from $T$ iff $y$ is chosen from $T$.
(3) Condition $\gamma$: If $S = \bigcup_{i \in I} S_i$ and $x \in C(S_i)$ for all $i \in I$, then $x \in C(S)$. Its interpretation is: If $x$ is chosen from every subset $S_i$ of $S$ and the union of all $S_i$ is $S$ itself, then $x$ is chosen from $S$.
(4) Condition $\delta$: For any $S, T \in P(X)$ and $x, y \in C(S)$, if $S \subseteq T$, then $C(T) \neq \{x\}$. Its interpretation is: If both $x$ and $y$ are chosen from $S$, a subset of $T$, then neither of them can be uniquely chosen from $T$.

Condition $\alpha$ was firstly proposed by Chernoff [2], Condition $\beta$, $\gamma$ and $\delta$ were due to Sen [4, 5]. These rationality conditions formulate the behavior of a choice function when a choice set is expanded or contracted, and thus they are frequently called expansion-contraction conditions. In [1], Arrow also put forward some similar conditions. For example, Arrow's C2 or C3 is just Condition $\alpha$ in a different form and the condition C4 combines $\alpha$ and $\gamma$. It is shown in [4] that Condition $\alpha$ and Condition $\gamma$ together are equivalent to the normality of the choice function.

In [6], a weakened version $\alpha_2$ of $\alpha$, a weakened version $\gamma_2$ of $\gamma$ and two strengthened versions $\beta^+$ and $\beta'$ of $\beta$ can be found. For other expansion-contraction conditions, the reader may refer to [17] for WD, ND, WND, SS and MR, [24] for W1-W5, [16] for PI etc.

Besides expansion-contraction conditions, there exist other type of rationality conditions which are related to preferences derived from a choice function. The most famous of these are the Weak Congruence Axiom (WCA), the Weak Axiom of Revealed Preference (WARP) and their strengthened versions: the Strong Congruence Axiom (SCA) and the Strong Axiom of Revealed Preference (SARP), which are defined as follows.

(1) WCA: If $y \in C(S)$, $x \in S$ and $xRy$, then $x \in C(S)$.
(2) WARP: If $x\tilde{P}y$, then $yR^c x$.
(3) SCA: If $y \in C(S)$, $x \in S$ and $xt(R)y$, then $x \in C(S)$, where $t(R)$ stands for the transitive closure of $R$.
(4) SARP: If $xt(\tilde{P})y$, then $yR^c x$, where $t(\tilde{P})$ stands for the transitive closure of $\tilde{P}$.

SCA was introduced by Richter [3] and its weakened form WCA by Sen [4]. WARP and SARP were introduced in the classic consumer theory by

Samuelson [8] and Houthakker [12] respectively. An amazing conclusion is that all the four conditions are equivalent [4]. They are also equivalent to the combination of Condition $\alpha$ and $\beta$.

There are a number of other revealed-preference-based conditions in the literature in addition to the preceding conditions. The reader may refer to [1] for C1, [10] for HARP, [17] for QRPA and [18] for Axiom 1 – Axiom 4 etc.

Given a choice function $C$, one main concern by researchers is whether or not there exists a binary relation $Q$ such that $C(S) = \{x|\forall y \in S, xQy\}$ for all $S$ in $P(X)$. If such relation does exist, then $C$ is called rational. Since we define choice functions on the domain $P(X)$, it is easy to show that if a choice function is rational then $Q = R$, and thus rationality and normality are equivalent in our context. If additional conditions are imposed on $R$, then we have some special rationality characterizations. A choice function is called transitive rational if the choice function is rational and $R$ is transitive. Considering the equivalency of some rationality conditions [4, 9], each of the following conditions is a necessary and sufficient condition of transitive rationality: WCA, SCA, WARP, SARP, $\alpha$ and $\beta$ together, $R = \tilde{R}$, $R = R_*$ ($R_* = \{(x,y)|x \notin S$ or $x \in C(S)$ or $y \notin C(S)\}$, the transitivity of $\bar{R}$ together with the normality of the choice function, $\bar{R} = \tilde{R}$ together with the normality of the choice function etc.

Besides transitive rationality, the investigation of other rationalities are also visible in the literature among which we mention mainly quasi-transitive rationality, semi-transitive rationality, pseudo-transitive rationality and semi-order rationality. A fuzzy choice function is called quasi-transitive rational if the fuzzy choice function is rational and $P_R$ is transitive. where $P_R$ stands for the strict relation of $R$, defined by $P_R = R \cap R^d$ ($R^d = \{(x,y)|yR^c x\}$). Concerning quasi-transitive rationality, a main conclusion is: for a normal choice function $C$, $R$ is quasi-transitive iff Condition $\delta$ is satisfied [4]. Other necessary and sufficient conditions can be found in [18, 19, 21, 24]. A fuzzy choice function is called semi-transitive rational if the fuzzy choice function is rational and $R$ is semi-transitive, i.e., $\forall a, b, c, d \in X$, $aRb$ and $bRc$ imply $aRd$ or $dRc$. For the characterization of semi-transitive rationality, the reader may refer to [19]. As for pseudo-transitive rationality and semi-order rationality, see [18, 19, 21, 24] for details.

Among choice functions, an important special family is that whose choice sets are based on a preference relation on $X$. Two famous preference-based choice sets are the set of best elements and the set of maximal elements [5]. Given a relation $R$ on $X$ and $S \in P(X)$, an element $x$ in $S$ is called a "best" (or "greatest") element of $S$ w.r.t. $R$ if $\forall y \in S$, $xRy$, and the set of all best elements in $S$ is denoted by $G_R(S)$, i.e.,

$$G_R(S) = \{x|x \in S \text{ and } \forall y \in S, xRy\}.$$

An element $x$ is called a maximal element of $S$ w.r.t. $R$ if $\forall y \in S$, $yP_R^c x$, where $P_R$ is the strict relation of $R$

and the set of all maximal elements in $S$ is denoted by $M_R(S)$, i.e.,

$$M_R(S) = \{x | x \in S \text{ and } \forall y \in S, yP^c x\}.$$

Clearly, $G_R(S) \subseteq M_R(S)$ for any $S \in P(X)$. Unfortunately, $G_R(S)$ or $M_R(S)$ may be empty, and either $G_R$ or $M_R$ is not necessarily a choice function as a result. It is shown in [5] that if $R$ is reflexive and complete then $G_R(S) \neq \emptyset$ iff $R$ is acyclic on $X$, i.e.,

$$\forall x_1, x_2, \cdots, x_n \in X, x_1 P x_2 P \cdots P x_n \text{ implies } x_1 R x_n.$$

It is pointed out in [23] that $\forall S \in P(X), G_R(S) = M_R(S)$ if $R$ is reflexive and complete. As a consequence, the acyclicity of $R$ is also a necessary and sufficient condition for $M_R(S) \neq \emptyset$ provided that $R$ is reflexive and complete.

In [5, 7], the properties of $G_R$ are investigated. It is verified that (1) $G_R$ satisfies Condition $\alpha$ and $\gamma$ but not necessarily Condition $\beta$; (2) Under the assumption that $G_R$ is a choice function, $G_R$ satisfies Condition $\beta$ iff $R$ satisfies $P \circ I \subseteq P$, where $I$ is the indifference part of $R$ defined by $xIy \iff xRy$ and $yRx$.

## 2 Banerjee Fuzzy Choice Functions

In 1995, Banerjee [44] presented the definition of a fuzzy choice function, which is called the Banerjee fuzzy choice function [31]. Let $F(X)$ denote the set of all non-empty fuzzy subsets of $X$. A Banerjee fuzzy choice function is a mapping $C : P(X) \to F(X)$ such that $\forall S \in P(X)$, supp$C(S) \subseteq S$, where supp $C(S)$ denotes the support of the fuzzy choice set $C(S)$. It is easily checked that supp$C(S) \subseteq S$ is equivalent to $C(S) \subseteq S$. So supp$C(S) \subseteq S$ can be replaced by $C(S) \subseteq S$ which is in the same form as in the crisp case. In this section, we assume that every choice set is normal, i.e., $\forall S \in P(X)$, $\exists x \in S$ such that $C(S)(x) = 1$.

Then, he carried out an investigation into the characterization of rationality of fuzzy choice functions in a general sense. To examine the rationality of a choice function $C$, Banerjee introduced some concepts related to a fuzzy choice function $C$, among which the most important ones include fuzzy revealed preference relation $R$ defined by

$$R(x, y) = \bigvee_{\{S | x, y \in S\}} C(S)(x),$$

and normality:

$$\forall S \in P(X), \forall x \in S, C(S)(x) = \bigwedge_{y \in S} R(x, y).$$

Clearly, (1) $\forall x, y \in X, C(S)(x) \leq R(x,y)$; (2)If every choice set is normal, i.e., $\forall S \in P(X)$, $\exists x \in S$ such that $C(S)(x) = 1$, then $R$ is reflexive and complete. Afterwards, Banerjee proposed three fuzzy congruence conditions FC1, FC2, FC3.

(1) FC1: $\forall S \in P(X)$, $y$ is dominant in $S$, i.e., $C(S)(y) \geq C(S)(x)$ for all $x \in S$, implies $C(S)(x) = R(x,y)$ is valid for all $x \in S$.
(2) FC2: $\forall S \in P(X)$ and $\forall x, y \in S$, $y$ is dominant in $S$ and $R(x,y) \geq R(y,x)$ imply that $x$ is dominant in $S$.
(3) FC3: $\forall S \in P(X), \forall \alpha \in (0,1]$ and $\forall x, y \in S, C(S)(y) \geq \alpha$ and $R(x,y) \geq \alpha$ imply $C(S)(x) \geq \alpha$.

It is easily checked that all the three conditions are fuzzy versions of WCA in the crisp case. Banerjee showed that FC1, FC2 and FC3 are necessary and sufficient conditions for characterizing the rationality of a fuzzy choice function. Meanwhile, Banerjee pointed out that the three congruence conditions are independent. This is an incorrect assertion. As a matter of fact, the following dependencies are verified in [45]:

(1) FC1 implies FC2;
(2) FC3 implies FC2.

Hence FC1 and FC3 are sufficient to characterize the rationality of a Banerjee fuzzy choice function.

As in the crisp case, various preference relations can be derived from a fuzzy choice function. Besides the fuzzy revealed preference relation $R$, some other preferences $\bar{R}, \tilde{R}_n, R_*$ derived from a fuzzy choice function $C$ are also defined in different frameworks [34, 35, 51]. Here we assume that $(T, S, n)$ is a De Morgan triple.

(1) $\bar{R}(x,y) = C(\{x,y\})(x)$;
(2) $\tilde{R}_n(x,y) = n(\tilde{P}_n(y,x))$, where $\tilde{P}_n(x,y) = \bigvee_{\{S|x,y \in S\}} T(C(S)(x), n(C(S)(y)))$;
(3) $R_*(x,y) = \bigwedge_{\{S|x,y \in S\}} I_T(C(S)(y), C(S)(x))$, where $T$ is a $t$-norm,

$$I_T(a,b) = \sup\{c | T(a,c) \leq b\}$$

is the $R$-implication associated with $T$, $n$ a negation.

In[51], under the condition that $n$ is a strong negation, the relationships between these preferences are investigated in detail and the main results include:

(1) $\tilde{R}_n \subseteq \bar{R} \subseteq R$;
(2) $R_* \subseteq \bar{R}$;
(3) $R = \tilde{R}_n$ iff the choice function is normal and $\bar{R} = \tilde{R}_n$;
(4) $R_* = \tilde{R}_n$ under a strong De Morgan triple;
(5) If $T$ is a continuous $t$-norm, then $R = R_*$ iff $\bar{R} = R_*$ and the choice function is normal.

*Remark 1.* Georgescu proved in [31] that the normality of the choice function and $\bar{R} = \tilde{R}_n$ imply $R = \tilde{R}_n$ in a different framework.

In the study of fuzzy choice functions, some important rationality conditions are fuzzified. As in the crisp case, they can be divided into revealed-preference-based conditions and expansion-contraction conditions. The former includes Weak Fuzzy Congruence Axiom (WFCA), Weak Axiom of Fuzzy Revealed Preference (WAFRP) and their strengthened versions: Strong Fuzzy Congruence Axiom (SFCA) and Strong Axiom of Fuzzy Revealed Preference (SAFRP). Let $(T, S, n)$ be a De Morgan triple. Then these rationality conditions are defined as follows.

(1) WFCA: $\forall S \in P(X), \forall x, y \in S, T(R(x,y), C(S)(y)) \leq C(S)(x)$;
(2) WAFRP: $\forall x, y \in X, \tilde{P}_n(x,y) \leq n(R(y,x))$;
(3) SFCA: $\forall S \in P(X), \forall x, y \in S, T(t(R)(x,y), C(S)(y)) \leq C(S)(x)$;
(4) SAFRP: $\forall x, y \in X, t(\tilde{P}_n)(x,y) \leq n(R(y,x))$.

It is easily checked that $WFCA$ is equivalent to FC3 if $T = \min$. WFCA and WAFRP are not generally equivalent although their crisp counterparts WCA and WARP are equivalent. However, the equivalency can be obtained again by imposing the condition that $(T, S, n)$ is a strong De Morgan triple [52].

The most important expansion-contraction conditions include $F\alpha, F\beta, F\gamma, F\delta$ [48, 53].

(1) $F\alpha$ : If $S_1 \subseteq S_2$ then $C(S_2)(x) \leq C(S_1)(x)$ for any $x \in S_1$.
(2) $F\beta$ : If $S_1 \subseteq S_2$ and $x, y \in S_1$, then

$$T(C(S_1)(x), C(S_1)(y), C(S_2)(x)) \leq C(S_2)(y).$$

(3) $F\gamma$: For any $x \in S_1 \cap S_2$,

$$T(C(S_1)(x), C(S_2)(x)) \leq C(S_1 \cup S_2)(x).$$

(4) $F\delta$: If $S_1 \subseteq S_2$, $x, y \in S_1$ and $x \neq y$, then

$$T(C(S_1)(x), C(S_1)(y)) \leq n(T(C(S_2)(x), \underset{t \neq x}{T}(n(C(S_2)(t))))).$$

It can be checked that they are indeed fuzzy versions of the conditions $\alpha, \beta, \gamma, \delta$ respectively. Concerning the connections between the above mentioned conditions, it is shown that

(1) If $n$ is a strong negation, then $F\beta$ implies $F\delta$ [53].
(2) If $T = \min$, then $F\alpha$ and $F\gamma$ are equivalent to the normality of the choice function [48].
(3) If $T = \min$, then $F\alpha$ and $F\beta$ are equivalent to WFCA [33].
(4) The normality of the choice function and $\bar{R} = \tilde{R}_n \iff WAFRP \iff SAFRP \iff R = \tilde{R}_n$ if the involved negation is a strong negation[51].

(5) $T$-regularity(reflexivity, completeness and $T$-transitivity) of $R$ and the normality of the choice function $\Longleftrightarrow$ $T$-regularity of $\bar{R}$ and the normality of the choice function $\Longleftrightarrow$ $WFCA$ $\Longleftrightarrow$ $SFCA$[34, 51].

In addition to these popular rationality conditions, many other crisp rationality conditions employed by Suzumura [9], Schwartz [24], Arrow [1] and Bandyopadhyay [17] are also extended to the fuzzy case and the relationships between the fuzzified conditions are investigated in [49, 54, 56].

Another important research topic is the rationalization of fuzzy choice functions. If there exists a reflexive and complete fuzzy relation $Q$ on $X$ such that $\forall S \in P(X), \forall x \in S, C(S)(x) = \bigwedge_{y \in S} Q(x,y)$, then we call the choice function $C$ rationalized by $Q$, or simply the choice function $C$ rational. It is interesting that rationality and normality are equivalent if every choice set is normal. Firstly, it is clear that normality implies rationality. On the other hand, if $C$ is rationalized by $Q$, then for any $x, y \in X$,

$$R(x,y) = \bigvee_{\{S|x,y \in S\}} C(S)(x) = \bigvee_{\{S|x,y \in S\}} (\bigwedge_{z \in S} Q(x,z)) \leq Q(x,y).$$

Thus $C(S)(x) = \bigwedge_{y \in S} Q(x,y) \geq \bigwedge_{y \in S} R(x,y)$. On the other hand, $C(S)(x) \leq R(x,y)$ for all $x, y \in X$, which means $C(S)(x) \leq \bigwedge_{y \in S} R(x,y)$. Hence, $C(S)(x) = \bigwedge_{y \in S} R(x,y)$ for all $x, y \in X$, i.e., $C$ is normal.

There are various rationality characterizations dependent on the properties of the binary relations involved, of which $T$-transitive rationality is the most extensively studied one. A fuzzy choice function is called $T$-transitive rational if the fuzzy choice function is rational and $R$ is $T$-regular. Each of the following statements is a necessary and sufficient condition of $T$-transitive rationality provided that $T$ is continuous [34, 51].

(1) $\bar{R}$ is $T$-regular and the choice function is normal;
(2) The choice function satisfies $WFCA$;
(3) The choice function satisfies $SFCA$;
(4) The choice function satisfies $F\alpha$ and $F\beta^+$, where $F\beta^+$ means that $S_1 \subseteq S_2$ and $x, y \in S_1$, imply $T(C(S_1)(x), C(S_2)(y)) \leq C(S_2)(x)$.
(5) $R = R_*$;
(6) $\bar{R} = R_*$ and the choice function is normal.

Furthermore, if $(T, S, n)$ is a strong De Morgan triple, then the following statements are equivalent [51] and thus every one can serve as a necessary and sufficient condition of $T$-transitive rationality.

(1) $R = \tilde{R}_n$;
(2) The choice function satisfies $WAFRP$;
(3) The choice function satisfies $SAFRP$;
(4) The choice function is normal and $\bar{R} = \tilde{R}_n$.

In addition, if $T = \min$, $T$-transitive rationality is equivalent to $F\alpha$ and $F\beta$ [34].

Besides $T$-transitive rationality, the investigation of other rationalities are also visible in the literature among which we mention mainly $T$-quasi-transitive rationality and acyclic rationality. A fuzzy choice function is called $T$-quasi-transitive rational if the fuzzy choice function is rationalized by a reflexive, complete fuzzy relation $Q$ such that $P_Q$ is $T$-transitive, where $P_Q$ stands for the strict relation of $Q$. It was shown in [54] if a fuzzy choice function satisfies $F\alpha$ and $F\delta$, then $R$ is $T$-quasi-transitive rational. In [49], two necessary conditions of $T$-quasi-transitive rationality are presented by the fuzzification of the Schwartz's crisp rationality conditions [24]. In [57], by using fuzzy counterparts of Aizerman's [14] and Nehring's [13] rationality conditions, several necessary conditions of $T$-quasi-transitive rationality are also proposed. Unfortunately, these necessary conditions are not sufficient to assure $T$-quasi-transitive rationality although all of them are sufficient in the crisp case.

If a fuzzy choice function $C$ is rational and rationalized by an acyclic relation $Q$ then $C$ is called acyclic rational. A fuzzy relation $Q$ on $X$ is acyclic if, for any $x_1, x_2, \cdots, x_n \in X$, $Q(x_1, x_2) > Q(x_2, x_1)$, $Q(x_2, x_3) > Q(x_3, x_2)$, $\cdots$, $Q(x_{n-1}, x_n) > Q(x_n, x_{n-1})$ imply $Q(x_1, x_n) \geq Q(x_n, x_1)$. As we know, a choice function is normal if and only if it is rational. Moreover, it is shown that if every choice set of a fuzzy choice function is normal, then the revealed preference is acyclic [49]. Hence we can draw the conclusion that every rational fuzzy choice function is acyclic rational provided that every choice set is normal. In [56], a necessary and sufficient condition is presented to characterize acyclic rationality by fuzzifying the Bandyopadhyay's rationality conditions in [17].

As far as $T$-pseudo-transitive rationality and $T$-semi-transitive rationality are concerned, some elementary results can be found in [50].

*Remark 2.* It is seen that relations and their properties are extensively employed in the study of choice functions, particularly in the characterization of choice functions. They are deeply investigated in preference modeling theory [58]. In this theory, relations are employed to model pairwise comparison results between alternatives including strict preference, indifference, incomparability and large preference. Some properties of relations, e.g., reflexivity, completeness, transitivity, semi-transitivity, Ferrers property are among the most important concepts for modeling particular preference structures such as weak order, partial order, interval order, semi-order etc. As relations are fuzzified, their properties are also fuzzified, and as a result fuzzy preference structures are correspondingly investigated. For the literature on fuzzy preference structures, we recommend the references [29, 30, 55, 60, 61, 62].

Finally, we point out that the definition of rationality is adopted from [31]. There is an alternative definition in [44].

## 3 Choices Based on Fuzzy Preferences

### 3.1 Exact Choices Based on Fuzzy Preferences

In fuzzy environment, the study of choice functions was essentially initiated by Orlovsky when modeling fuzzy preferences in 1978 [37]. Orlovsky argued that a decision-maker may have a vague idea of preferences between alternatives, i.e. he find it difficult to state definitely that an alternative is preferred to another. In such cases, a number in $[0, 1]$ may be readily employed to reflect the degree of preference. From then on, fuzzy preference relations have been extensively utilized in decision-making analysis including the research of choice functions.

Let $\mathcal{R}(X)$ the set of some fuzzy relations on $X$. The following definition is adopted from [39]. An $\mathcal{R}$-preference-based choice function (PCF) is a mapping $C : P(X) \times \mathcal{R}(X) \to P(X)$ such that

$$\forall S \in P(X), \forall R \in \mathcal{R}(X), C(S, R) \subseteq S.$$

For any $S \in P(X)$, $R \in \mathcal{R}(X)$ and $x \in X$, let

$$OV(x, S, R) = \min_{y \in S} \min\{1 - R(y, x) + R(x, y), 1\}.$$

Then a PCF is defined by

$$C_{OV}(S, R) = \{x \in S | \forall y \in S, OV(x, S, R) \geq OV(y, S, R)\},$$

which is called the Orlovsky choice function [38].

*Remark 3.* Take the notation

$$PD(S, R) = \{x \in S | \forall y \in S, R(x, y) \geq R(y, x)\}.$$

It is shown that $C_{OV} = PD$ provided $R$ satisfies acyclicity. Acyclicity is also a necessary and sufficient condition for $PD(S, R) \neq \emptyset$ (or equivalently, $C_{OV}(S, R) \neq \emptyset$)[23].

Besides the Orlovsky choice function, some other crisp choice functions based on fuzzy preferences can be found in the literature. Dutta et al. suggested four choice functions in [25] and Barrett et al. proposed nine unambiguous choice functions in [42]. In addition, Basu [26], Switalski [27] and Roubens [43] also suggested some ways to choose alternatives when confronted with fuzzy preferences. Certainly, the use of different choice function reflects the decision-maker's different angle of viewing alternatives and may lead to different choice results.

The Orlovsky choice function attracted much attention among which the most important and fruitful research is the characterization of the Orlovsky choice function. In 1993, Banerjee firstly presented a characterization of the Orlovsky choice function for the so-called Banerjee transitive class of fuzzy relations [38]. By a Banerjee transitive fuzzy relation, we mean the fuzzy relation $R$ on $X$ such that $\forall x, y, z \in X$,

$$R(x, y) \geq R(y, x) \text{ and } R(y, z) \geq R(z, y) \text{ imply } R(x, z) \geq R(z, x).$$

This property was firstly proposed by Ponsard [59]. We adopted the name from [39]. Bouyssou pointed out that the contraction condition $\alpha$ is mistakenly taken for granted [39], and the characterization is incorrect as a result. Then he reformulated the characterization for the same class of fuzzy relations. Meanwhile, Bouyssou proposed a characterization of the Orlovsky choice function for the acyclic class of fuzzy relations. Afterwards, Sengupta characterized the Orlovsky choice function for the weak transitive class of fuzzy relations [41]. A fuzzy relation $R$ on $X$ is called weakly transitive if $\forall x, y, z \in X$,

$$R(x, y) > R(y, x) \text{ and } R(y, z) > R(z, y) \text{ imply } R(x, z) > R(z, x).$$

In addition to characterizing the Orlovsky choice function, Bouyssou characterized the "sum of differences" choice function by using Neutrality, Monotonicity and Independence of circuits [40].

Another research topic is the rationality properties of crisp preference-based choice functions. In [42], Barrett et al. proposed two classes of rationality conditions: conditions for choosing an alternative and conditions for rejecting an alternative. The first class of conditions includes:

(1) *reward for pairwise weak dominance*:

$$\forall y \in S, R(x, y) \geq R(y, x) \Rightarrow x \in C(S, R);$$

(2)*reward for pairwise strict dominance*:

$$\forall y \in S \setminus \{x\}, R(x, y) > R(y, x) \Rightarrow x \in C(S, R).$$

The second class of conditions includes:
(1) *strict rejection condition*:

$$\forall x \in S, R(y, x) \geq R(x, y) \text{ and } \exists y \in S, R(y, x > R(x, y) \text{ implies } x \notin C(S, R);$$

(2)*weak rejection condition*:

$$\forall y \in S \setminus \{x\}, R(y, x) > R(x, y) \text{ implies } x \notin C(S, R).$$

Clearly, these conditions are all based on pairwise comparisons. In addition, they presented faithfulness (upper faithfulness, lower faithfulness) as a property to assess the rationality of preference-based choice functions. Afterwards, nine choice functions are examined based on these conditions. The results show that $\max - mD$ satisfies most rationality properties, where $\max - mD$ is defined by

$$C_{mD}(S, R) = \{x | \forall y \in S, mD(x, S, R) \geq mD(y, S, R)\}$$

and for any $S \in P(X)$, $x \in S$,

$$mD(x, S, R) = \min_{y \in S \setminus \{x\}} (R(x, y) - R(y, x)).$$

## 3.2 *Fuzzy Choices Based on Fuzzy Preferences*

In 1993, Roubens defined several fuzzy choice set based on a fuzzy relation $R$ on $X$, among which the following are typical ones [43]. Let $P$ be the strict part of $R$ defined by $P(x, y) = T(R(x, y), 1 - R(y, x))$ with $T$ a $t$-norm. Let $S \in P(X)$ and $y \in S$.

(1) $D^R(y) = \bigwedge_{x \in S \setminus \{y\}} R(y, x)$ is the degree to which $y$ is preferred to any element of $S$.
(2) $N^R(y) = 1 - \bigvee_{x \in S \setminus \{y\}} R(x, y)$ is the degree to which any element of $S$ is not preferred to $y$.
(3) $SD^R(y) = \bigwedge_{x \in S \setminus \{y\}} P(y, x)$ is the degree to which $y$ is strictly preferred to any element of $S$.
(4) $SN^R(y) = 1 - \bigvee_{x \in S \setminus \{y\}} P(x, y)$, is the degree to which any element of $S$ is not strictly preferred to $y$.

Then he proposed some rationality properties to examine the rationality of these choice actions which include *heritage property, condordance property, independence of irrelevant alternatives and reward for strict dominance.* It can be easily checked heritage property and condordance property are just $F\alpha$ and $F\gamma$ respectively while reward for strict dominance is similar to reward for pairwise strict dominance proposed by Barrett et al [42]. Independence condition means that if $x$ is better than $y$ in a set, then the same is true for a larger set of alternatives. Afterwards, Roubens showed the fulfilment of these properties for the above mentioned preference-based fuzzy choices under some conditions.

As a matter of fact, exact choices and fuzzy choices are closely related to each other if a fuzzy preference is given. For instance, Barrett et al. [42] proposed nine choice functions in order to make unambiguous choices. However, we can see a fuzzy choice function behind every exact choice function.

Take their $C_{mD}$ as an example. In this case, $C_{mD}$ is the final exact choice function. In forming this choice function, $mD(x, S, R)$ is employed, which may be regarded as a fuzzy choice set and thus a fuzzy choice function when $S$ varies in $P(X)$. Conversely, a fuzzy choice function can be also employed to make crisp choices. More specifically, assume that $C(S, R)$ $(S \in P(X))$ is a fuzzy choice related to a fuzzy relation $R$. Then by simply letting $C'(S, R) = \{x | \forall y \in S, C(S, R)(x) \geq C(S, R)(y)\}$, we have the exact choice function $C'$. Particularly, if $C(S, R)$ is normal for every $S \in P(X)$, then we may make exact choice $C'(S, R) = \{x | C(S, R)(x) = 1\}$ for any $S$, which is called the set of non-fuzzy non-dominated alternatives in [37]. Similar practice can be found in [43] for $D^R, N^R, SD^R$ and $SN^R$.

For a survey of preference-based choice functions, the reader may refer to [63].

## 4 Georgescu Fuzzy Choice Functions

In the definition of Banerjee fuzzy choice function, the available set of alternatives are crisp sets, i.e, we make the fuzzy choice $C(S)$ from the *crisp* set $S$. In other words, the domain of a Banerjee choice function is crisp although its range is fuzzy. Georgescu argued that "If a fuzzy subset $S$ of $X$ will represent a vague criterion and $x$ is an alternative, then the real number $S(x)$ means the degree to which the alternative $x$ verifies the criterion $S$. $S(x)$ will be called the availability degree of $x$" ([31], Page 87). Hence, she presented a new definition of fuzzy choice function by extending the domain from crisp sets to fuzzy sets. A Georgescu fuzzy choice function is a mapping $C$: $F(X) \to F(X)$ such that $C(S) \subseteq S$ is valid for any $S \in F(X)$. This is the most general definition of a fuzzy choice function in the literature. In this section, a fuzzy choice function means a Georgescu fuzzy choice function and the involved work is mainly done by Georgescu [31, 32, 33, 34, 35] in the framework $(*, \to, \neg)$, where $*$ is a continuous $t$-norm, $x \to y = \sup\{z | x * z \leq y\}$ and $\neg x = x \to 0$. We assume that every choice set is normal, i.e., $\forall S \in F(X)$, $\exists x \in [0, 1]$, $C(S)(x) = 1$.

*Remark 4.* Fuzzy logic connectives are fundamental tools for the research on Georgescu fuzzy choice functions. For their detailed discussion, the reader may refer to [28, 46, 47, 65, 66, 67].

For a Georgescu fuzzy choice function $C$, some main fuzzy preferences derived from $R$ are defined as follows.

(1) $R(x, y) = \bigvee_{S \in F(X)} (C(S)(x) * S(y))$.
(2) $\bar{R}(x, y) = C(\{x, y\})(x)$ and $\bar{P}(x, y) = \bar{R}(x, y) * \neg\bar{R}(y, x)$.
(3) $\tilde{R}(x, y) = \neg\tilde{P}(y, x)$, where $\tilde{P}(x, y) = \bigvee_{S \in F(X)} (C(S)(x) * S(y) * \neg C(S)(y))$.

## 4.1 Rationality Characterization

Let $Q$ be a fuzzy relation on $X$. Define $M_Q$ and $G_Q$ by: $S \in F(X)$,

$$M_Q(x) = S(x) * \bigwedge_{y \in X} [(S(y) * Q(y,x)) \to Q(x,y)],$$

$$G_Q(x) = S(x) * \bigwedge_{y \in X} ((S(y) \to Q(x,y)).$$

Clearly, $G_Q(S) \subseteq M_Q(S)$ for any $S \in F(X)$. It is verified [31] that $G_Q(S) = M_Q(S)$ if $Q$ is reflexive and complete. A fuzzy choice function is called $G$-rational ($M$-rational respectively) if there exists a fuzzy relation $Q$ on $X$ such that $C = G_Q$ ($C = M_Q$ respectively). Particularly, $G$-rational is called $G$-normal and $M$-rational is called $M$-normal when $Q = R$. Concerning normality and rationality, there are the following conclusions [31]:

(1) $M$-rationality implies $G$-rationality;
(2) If $* = \min$, $G$-normality and $G$-rationality are equivalent;
(3) If $* = \min$, $M$-normality implies $G$-normality.

If additional conditions are imposed, then we have special rationalities. A fuzzy choice function $C$ is called full rational if there exists a fuzzy relation $Q$ on $X$ which is reflexive, $*$-transitive and strongly total ($\forall x, y \in X, x \neq y, Q(x,y) = 1$ or $Q(y,x) = 1$) such that $C = G_Q$. A fuzzy choice function $C$ is called strongly complete rational if there exists a fuzzy relation $Q$ on $X$ which is strongly complete (reflexive, strongly total) such that $C = G_Q$. It is verified that any $M$-normal fuzzy choice function $C$ is strongly complete rational under the $t$-norm min.

Compared with the rationality characterization of crisp choice functions, the results related to the fuzzy case in the sense of Georgescu are much fewer.

## 4.2 Rationality Conditions

Some important rationality conditions including WCA, SCA, WARP, SARP, etc. have been fuzzified by Georgescu. To distinguish them from the conditions in the sense of Banerjee, we take the notations WFCA(G), SFCA(G), WAFRP(G), SAFRP(G) etc.

(1) WAFRP(G): $\tilde{P}(x,y) \leq \neg R(y,x)$ for all $x, y \in X$.
(2) SAFRP(G): $t(\tilde{P})(x,y) \leq \neg R(y,x)$ for all $x, y \in X$, where $t(\tilde{P})$ is the $*$-transitive closure of $\tilde{P}$.
(3) WFCA(G): $R(x,y) * C(S)(y) * S(x) \leq C(S)(x)$ for all $S \in F(X)$ and $x, y \in X$.
(4) SFCA(G): $t(R)(x,y) * C(S)(y) * S(x) \leq C(S)(x)$ for all $S \in F(X)$ and $x, y \in X$, where $t(R)$ is the $*$-transitive closure of $R$.

Generally, WFCA(G) implies WAFRP(G), and SFCA(G) implies SAFRP(G) [31]. In addition, if we consider the following statements:

(i) $R$ is regular and $C$ is $G$-normal;
(ii) $\bar{R}$ is regular and $C$ is $G$-normal;
(iii) $C$ verifies WFCA(G);
(iv) $C$ verifies SFCA(G);
(v) $C$ verifies WAFRP(G);
(vi) $C$ verifies SAFRP(G);
(vii) $R = \tilde{R}$;
(viii) $\bar{R} = \tilde{R}$ and $C$ is $G$-normal,

then the following assertions are true [31]:

(1) (i) and (ii) are equivalent.
(2) (i) implies (iii); if $* = \min$, then (iii) implies (i).
(3) (iii) and (iv) are equivalent.
(4) If $*$ is the Lukasiewicz $t$-norm, then (iii), (v), (vi), (vii) are equivalent.
(5) (viii) implies (vii).

These assertions extend the results obtained by Sen in [4], where all the assertions are equivalent in the crisp case.

Another fuzzy version WAFRP$^0$ of WARP also plays an important role in the study of Georgescu fuzzy choice function. Let $A, B \in F(X)$. Take the notations: $a \leftrightarrow b = (a \to b) \wedge (b \to a)$;
$I(A,B) = \bigwedge_{x\in X} (A(x) \to B(x))$;
$E(A,B) = \bigwedge_{x\in X} (A(x) \leftrightarrow B(x))$.

Then we have WAFRP$^0$: For any $x, y \in X$ and $S_1, S_2 \in F(X)$,

$$(S_1(x) * C(S_2)(x)) * (S_2(y) * C(S_1)(y)) \leq E(S_1 \cap C(S_2), S_2 \cap C(S_1)).$$

WAFRP$^0$ implies $M$-normality and thus $G$-rationality and $M$-rationality as well. In addition, WAFRP$^0$ is equivalent to WFCA(G) under the $t$-norm min.

For a Georgescu fuzzy choice function, some expansion-contraction conditions have the following fuzzy versions:

(1) $F\alpha(G): \forall x \in X, \forall S, T \in F(X), I(S,T) * S(x) * C(T)(x) \leq C(S)(x)$.
(2) $F\beta(G): \forall x, y \in X, \forall S, T \in F(X)$,

$$I(S,T) * C(S)(x) * C(S)(y) \leq C(T)(x) \leftrightarrow C(T)(y).$$

(3) $FAA: \forall x \in X, \forall S, T \in F(X), I(S,T) * S(x) * C(T)(x) \leq E(S \cap C(T), C(S))$.

Clearly, normality implies $F\alpha(G)$. Concerning the relationships between these conditions, it is worth mentioning that, if the involved $t$-norm $*$ is min, then (1) $WFCA(G)$ is equivalent to the combination of $F\alpha(G)$ and $F\beta(G)$; (2) FAA is equivalent to WFCA(G) and can be employed to characterize the full rationality of the choice function.

## 4.3 Rationality Indicators

In the study of Georgescu fuzzy choice functions, a unique aspect is the measure of the degree of rationality and the degree of the satisfaction of rationality conditions. Instead of examining whether a fuzzy choice function is rational or not, a rationality degree is defined to measure the rationality of the choice function. Similarly, instead of checking whether or not a rationality property is satisfied or not, an indicator is defined to indicate the degree to which the property is satisfied. In order to achieve this aim, we firstly introduce a basic notion: the similarity of two fuzzy choice functions $C_1$ and $C_2$ defined by

$$E(C_1, C_2) = \bigwedge_{x \in X} \bigwedge_{S \in F(X)} (C_1(S)(x) \leftrightarrow C_2(S)(x)).$$

Let $\mathcal{R} = F(X \times X)$ and $R$ be the revealed preference. Firstly, we present the rationality and normality degrees:

$G$-rationality degree: $Rat_G(C) = \bigvee_{Q \in \mathcal{R}} E(C, G_R)$;
$M$-rationality degree: $Rat_M(C) = \bigvee_{Q \in \mathcal{R}} E(C, M_R)$;
$G$-normality degree: $Norm_G(C) = E(C, G_R)$;
$M$-normality degree: $Norm_M(C) = E(C, M_R)$.

Concerning the connections between these rationality indicators, the following are valid:

$$Rat_M(C) \leq Rat_G(C), Norm_M(C) \leq Norm_G(C) \text{ and } Rat_G(C) = Norm_G(C),$$

which extend the results related to the relationships between $M$-rationality, $M$-normality, $G$-rationality and $G$-normality. Moreover, Georgescu defined and discussed strongly complete rationality indicator.

As for rationality conditions, the following indicators are typical:

(1) $WAFRP(C) = \bigwedge_{x,y \in X} (\tilde{P}(x,y) \to \neg R(y,x))$;
(2) $SAFRP(C) = \bigwedge_{x,y \in X} (t(\tilde{P})(x,y) \to \neg R(y,x))$;
(3) $WFCA(C) = \bigwedge_{x,y \in X} \bigwedge_{S \in F(X)} ((S(x) * C(S)(y) * R(x,y)) \to C(S)(x))$;
(4) $SFCA(C) = \bigwedge_{x,y \in X} \bigwedge_{S \in F(X)} ((S(x) * C(S)(y) * t(R)(x,y)) \to C(S)(x))$.

If the involved $t$-norm is min, the equalities

$$WFCA(C) = SFCA(C) = E(C, G_R) \wedge Trans(R) = E(C, G_R) \wedge Trans(\bar{R})$$

can be found in [31], which extend the conclusions in [4] to degree descriptions. If $*$ is the Lukasiewicz $t$-norm, then $WFARP(C) = WFCA(C) = E(R, \tilde{R})$.

The rationality condition $WAFRP^0$ indicator was also investigated. Its definition is $WAFRP^0(C) =$

$$\bigwedge_{x,y\in X}\bigvee_{S,T\in F(X)}[(S(x)*C(T)(x)*(T(y)*C(S)(y)))\to E(S\cap C(T),T\cap C(S))].$$

It is verified in [64] that

$$WFCA(C) = WAFRP^0(C) \leq \min\{WAFRP(C), Norm_M(C), E(G_R, M_R)\}$$

if $* = \min$.

Important expansion-contraction condition indicators include $F\alpha(C)$, $F\beta(C)$, $FAA(C)$, $F\alpha_2(C)$, $F\beta_2(C)$, etc.

(1) $F\alpha(C) = \bigwedge_{x\in X}\bigwedge_{S,T\in F(X)}[(I(S,T)*S(x)*C(T)(x))\to C(S)(x)]$;
(2) $F\beta(C) = \bigwedge_{x,y\in X}\bigwedge_{S,T\in F(X)}[(I(S,T)*C(S)(x)*C(S)(y))\to (C(T)(x)\leftrightarrow C(T)(y))]$;
(3) $F\alpha_2(C) = \bigwedge_{x\in X}\bigwedge_{S\in F(X)}[(I(\{x,y\},S)*C(S)(x))\to C(\{x,y\})(x)]$;
(4) $F\beta_2(C) = \bigwedge_{x,y\in X}\bigwedge_{S\in F(X)}[(I(S\{x,y\},S)*C(\{x,y\})(x)*C(\{x,y\})(y))\to (C(S)(x)\leftrightarrow C(S)(y))]$;
(5) $FAA(C)= \bigwedge_{x\in X}\bigwedge_{S,T\in F(X)}[(I(S,T)*S(x)*C(T)(x))\to E(S\cap C(T),C(S)))]$.

If the involved $t$-norm is min, then $WFCA(C) = F\alpha(C)\wedge F\beta(C)$.

*Remark 5.* In [31], it was claimed that $WFCA(C) = F\alpha(C)\wedge F\beta(C)$ without the requirement $* = \min$. However, the established proof was based on this requirement.

Meanwhile, two transitive rationality indicators $TrRat_G(C)$ and $TrRat_M(C)$ were defined in [64].

$$TrRat_G(C) = \bigvee_{Q\in\mathcal{R}} E(C,G_Q)\wedge Trans(Q);$$

$$TrRat_M(C) = \bigvee_{Q\in\mathcal{R}} E(C,M_Q)\wedge Trans(Q).$$

Main conclusions include ($* = \min$)

(1) $F\alpha_2(C)\wedge F\beta_2(C) = E(C,G_R)\wedge Trans(R) \leq TrRat_G(C)$;
(2) $F\alpha_2(C)\wedge F\beta_2(C) = F\alpha(C)\wedge F\beta_2(C) = F\alpha_2(C)\wedge F\beta(C)$;
(3) $FAA(C) = E(C,G_R)\wedge Trans(R)$.

Recently, two acyclic rationality indicators $AcRat_G(C)$ and $AcRat_M(C)$ are put forward by Georgescu in [36] and the equality $Rat_G(C) = AcRat_G(C)$ is verified.

## 5 Concluding Remarks

This state-of-the-art survey presents a brief introduction to various types of choice functions in fuzzy environment. The involved topics mainly focus on revealed preferences, rationality properties and rationalization. It should be noted that all the results are heavily dependent on the context. In our context, the domain of a crisp choice function is assumed to be $P(X)$. This assumption largely simplifies our discussion. In this case, normality and rationality are equivalent, the revealed preference become reflexive, complete and acyclic. Meanwhile, some notions in the literature become indifferent. For example, rationality and acyclic rationality become identical. $M$-rationality and $G$-rationality defined by Suzumura [10] become the same notion etc. In addition, in the discussion of Banerjee fuzzy choice functions and Georgescu fuzzy choice functions, different frameworks are adopted in considering the current research state. For the former, the framework is a De Morgan triple and it is a residuated lattice $([0,1], \vee, \wedge, *, \rightarrow, 0, 1)$ for the latter. Hence, different fuzzy logic connectives may be employed in formulating rationalities although their intuitive interpretations are the same. This difference of framework may make the difference of the research results extremely subtle although some conclusions appear in almost the same or similar form. A Georgescu fuzzy choice function is the most general choice function since both the domain and range of the choice function are fuzzy. However, this generality often makes the investigation difficult. As a consequence, it is observed that many results related to Georgescu fuzzy choice functions are obtained with the condition $* = \min$. In this case, as we know, the corresponding negation should be the intuitionistic negation [29] with the definition $\neg a = 1$ if $a = 0$ and $\neg a = 0$ elsewhere, which makes many rationality properties associated with the negation become special.

**Acknowledgements.** One of the authors, Wang Xuzhu, wish to express his sincere gratitude to Prof. Kerre for the patient and professional guidance during Wang's Ph. D. work and for the successful cooperation after Wang left Gent University from which Wang benefited greatly.

## References

1. Arrow, K.J.: Rational choice functions and orderings. Economica 26, 121–127 (1959)
2. Chernoff, H.: Rational selection of decision functions. Econometrica 22, 423–443 (1954)
3. Richter, M.K.: Revealed preference theory. Econometrica 34, 635–645 (1966)
4. Sen, A.K.: Choice functions and revealed preference. Review of Economic Studies 38, 307–317 (1971)
5. Sen, A.K.: Collective Choice and Social walfare. Holden-Day, San Francisco (1970)
6. Sen, A.K.: Social choice theory: a re-examination, vol. 45, pp. 53–89 (1977)

7. Sen, A.K.: Quasi-transitivity, rational choice and collective decisions. Review of Economic Studies 386, 381–393 (1969)
8. Samuelson, P.A.: A note on the pure theory of consumers' behaviour. Economica, New Series 5, 353–354 (1938)
9. Suzumura, K.: Remarks on the theory of collective choice. Economica 43, 381–390 (1976)
10. Suzumura, K.: Rational choice and revealed preference. Review of Economic Studies 46, 149–158 (1976)
11. Suzumura, K.: Houthakker's axiom in the theory of rational choice. Jourmal of Economic Theory 14, 284–290 (1977)
12. Houthakker, H.S.: Revealed preference and utility functions. Economic 17, 159–174 (1950)
13. Nehring, K.: Choice and revealed preference without binariness. Social Choice and Welfare 14, 403–425 (1997)
14. Aizerman, M.A., Malishevski, A.V.: General theory of best variants choice: some aspects. IEEE Transactions on Automatic Control 26, 1030–1040 (1981)
15. Herzberger, H.G.: Ordinal preference and rational choice. Econometrica 41, 187–237 (1973)
16. Plott, C.R.: Path independence, rationality and social choice. Econometrica 41, 1075–1091 (1973)
17. Bandyopadhyay, T.: Revealed preference and the axiomatic foundations of intransitive indefference: the case of asymmetric subrelations. Journal of Mathematical Psychology 34, 419–434 (1990)
18. Bandyopadhyay, T., Sengupta, K.: Revealed preference axioms for rational choice. Economic Journal 101, 202–213 (1991)
19. Fishburn, P.C.: Semiorders and choice functions. Econometrica 43, 975–976 (1975)
20. Luce, R.D.: Semiorder and a theory of utility discriminations. Econometrica 24, 178–191 (1956)
21. Jamison, D.T., Lau, L.J.: Semiorders and the theory of choice. Economitrica 41, 901–912 (1973)
22. Matsumoto, Y.: Non-binary social choice: revealed preference interpretations. Economica 52, 185–194 (1985)
23. Montero, F.J., Tejada, J.: A necessary and sufficient condition for the existence of Orlovsky's choice set. Fuzzy Sets and Systems 26, 121–125 (1988)
24. Schwartz, T.: Choice functions, "rationality" conditions, and variations on the weak axioms of revealed preference. Journal of Economic Theory 13, 414–427 (1976)
25. Dutta, B., Panda, S.C., Pattanaik, P.K.: Exact choice and fuzzy preferences. Mathematical Social Sciences 11, 53–68 (1986)
26. Basu, K.: Fuzzy revealed preference theory. J. Economic Theory 32, 212–227 (1984)
27. Switalski, Z.: Choice functions associated with fuzzy preference relations. In: Kacprzyk, J., Roubens, M. (eds.) Nonconventioal Preference Relations in Decision Making. Springer, Berlin (1988)
28. Schweizer, B., Sklar, A.: Probabilistic Metric Spaces. North-Holland, Amsterdam (1983)
29. Fodor, J., Roubens, M.: Fuzzy Preference Modeling and Multicriteria Decision Support. Kluwer Academic Publishers, Dordercht (1994)

30. Fodor, J.: An axiomatic approach to fuzzy preference modeling. Fuzzy Sets and Systems 52, 47–52 (1992)
31. Georgescu, I.: Fuzzy Choice Functions. Springer, Berlin (2007)
32. Georgescu, I.: Rational and congruous fuzzy consumers. In: Proceeding of the International Conference on Fuzzy Information Processing, Theory and Applicatinos, pp. 133–137 (2003)
33. Georgescu, I.: Consistency conditions in fuzzy consumers theory. Fundamenta Informaticae 61, 1–23 (2004)
34. Georgescu, I.: On the axioms of revealed preference in fuzzy consumer theory. Journal of Systems Sciences and Systems Engineering 13, 279–296 (2004)
35. Georgescu, I.: Revealed preference, congruence and rationality: a fuzzy approach. Fundamenta Informaticae 63, 1–22 (2004)
36. Georgescu, I.: Acyclic rationality indicators of fuzzy choice functions. Fuzzy Sets and Systems 160, 2673–2685 (2009)
37. Orlovsky, S.A.: Decision-making with a fuzzy preference relation. Fuzzy Sets and Systems 1, 155–167 (1978)
38. Banerjee, A.: Rational choice under fuzzy preference: The Orlovsky choice function. Fuzzy Sets and Systems 53, 295–299 (1993)
39. Bouyssou, D.: Acyclic fuzzy preference and The Orlovsky choice function: A note. Fuzzy Sets And Systems 89, 107–111 (1997)
40. Bouyssou, D.: A note on the sum of difference choice function for fuzzy preference relations. Fuzzy Sets And Systems 47, 197–202 (1992)
41. Sengupta, K.: Fuzzy preference and orlovsky choice procedure. Fuzzy Sets And Systems 93, 231–234 (1998)
42. Barrett, C.R., Pattanaik, P.K., Salles, M.: On choosing rationally when preferences are fuzzy. Fuzzy Sets and Systems 34, 197–212 (1990)
43. Roubens, M.: Some properties of choice functions based on valued binary relation. European Journal of Operation Research 40, 309–321 (1989)
44. Banerjee, A.: Fuzzy choice functions, revealed preference and rationality. Fuzzy Sets and Systems 70, 31–43 (1995)
45. Wang, X.: A note on congruence conditions of fuzzy choice functions. Fuzzy Sets and Systems 145, 355–358 (2004)
46. Wang, X., Ruan, D., Kerre, E.: Mathematics of Fuzziness – Basic Issues. Springer, Heidelberg (2009)
47. Klement, E.P., Mesiar, R., Pap, E.: Triangular Norms, Trends in Logic. Kluwer Academic Publishers, Dordrecht (2000)
48. Guo, Y., Zheng, G., Wang, X.: The consistency and rationality of fuzzy choice function. Fuzzy Systems and Mathematics 20, 53–56 (2006) (in chinese)
49. Zhang, H., Wang, X., Ma, X.: The characterization of rationality and quasi-transitive rationality of fuzzy choice functions. Fuzzy Systems and Mathematics 20, 68–73 (2006) (in chinese)
50. Zhang, H., Zhang, Q., Wang, X.: The characterization of $W_\varphi$-pseudotransitive rationality and $W_\varphi$-semiorder rationality of fuzzy choice functions. Fuzzy Systems and Mathematics 23, 36–40 (2009) (in chinese)
51. Wu, C., Wang, X., Hao, Y.: An investigation into some revealed preference relations of fuzzy choice functions. Fuzzy Systems and Mathematics 24, 84–91 (2010) (in chinese)
52. Wu, C., Wang, X., Jiang, W.: Some equivalencies of rationality conditions of fuzzy choice functions under a strong De Morgan triple. J. of Taiyuan University of Technology 2, 203–206 (2009) (in chinese)

53. Wu, C., Wang, X., Hao, Y.: A further study on rationality conditions of fuzzy choice functions, submitted to Fuzzy Sets and Systems
54. Wu, C., Wang, X.: Fuzzification of Suzumura's Results on Choice Functions. To appear in the Proceeding of the 7th International Conference on Fuzzy Systems and Knowledge Discovery (FSKD 2010) (2010)
55. Bufardi, A.: On the construction of fuzzy preference structures. Journal of Multi-criteria Decision and Analysis 7, 169–175 (1998)
56. Hao, Y.: The Characterization of $W_\varphi$-transitive rationality and acyclic rationality of fuzzy choice functions. To appear in Fuzzy Information and Engineering
57. Ma, X., Wang, X.: The characterization of two classes of quasi-transitive rationalities of fuzzy choice functions. J. of Taiyuan University of Science and Technology 29, 36–39 (2008)
58. Roubens, M., Vincke, P.: Preference Modeling. Springer, Berlin (1985)
59. Ponsard, C.: An application of fuzzy subsets theory to the analysis of the consumer's spatial preference. Fuzzy Sets and Systems 5, 235–244 (1981)
60. De Baets, B., Van de Walle, B.: Weak and strong fuzzy interval orders. Fuzzy Sets and Systems 79, 213–225 (1996)
61. De Baets, B., Van de Walle, B., Kerre, E.: Fuzzy preference structures without incomparability. Fuzzy Sets and Systems 76, 333–348 (1995)
62. Llamazares, B.: Characterization of fuzzy preference structures through Lukasiewicz triplets. Fuzzy Sets and Systems 136, 217–235 (2003)
63. Kulshreshtha, P., Shekar, B.: Interraltionships among fuzzy preference-based choice functions and significance of rationality conditions: a taxonomic and intuitive perspective. Fuzzy Sets and Systems 109, 429–445 (2000)
64. Wu, X.: A further study on rationality indicators of fuzzy choice functions, MS degree thesis, Taiyuan University of Technology (2010) (in chinese)
65. Ruan, D., Kerre, E.: Fuzzy implication operators and generalized fuzzy method of cases. Fuzzy Sets and Systems 54, 23–37 (1993)
66. Shi, Y., Van Gasse, B., Ruan, D., Kerre, E.: On the first place antitonicity in QL-implications. Fuzzy Sets and Systems 159, 2988–3013 (2008)
67. Hájek, P.: Methematics of Fuzzy Logic. Kluwer Academic Publishers, Dordrecht (1998)

# A Fuzzy Extension to Compact and Accurate Associative Classification

Guoqing Chen, Yixi Xiong, and Qiang Wei

**Abstract.** Classification has been one of the focal points in data mining research and applications. With an effective approach to building compact and accurate associative classification (namely GARC – Gain-based Association Rule Classification (Chen, Liu, Yu, Wei, & Zhang, 2006)) in forms of association rules, this chapter explores a way of fuzzy extension to GARC in dealing with the problem caused by crisp partitions for continuous attribute domains in data. Concretely, the sharp boundaries of the partitioned intervals are smoothened using fuzzy sets (or often conveniently labeled in linguistic terms) so as to reflect a variety of fuzziness on the domains (parameterized in $f_2$), giving rise to a fuzzy associative classifier (i.e., $GARC_{f2}$). Furthermore, due to the fuzziness involved, the notions of information gain, rule redundancy and conflicts are extended, aimed at providing the desirable features of GARC in the fuzzy extension context for accuracy and compactness. Moreover, data experiments on benchmarking datasets as well as a real-world application illustrate the effectiveness of the proposed fuzzy associative classifier.

**Keywords:** Associative Classification, Information Gain, Fuzzy Partitions, Credit Rating.

In recent decades, there have been numerous research efforts as well as a wide range of applications in both classification and association rule mining (Fayyad & Uthurusamy, 1994; Han, Kamber, & Pei, 2005; Weiss & Kulikowski, 1991). In the context of massive data, association rule mining is to discover associative patterns between data items in a manner they occur together frequently, while classification is to build a classifier usually based on training/testing datasets, which enables to

Guoqing Chen · Yixi Xiong · Qiang Wei
School of Economics and Management, Tsinghua University
Beijing 100084 China

C. Cornelis et al. (Eds.): 35 Years of Fuzzy Set Theory, STUDFUZZ 261, pp. 171–193.
springerlink.com 

classify the newly coming data to a specific group (class). Though one may find a number of existing approaches that could build classifiers effectively, such as Decision Tree, Bayesian networks, Support Vectors Machine, Neural Network, etc. (Berzal, Cubero, Sanchez, & Serrano, 2004; Friedman, Geiger, & Goldszmidt, 1997; Quinlan, 1993; Shafer, Agrawal, & Mehta, 1996; Sharpe & Glover, 1999; Shie & Chen, 2008), associative classification, which results from a combined perspective of classification and association rule mining, has attracted an increasing amount of attention of academia and practitioners due to its accuracy and understandability (Ali, Manganaris, & Srikant, 1997; Berzal et al., 2004; Chen, Liu, Yu, Wei, & Zhang, 2006; Chen, Zhang, & Yu, 2006; Dong, Zhang, Wong, & Li, 1999; Li, Han, & Pei, 2001; Liu, Hsu, & Ma, 1998; Liu, Ma, & Wong, 2000; Wang, Zhou, & He, 2000; Yin & Han, 2003; Zhang, Chen, & Wei, 2009).

Concretely, an association rule is a link between data items in forms of $X \Rightarrow Y$, where $X$ and $Y$ are itemsets (i.e., set of items/attributes) with $X \cap Y = \varnothing$, expressing that the event for $X$ is associated with the event for $Y$. For example, a rule like "*Milk⇒Bread*" means that the customers who buy milk may also tend to buy bread at the same time. There are two basic measures to evaluate a rule, namely, Degree of Support (Dsupp) and Degree of Confidence (Dconf). Dsupp($X \Rightarrow Y$) is the percentage of transactions containing both $X$ and $Y$ with respect to the whole transaction set (dataset), while Dconf($X \Rightarrow Y$) is the percentage of transactions containing $Y$ among those transactions (i.e., tuple/record) containing $X$. That is, for itemsets $X$ and $Y$, Dsupp($X \Rightarrow Y$) = $\|X \cup Y\| / |D|$, and Dconf($X \Rightarrow Y$) = $\|X \cup Y\| / \|X\|$, where $|D|$ is the number of transactions in dataset D, and $\|X\|$ and $\|X \cup Y\|$ are the numbers of transactions that contain, respectively, $X$ and $X \cup Y$. Given two thresholds for degrees of support and confidence (i.e., minimal support $\alpha$ and minimal confidence $\beta$, with $\alpha$, $\beta \in [0,1]$), an itemset $X$ is called frequent, if Dsupp($X$) = $\|X\| / |D| \geq \alpha$. Furthermore, for any rule $X \Rightarrow Y$, it is called a valid association rule if Dsupp($X \Rightarrow Y$) $\geq \alpha$ and Dconf($X \Rightarrow Y$) $\geq \beta$. To find such valid rules, many approaches have been proposed (Agrawal, Imielinski, & Swami, 1993; Agrawal & Srikant, 1994; Chen, Wei, Liu, & Wets, 2002; Mannila, Toivonen, & Verkamo, 1994; Srikant & Agrawal, 1996; Srikant, Vu & Agrawal, 1997), among which the Apriori method is usually deemed as a classical algorithm (Agrawal & Srikant, 1994).

In the spirit of association rules, associative classification can be viewed as a special case of association rule mining, resulting in a classifier composed of a number of class association rules (CARs) such as CBA (Liu et al., 1998; Liu et al., 2000). A CAR is of the form $X \Rightarrow C_i$, where $X$ is a set of items and $C_i$ is a predetermined class label. Let $t$ be a transaction in dataset D, $t$ is called to satisfy $X \Rightarrow C_i$ if $X \subseteq t$ and $C_i$ is the predetermined class for $t$. Associative classification usually searches for all the CARs that satisfy given minimum support and minimum confidence thresholds. Notably, associative classification has been considered effective in many cases (Berzal et al., 2004; Chen et al., 2006; Chen et al., 2006; Dong et al., 1999; Li et al., 2001; Liu et al., 1998; Liu et al., 2000; Wang et al, 2000; Yin & Han, 2003; Zhang et al., 2009).

While classifiers in forms of association rules are usually appealing for use and explanation by human users, one may however suffer from the fact that the number of discovered CARs is huge, thus hard to manage in understanding the rules as well as in resolving possible rule redundancy and conflicts. Therefore, the construction of an effective classifier to cope with the size of rules and redundant/conflict rules is desirable. In this regard, an approach, namely GARC (i.e., Gain-based Association Rule Classification) (Chen et al., 2006), has been introduced that possesses satisfied accuracy for classification as well as compactness in size of the resultant classifier.

In GARC, only those "informative" rules are generated. The characteristic of this approach is threefold (Chen et al., 2006): First, GARC employs information gain measure to select the best-split attribute for 1-itemsets (i.e., the itemsets each containing a single item), which is to be included in the generation of candidate k-itemsets (i.e., the itemsets each containing k items, $k \geq 2$). Second, both *frequent itemsets* and *excluded itemsets* (i.e., infrequent itemsets) are used in the generation of candidate itemsets. Third, GARC defines rule redundancy and rule conflicts as follows: in a set of classification rules $\varphi$, if rule $X \Rightarrow C_i$ exists, then (1) any rule of the form $Z \Rightarrow C_i$ ($Z \supset X$) is called *redundant*, that is, $Z \Rightarrow C_i$ would not be regarded semantically necessary from the perspective of classification; and (2) for $i \neq j$, any rule of the form $Z \Rightarrow C_j$ ($Z \supseteq X$) is called *conflicting* if $Z \Rightarrow C_j$ does not precedes $X \Rightarrow C_i$ (i.e., (i) Dconf($X \Rightarrow C_i$) > Dconf($Z \Rightarrow C_j$), or (ii) Dconf($X \Rightarrow C_i$) = Dconf($Z \Rightarrow C_j$) while Dsupp($X \Rightarrow C_i$) > Dsupp($Z \Rightarrow C_j$)), which may lead to misclassification in the decision making process. A nonempty classification rule set is called a compact set of classification rules if neither rule redundancy nor rule conflicts exists. Furthermore, GARC applies two strategies in avoiding rule conflicts and redundancy: (1) If $\beta > 50\%$, then rules $X \Rightarrow C_i$ and $X \Rightarrow C_j$ will not hold simultaneously. (2) For a classification rule $X \Rightarrow C_i$, if $1 - \text{Dconf}(X \Rightarrow C_i) < \alpha$ or $\text{Dsupp}(X) < 2\alpha$, then any itemset like $Z \Rightarrow C_j$ ($Z \supseteq X$, $i \neq j$) is an excluded itemset. As a result, GARC constructs a more condensed and understandable (in terms of much fewer rules than CBA) classifier while the classification accuracy is satisfactory compared with other major approaches (Chen et al., 2006).

It is worth mentioning that classical association rule mining methods and related associative classification methods were originally designed for binary databases in that the values of attributes (items) are either 1 (true) or 0 (false), reflecting whether or not a corresponding event takes place. In other words, for itemsets $X$ and $Y$, the semantics of rule $X \Rightarrow Y$ could be explained in a way as whether the events for $X$ and $X \cup Y$ occur or not. However, in many real world applications, not only shall the occurrence of events be considered important, but also the quantity involved in the events. That is, we often encounter the cases where the values of attributes (such as Age, Income, Price, etc.) are continuous in the domains. These cases are usually dealt with in light of quantitative association rules (QAR) (such as "*Milk*[6,12] ⇒*Bread*[2,6]", meaning that the customers who buy 6 to 12 bottles of milk may also tend to buy 2 to 6 pieces of bread at the same time) (Aumann & Lindell, 1999; Dougherty, Kohavi, & Sahami, 1995; Srikant & Agrawal, 1995). In doing so, typical approaches use sharp partitioning for the respective domains to generate several intervals, each then treated as a new attribute. Then, the original dataset with continuous attribute values could be transformed into a new dataset with binary attribute values, on which the classical Apriori method can be used in finding such QARs. Another example of the sharp partitioning is on an *Age* domain (0,100) to form such new attributes as *Age*(0, 30], *Age*(30, 60], *Age*(60, 100).

Notably, though sharp partitioning is considered effective, it may suffer from the "boundary problem", leading to misclassification (Bosc & Pivert, 2001; Chen, Wei, & Kerre, 2006; Chen, Yan, & Kerre, 2004; Chen, Yan, & Wei, 2009). That is to say, the intervals with sharp boundaries may result in over-sensitive grouping of attribute values near the boundaries, which then affect the classification outcomes. For instance, a transaction $t$ with age of 31 will have the values 0, 1, 0 for attributes *Age*(0, 30], *Age*(30, 60], *Age*(60, 100), respectively, meaning that the event relating to *Age*(30,60] is relevant to transaction $t$. On the other hand, age of 30 will be grouped to another attribute *Age*(0,30), though semantically age of 30 or 31 may not look like so different. In other words, the intervals resulting from sharp partitioning may amply (or over-emphasize) the difference between the values at boundaries.

In this regard, fuzzy logic (Zadeh, 1965) is considered useful to smooth the differences with gradual grades for the values around boundaries. This will extend the belongings of these values to attributes from {0,1} to [0,1]. An approach that is considered effective is to define fuzzy sets upon the sharply partitioned intervals (e.g., the crisp intervals used in classical classification methods such as CBA and GARC). Whereas there are different ways to define the fuzzy sets (e.g., parameterized by $f_1$ as the degree of fuzziness in (Xiong, 2010)), this chapter presents a fuzzy extension to GARC in that fuzzy sets are defined on the boundaries of the crisp intervals in consideration of a boundary value possessing the greatest degree of fuzziness. With the degree of fuzziness as parameterized by $f_2$, the fuzzy extension is then referred to as $\text{GARC}_{f2}$.

The chapter is organized as follows. Section 1 introduces some preliminary notions on fuzzy sets. Section 2 elaborates on the steps of $\text{GARC}_{f2}$ in constructing classifiers including the respective algorithm. Section 3 shows the experimental results and analysis on benchmarking datasets. Finally, section 4 illustrates the advantages of $\text{GARC}_{f2}$ through a real-world application.

## 1 Preliminary Notions

Fuzzy sets theory provides a means to deal with imprecision and uncertainty (Zadeh, 1965). Let $U$ be a domain, a fuzzy set A (sometimes labeled in a linguistic term) on $U$ is characterized by a membership function $\mu_A(u)$: $U \rightarrow [0,1]$, where, for any $u$ in $U$, $\mu_A(u)$ is the grade of membership of $u$ for A, representing the degree of $u$ belonging to A. For example, *Young-Age*, *Middle-Age*, *Old-Age* could be defined on Domain(*Age*) with the corresponding membership functions being trapezoid as shown in Figure 1.

Furthermore, fuzzy partitioning (Bezdek, 1981; Dunn, 1974) is used for transforming quantitive datasets to fuzzy datasets. Generally, suppose there are $n$ transactions $\{t_1, t_2, \ldots, t_m\}$ in an original dataset $D$ with schema $\mathbf{R}(\mathbf{A})$, where $\mathbf{A}$ is a set of $m$ attributes, i.e., $\mathbf{A} = \{A_1, A_2, \ldots, A_m\}$, each $A_k$ ($1 \leq k \leq m$) can be associated

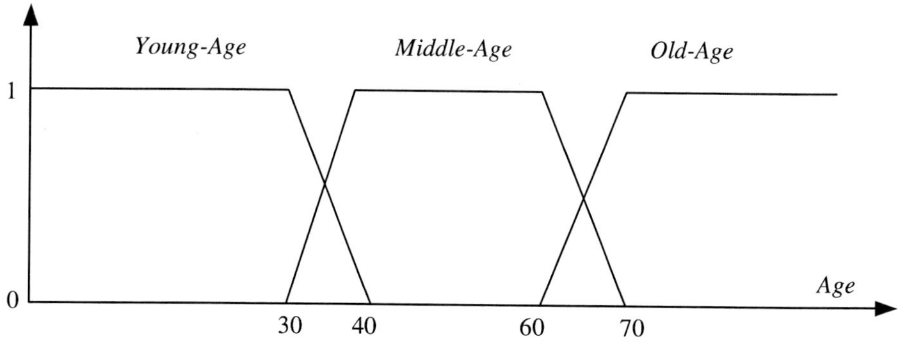


**Fig. 1** Membership functions for *Young-Age*, *Middle-Age* and *Old-Age*

with $q_k$ sets (i.e., intervals) defined on domain($A_k$), denoted as $\{A_k^1, \ldots, A_k^{q_k}\}$. Traditionally, for each value $t_i(A_k)$ $(1 \le i \le n)$, i.e., value of $t_i$ for attribute $A_k$, it will totally belong to one of the sets $\{A_k^1, \ldots, A_k^{q_k}\}$. The result can be described by an $n \times q_k$ matrix $X_k$, where each element $x_{ij}$ values 1 if $x_{ij} \in A_k^j$ $(1 \le j \le q_k)$ or 0 otherwise. In other words, the matrix $X_k$ has the following properties: (1) $x_{ij} \in \{0,1\}$, (2) $\sum_{j=1}^{q_k} x_{ij} = 1, \forall i$, (3) $0 < \sum_{i=1}^{n} x_{ij} < n, \forall j$, then $X_k$ is called to be a *sharp partition* of attribute $A_k$. If the value of $x_{ij}$ is not constrained in Boolean value $\{0,1\}$ but in $[0,1]$, which means that $t_i(A_k)$ can partially belong to one of the sets $\{A_k^1, \ldots, A_k^{q_k}\}$, then $X_k$ is called to be a *fuzzy partition* of attribute $A_k$ if (1) $x_{ij} \in [0,1]$, (2) $\sum_{j=1}^{q_k} x_{ij} = 1, \forall i$, (3) $0 < \sum_{i=1}^{n} x_{ij} < n, \forall j$.

For instance, given a dataset D with continuous attributes as shown in Table 1, the resultant datasets based on sharp partitioning and fuzzy partitioning are shown as Table 2 and Table 3.

**Table 1** Original Dataset D

| *D* | *Age* | *Income* | *Class* |
|---|---|---|---|
| #1 | 33 | 64000 | $C_1$ |
| #2 | 25 | 26000 | $C_2$ |
| #3 | 46 | 39000 | $C_3$ |
| #4 | 62 | 22000 | $C_2$ |
| #5 | 36 | 52000 | $C_1$ |

**Table 2** Sharp-Partitioned Dataset $D_s$

| $D_s$ | *Age* (0,35] | *Age* (35,65] | *Age* (65,100] | *Income* [0,25000] | *Income* (25000,60000] | *Income* (60000,∝) | *Class* |
|---|---|---|---|---|---|---|---|
| #1 | 1 | 0 | 0 | 0 | 0 | 1 | $C_1$ |
| #2 | 1 | 0 | 0 | 0 | 1 | 0 | $C_2$ |
| #3 | 0 | 1 | 0 | 0 | 1 | 0 | $C_3$ |
| #4 | 0 | 0 | 1 | 1 | 0 | 0 | $C_2$ |
| #5 | 0 | 1 | 0 | 0 | 1 | 0 | $C_1$ |

**Table 3** Fuzzy-Partitioned Dataset $D_f$

| $D_f$ | *Young-Age* | *Middle-Age* | *Old-Age* | *Low-Income* | *Middle-Income* | *High-Income* | *Class* |
|---|---|---|---|---|---|---|---|
| #1 | 0.7 | 0.3 | 0 | 0 | 0.3 | 0.7 | $C_1$ |
| #2 | 1 | 0 | 0 | 0.4 | 0.6 | 0 | $C_2$ |
| #3 | 0 | 1 | 0 | 0 | 1 | 0 | $C_3$ |
| #4 | 0 | 0.2 | 0.8 | 0.8 | 0.2 | 0 | $C_2$ |
| #5 | 0.4 | 0.6 | 0 | 0 | 0.9 | 0.1 | $C_1$ |

Note that the fuzzy sets on the domain of *Income* for Table 3 are shown as defined in Figure 2.

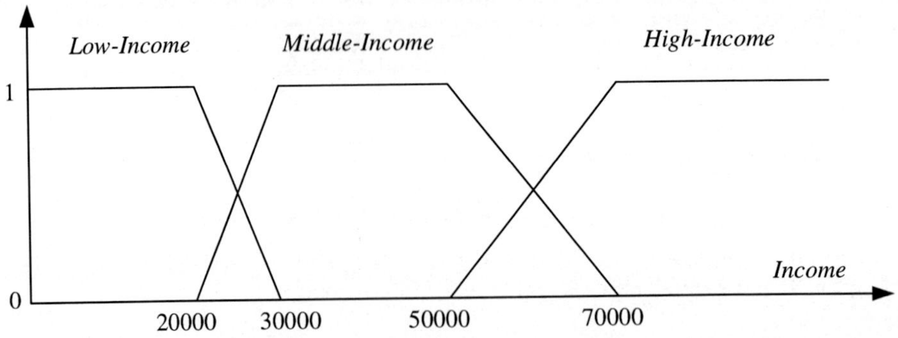


**Fig. 2** Membership functions for *Low-Income*, *Middle-Income* and *High-Income*

## 2 Gain Based Associative Classification with Fuzzy Partitioning

This section will introduce GARC$_{f2}$ (Gain based Associative Classification with Fuzzy Partitioning in $f_2$) in three steps: (1) transforming the traditional datasets with continuous attributes to fuzzy-partitioned datasets, (2) mining a compact set of fuzzy class association rules, and (3) building a classifier based on the discovered rules.

## 3 Fuzzy Partitioning with Trapezoidal Membership Functions

As mentioned in Section 2, fuzzy partitioning is to define a number of fuzzy sets, usually labeled by linguistic terms, upon each attribute domain. In this chapter, trapezoidal membership functions will be adopted, since they are not only easy to construct, but also convenient to illustrate the values close to the boundaries (Chen, 1998; Klir & Yuan, 1995). In addition, though represented in linear segments, they are sufficient for use in reflecting a wide variety of real-world semantics.

Concretely, a membership function concerned is of the form: trapezia($a$, $b$, $c$, $d$) (denoted as $\pi$ in brief), where $a$, $b$, $c$ and $d$ are four parameters. Furthermore, the membership functions are assumed to be normalized, i.e., for any $\mu: \pi(a, b, c, d)$, $\mu(b) = \mu(c) = 1$. Moreover, given an attribute A in D, suppose Domain(A) is fuzzy-partitioned into a series of $v$ attributes $A_1, A_2, \ldots, A_v$ in $D_f$, where $A_q$ ($q = 1, 2, \ldots, v$) is a fuzzy set with a trapezoidal membership function $\mu_{Aq}: \pi(a_q, b_q, c_q, d_q)$, then $A_1, A_2, \ldots, A_v$ could be semantically viewed as in an increasing order across the value range in Domain(A), e.g., denoted as $A_q \prec A_{q+1}$ (where $c_q \le b_{q+1}$ and $\mu_{Aq}(c_q) = \mu_{Aq}(b_{q+1}) = 1$). For instance, *Low-Income* $\prec$ *High-Income*, and *Young-Age* $\prec$ *Middle-Age*. This represents a gradual smoothing around the juncture of two adjacent fuzzy sets (attributes/labels). To make $\{A_1, A_2, \ldots, A_v\}$ a fuzzy partition on Domain(A), we shall also have $a_q = c_{q-1}$, $b_q = d_{q-1}$, $(q = 2, 3, \ldots, v)$. Thus, importantly, for any $u$ in Domain(A), there exist two membership functions $\mu_{Aq}$ and $\mu_{Aq+1}$ defined on Domain(A) such that $\mu_{Aq} \prec \mu_{Aq+1}$ and $\mu_{Aq}(u) + \mu_{Aq+1}(u) = 1$.

In particular, given a domain $[p_0, p_v]$ with a crisp partitioning of $v$ intervals $[p_0, p_1)$, $[p_1, p_2)$, …, $[p_{v-1}, p_v]$, if a fuzzy partitioning with $v$ attributes is still desired, then a trapezoidal membership function $\pi(a_q, b_q, c_q, d_q)$ could be defined for each attribute $A_q$ ($q = 1, 2, \ldots, v$) in $D_f$ with respect to interval $[p_{q-1}, p_q]$. Moreover, a single parameter $f_2$ could be defined, as follows, to represent the degree of smoothing around the boundary of each crisp interval. In other words, $f_2$ could be viewed as the degree of fuzziness that $A_q$ reflects for $[p_{q-1}, p_q]$ (See Figure 3).

$$f_2 = \frac{p_{q-1} - a_q}{\min\left(p_{q-1} - p_{q-2}, p_q - p_{q-1}\right)} = \frac{b_q - p_{q-1}}{\min\left(p_{q-1} - p_{q-2}, p_q - p_{q-1}\right)}$$

$$= \frac{p_q - c_q}{\min\left(p_q - p_{q-1}, p_{q+1} - p_q\right)} = \frac{d_q - p_q}{\min\left(p_q - p_{q-1}, p_{q+1} - p_q\right)}$$

where $q = 2, 3, \ldots, v-1$. It is easy to see that $(a_q+b_q)/2=p_{q-1}$, $(c_q+d_q)/2=p_q$.

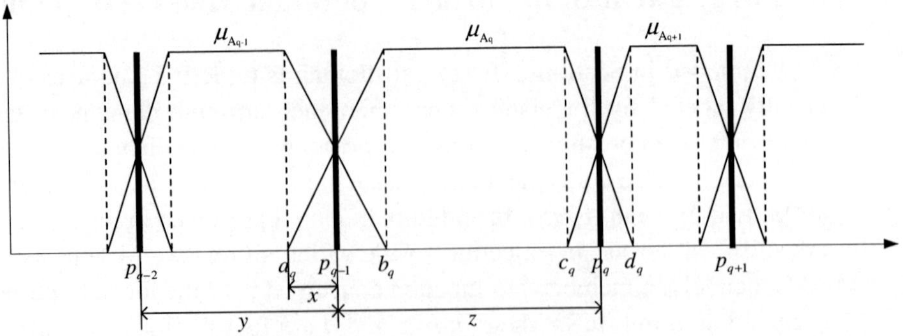


**Fig. 3** Sharp and Fuzzy Partitions with $f_2$ ($f_2 = 0.25$)

For example, for the fuzzy set $A_q$ defined on the domain of A, of which trapezoidal membership function is showed as $\mu_{Aq}$ : $\pi(a_q, b_q, c_q, d_q)$, $x$ is the distance between $a_q$ and $p_{q-1}$, while $y$ and $z$ are the length of intervals $[p_{q-2}, p_{q-1}]$ and $[p_{q-1}, p_q]$, respectively (See Figure 3). Given parameter $f_2 = 0.25$, then we have $x$ / min($y$, $z$) = $x$ / $y$ = 0.25.

It is worth mentioning that crisp intervals is a special case of trapezoidal membership functions, i.e., $\pi(b, b, c, c)$ with $b \neq c$. Apparently, the higher $f_2$ is, the fuzzier $A_q$ is. The range of $f_2$ is from 0 to 0.5 expressing an increasing degree of fuzziness, where $f_2 = 0$ represents a crisp interval (i.e., $a = b$ and $c = d$). An advantage of using $f_2$ is to only set a single parameter instead of setting many ones (e.g., 4×$v$)), which is practically appealing.

# 4 Generating a Fuzzy Associative Classifier (GARC$_{f2}$)

## *4.1 Fuzzy Class Association Rules (Fuzzy CARs)*

The fuzzy partitioned dataset gives rise to fuzzy association rules. In classification, it results in fuzzy class association rules. For example, compared with a classical CAR: (*Age*(0,35], *Income*(25000, 60000]) ⇒ *Class* 1 (meaning that the customers at ages of (0, 35] and with income of (25000, 60000] will be classified into Class 1), a fuzzy CAR may look like (*Young-Age*, *Middle-Income*) ⇒ *Class* 1 (meaning that the customers who are *young* and have *middle income* will be classified into Class 1), which is more general and is of natural-language nature.

Generally, for the original set of items/attributes $\boldsymbol{I} = \{I_1, I_2, \ldots, I_m\}$ and database $D$ with schema $\boldsymbol{R}(\boldsymbol{I})$, each $I_k$ ($1 \leq k \leq m$) can be associated with $q_k$ fuzzy sets defined on domain($I_k$), and usually labeled as $q_k$ new attributes in $D_f$. That is, the new database $D_f$ is with respect to schema $\boldsymbol{R}(\boldsymbol{I}_f)$ where $\boldsymbol{I}_f = \{ I_1^1, \ldots, I_1^{q_1}, \ldots, I_k^1, \ldots, I_k^{q_k}, \ldots, I_m^1, \ldots, I_m^{q_m} \}$. Let G={$C_1, C_2, \ldots, C_g$} be a set of class labels, then a *fuzzy rule item* is of the form $X \Rightarrow C$, where $X \subseteq \boldsymbol{I}_f$, and $C \in$ G.

In fuzzy context, the notions of degrees of support and confidence can be extended as follows (Chen &Wei, 2002; Chen et al., 2006). For any $X=\{x_1, x_2, \ldots, x_p\} \subseteq \boldsymbol{I}_f \cup \{C_j\}$ (j = 1,2,…,g) and transaction $t$ in $D_f$,

$$\mathrm{Dsupp}_t(X) = \mu_X(t) = \min(t(x_1), t(x_2), \ldots, t(x_p)),$$

$$\mathrm{Dsupp}(X) = \frac{\| X \|}{|D_f|} = \frac{\sum_t \mathrm{count}(\mu_X(t))}{|D_f|},$$

$$\mathrm{Dsupp}(X \Rightarrow C_i) = \mathrm{Dsupp}(XC_i) = \frac{\| XC_i \|}{|D_f|} = \frac{\sum_{t|t.class=C_i} \mathrm{count}(\mu_X(t))}{|D_f|},$$

$$\mathrm{Dconf}(X \Rightarrow C_i) = \frac{\| XC_i \|}{\| X \|} = \frac{\sum_{t|t.class=C_i} \mathrm{count}(\mu_X(t))}{\sum_t \mathrm{count}(\mu_X(t))},$$

where $t(x_i)$ ($i = 1, 2, \ldots, p$) is the corresponding membership degree of $t$ for attribute $x_i$, $|D_f|$ is the cardinality of $D_f$ (i.e., the number of transactions in $D_f$), and $\|X\|$ is the fuzzy cardinality of $X$ calculated using $\Sigma$count operator. Note that t-norm minimum is used in the definition of $\mathrm{Dsupp}_t(X)$ due to its semantics. Other t-norms may also be considered under specific application contexts.

Take Table 3 for example, $\mathrm{Dsupp}_{\#1}$(*Young-Age*, *Middle-Income*) = min(0.7, 0.3) = 0.3. Moreover, Dsupp(*Young-Age*, *Middle-Income*) = (min(0.7, 0.3) + min(1, 0.6) + min(0, 1) + min(0, 0.2) + min(0.4, 0.9))/5 = 1.3/5 = 0.26, Dsupp((*Young-Age*, *Middle-Income*)$\Rightarrow C_1$) = (min(0.7, 0.3) + min(0.4, 0.9))/5 = 0.7/5 = 0.14, then Dconf((*Young-Age*, *Middle-Income*)$\Rightarrow C_1$) = 0.14/0.26 = 0.54.

Then, given minimal support $\alpha$ and minimal confidence $\beta$, with $\alpha, \beta \in [0, 1]$, a fuzzy rule item $X \Rightarrow C_i$ is called *frequent* if $\mathrm{Dsupp}(X \Rightarrow C_i) \geq \alpha$. Importantly, if fuzzy rule item $X \Rightarrow C_i$ is not frequent, then any fuzzy rule item of the form $Z \Rightarrow C_i$ ($Z \supset X$) is also infrequent, this is because $\mu_Z(t) \leq \mu_X(t)$ for each $t$, and consequently we have:

$$\mathrm{Dsupp}(Z \Rightarrow C_i) = \frac{\sum_{t|t.class=C_i} \mathrm{count}(\mu_Z(t))}{|D_f|} \leq \frac{\sum_{t|t.class=C_i} \mathrm{count}(\mu_X(t))}{|D_f|} = \mathrm{Dsupp}(X \Rightarrow C_i)$$

A fuzzy rule item $X \Rightarrow C_i$ is called a valid *fuzzy class association rule* (fuzzy CAR) if $\mathrm{Dsupp}(X \Rightarrow C_i) \geq \alpha$ and $\mathrm{Dconf}(X \Rightarrow C_i) \geq \beta$. It is worthwhile to indicate that, since crisp intervals are special cases of fuzzy sets (e.g., an interval $[a, b]$ can be represented by a fuzzy set $\int_{a \leq u \leq b} 1.0/u$ ), quantitative association rules are special cases of fuzzy association rules (with fuzzy partitioning), and therefore, CARs are special cases of fuzzy CARs.

## 5 Fuzzy Information Gain

Information gain is one of the measures used in decision tree classification methods for selecting best split attributes (Breiman, 1984; Weiss & Kulikowski, 1991). It is also used as a measure to reduce the number of itemsets in some associative classification (e.g. GARC).

In GARC, information gain measure is used to reduce the search space in generating candidate itemsets as well as rules. That is to say, only those candidate itemsets containing the so-called best split attribute value will be generated. This will significantly help improve the rule set without loss of accuracy (Chen et al., 2006). In the fuzzy context with $f_2$, we extend the fuzzy information entropy of a fuzzy sub-dataset S in $D_f$ with class labels as follows (De Luca & Termini, 1972):

$$\text{info}(S) = \sum_{p=1}^{g} \frac{\sum\text{count}(C_p, S)}{|S|} \times \log\left(\frac{\sum\text{count}(C_p, S)}{|S|}\right)$$

where $\Sigma\text{count}(C_p, S)$ represents the fuzzy cardinality of transactions in S with class label $C_p$. In addition, given an attribute A in D, without loss of generality, suppose Domain(A) is partitioned into v attributes each being a fuzzy set on Domain(A), e.g., $A_1, A_2, \ldots, A_v$, and $D_f^{A_q}$ represents the fuzzy sub-dataset in $D_f$ in which each transaction t has $\mu_{A_q}(t) > 0$, for $q = 1, 2, .., v$, then the fuzzy information entropy of A could be calculated below:

$$\begin{aligned}\text{info}_A(D_f) &= \sum_{q=1}^{v} \frac{|D_f^{A_q}|}{|D_f|} \times \text{info}\left(D_f^{A_q}\right)\\ &= -\sum_{q=1}^{v} \frac{\|A_q\|}{|D_f|} \times \left(\sum_{p=1}^{g} \frac{\sum\text{count}(C_p, D_f^{A_q})}{\|D_f^{A_q}\|} \times \log\left(\frac{\sum\text{count}(C_p, D_f^{A_q})}{\|D_f^{A_q}\|}\right)\right)\\ &= -\sum_{q=1}^{v} \text{Dsupp}(A_q) \times \left(\sum_{k=1}^{g} \text{Dconf}(A_q \Rightarrow C_p) \times \log\left(\text{Dconf}(A_q \Rightarrow C_p)\right)\right)\end{aligned}$$

Hence, the fuzzy information gain for attribute A is obtained as follows:

$$\text{gain}(A) = \text{info}(D_f) - \text{info}_A(D_f)$$

## 6 Pruning Strategies

As indicated previously, building an associative classifier needs to deal with the situation where there exist redundant and conflicting rules. This is also the case in generating fuzzy CARs. For example, suppose we already have *High_Income* $\Rightarrow$ $C_2$, then (*Young_Age*, *High_Income*) $\Rightarrow$ $C_2$ might be redundant; and *High_Income* $\Rightarrow$ $C_3$ might be conflicting. Rather than filtering these redundant and conflicting rules after generating all fuzzy CARs, a more effective and desirable way is to incorporate some pruning strategies in the process of generating itemsets as well as CARs, resulting in fewer itemsets (therefore rules) to be generated.

First of all, as in GARC, the strategy could be applied that if $X \Rightarrow C_i$ holds, then any candidate itemsets containing $XC_i$ need not to be generated and tested. In this way, the number of candidate itemsets to be generated is reduced. Moreover, the following strategies would be helpful in pruning candidate rule items.

**Strategy 1.** Given a rule item $X \Rightarrow C_i$ in $D_f$, if the condition

$$\text{Dsupp}(X) \times \left(1 - \text{Dconf}\left(X \Rightarrow C_i\right)\right) < \alpha$$

is satisfied, then rule items of the form $Z \Rightarrow C_j$ ($Z \supseteq X$, $j \neq i$) will not hold (i.e., not valid) in $D_f$.

**Proof.** Since $\text{Dsupp}(Z \Rightarrow C_j) \leq \text{Dsupp}(X \Rightarrow C_j)$, we only need to prove that $X \Rightarrow C_j$ ($j \neq i$) is infrequent.

As each $t$ in $D_f$ will be assigned with a unique class label $C_i$, $i = 1,2,\ldots,g$, we have $\| X \| = \sum_t \text{count}(\mu_X(t)) = \sum_{t|t.class=C_i} \text{count}(\mu_X(t)) + \sum_{t|t.class \neq C_i} \text{count}(\mu_X(t)), \forall i$, which means:

$$\| X \| = \sum_{t|t.class=C_i} \text{count}(\mu_X(t)) + \sum_{\substack{j=1 \\ j \neq i}}^{g} \left( \sum_{t|t.class=C_j} \text{count}(\mu_X(t)) \right) = \left\|XC_i\right\| + \sum_{\substack{j=1 \\ j \neq i}}^{g} \left( \left\|XC_j\right\| \right).$$

Further, from $\dfrac{\| X \|}{| D_f |} = \dfrac{\| XC_i \|}{| D_f |} + \sum_{\substack{j=1 \\ i \neq j}}^{g} \dfrac{\| XC_j \|}{| D_f |}$, we have:

$$\sum_{\substack{j=1 \\ j \neq i}}^{g} \text{Dconf}(X \Rightarrow C_j) = 1 - \text{Dconf}(X \Rightarrow C_i).$$

Thus, $\text{Dconf}(X \Rightarrow C_j) \leq 1 - \text{Dconf}(X \Rightarrow C_i), \forall j \neq i$.

Since $\text{Dsupp}\left(X \Rightarrow C_j\right) = \dfrac{\left\|XC_j\right\|}{\left|D_f\right|} = \dfrac{\left\|X\right\|}{\left|D_f\right|} \times \dfrac{\left\|XC_j\right\|}{\left\|X\right\|} = \text{Dsupp}\left(X\right) \times \text{Dconf}\left(X \Rightarrow C_j\right)$,

we have $\text{Dsupp}\left(X \Rightarrow C_j\right) \leq \text{Dsupp}\left(X\right) \times \left(1 - \text{Dconf}\left(X \Rightarrow C_i\right)\right) < \alpha$, which means $X \Rightarrow C_j$ is infrequent, and therefore will not hold in $D_f$. □

**Strategy 2.** Suppose rule $X \Rightarrow C_i$ holds in $D_f$, if the condition

$$\text{Dsupp}\left(X\right) < \max\left(\frac{1}{1-\beta}, 2\right) \times \alpha$$

is satisfied, then rule items of the form $Z \Rightarrow C_j$ ($Z \supseteq X$, $j \neq i$) will not hold in $D_f$.

**Proof.** Similarly, we only need to prove that $X \Rightarrow C_j$ ($j \neq i$) is infrequent. The following proof consists of two situations:

(1) If $\beta \le 0.5$, then $\frac{1}{1-\beta} \le 2$. In this situation, the condition is equivalent to

$\mathrm{Dsupp}(X) < 2\alpha$.

Since $\| X \| = \sum_{t|t.class=C_i} \mathrm{count}(\mu_X(t)) + \sum_{\substack{j=1 \\ j\neq i}}^{g} \left( \sum_{t|t.class=C_j} \mathrm{count}(\mu_X(t)) \right) = \|XC_i\| + \sum_{\substack{j=1 \\ j\neq i}}^{g} (\|XC_j\|)$, then

$$\frac{\| X \|}{\| X \|} = \frac{\| XC_i \|}{\| X \|} + \sum_{\substack{j=1 \\ i\neq j}}^{g} \frac{\| XC_j \|}{\| X \|}.$$

Thus, we have $\sum_{\substack{j=1 \\ j\neq i}}^{g} \mathrm{Dsupp}(X \Rightarrow C_j) = \mathrm{Dsupp}(X) - \mathrm{Dsupp}(X \Rightarrow C_i)$.

Since $X \Rightarrow C_i$ holds in $D_f$, then $\mathrm{Dsupp}(X \Rightarrow C_i) \ge \alpha$. Further, we have $\mathrm{Dsupp}(X \Rightarrow C_j) \le \mathrm{Dsupp}(X) - \mathrm{Dsupp}(X \Rightarrow C_i) < 2\alpha - \alpha = \alpha, \forall j \neq i$. So $X \Rightarrow C_j$ is infrequent, thus will not hold in $D_f$.

(2) If $\beta > 0.5$, then $\frac{1}{1-\beta} \le 2$. In this situation, the condition is equivalent to

$(1-\beta)\mathrm{Dsupp}(X) < \alpha$.

Since $X \Rightarrow C_i$ holds in $D_f$, then $\mathrm{Dconf}(X \Rightarrow C_i) \ge \beta$. Consequently, we have $\mathrm{Dsupp}(X) \times (1 - \mathrm{Dconf}(X \Rightarrow C_i)) \le (1-\beta)\mathrm{Dsupp}(X) < \alpha$, which means the condition in Strategy 1 is satisfied. Therefore, $X \Rightarrow C_j$ is infrequent, and will not hold in $D_f$. □

Importantly, these strategies will be incorporated into the mining process as pruning strategies (rather than post-process filtering strategies) to reduce the number of candidate itemsets (rule items) generated.

Finally, with the built classifier, a newly coming transaction t could be classified using a particular fuzzy CAR: $X \Rightarrow C_i$, chosen in the classifier. That requires a match between $t$ and $X$, which can be determined with a measure called weighted confidence, denoted as $\mathrm{Wconf}_t(X \Rightarrow C_i)$ (Chen & Chen, 2008): $\mathrm{Wconf}_t(X \Rightarrow C_i) = \mathrm{Dsupp}_t(X) \times \mathrm{Dconf}(X \Rightarrow C_i)$. For each fuzzy CAR, there is a Wconf value for $t$, then the fuzzy CAR with the highest Wconf value will be used to classify $t$.

## 7 The $\mathrm{GARC}_{f2}$ Algorithm

The algorithmic detail of the fuzzy extension to GARC with $f_2$ (i.e., $\mathrm{GARC}_{f2}$) is provided in its general form as shown in Figure 4.

1. $D_f$ = Trapezia_Fuzzy(D); // Fuzzy partitioning with $f_2$
2. $fCAR_1 = \{ r \mid r$ is an 1-itemset, $Dsupp(r) \geq \alpha$ and $Dconf(r) \geq \beta \}$;
   // Initial 1-itemset rule set
3. $prfCAR_1$ = prunRules($fCAR_1$); // Deleting redundant rules and conflicting rules
4. $F_1 = \{p \mid p$ is an 1-itemset, $Dsupp(p) \geq \alpha$ and $p$ is not in $prfCAR_1$ set$\}$ // Strategy 1
5. $bestattr = gain()$; // function $gain()$ returns best attribute with maximal gain
6. for $k$ from 2 to $m$ do
7. $C_k = CandidateGen(F_{k-1}, bestattr)$; // Strategy 2
8. $fCAR_k = \{r \mid r \in C_k, Dsupp(r) \geq \alpha, Dconf(r) \geq \beta\}$;
9. $prfCAR_k$ = prunRules($fCAR_k$); // Deleting redundant rules and conflicting rules
10. $F_k = \{p \mid p$ is a $k$-itemset, $Dsupp(p) \geq \alpha$ and $p$ is not in $prfCAR_k$ set$\}$ // Strategy 1
11. end for;
12. $CompactfCARs = \cup_k(prfCAR_k)$; // Generating a Compact Set of Fuzzy CARs
13. Sort($CompactfCARs$);

**Fig. 4** The GARC$_{f2}$ algorithm

In Figure 4, line 1 performs the fuzzy partitioning operation based on a given parameter $f_2$, so that the original dataset D is then transformed to the fuzzy-partitioned dataset $D_f$. Lines 2-5 scan the dataset to measure all 1-itemsets rule items and then to get the best split attribute by fuzzy information gain calculation. Rule items that satisfy the minimum support and minimum confidence criterions are written into the set of fuzzy CARs (line 2). The generated 1-itemset rule set (called $fCAR_1$) is subject to a pruning operation (line 3), which deletes the redundant rules and conflicting rules. The pruning operation is also done in each subsequent pass of $fCAR_k$ (line 9). Lines 6-11 consecutively run operations of generation, scan and filtering to discover fuzzy CARs without rule redundancy and conflicts. For each subsequent pass, say pass $k$, frequent ($k$-1)-itemsets that are not in the pruned rule set $F_{k-1}$ (line 4 and line 10) are used to generate candidate rule items in the next loop, according to Strategy 1, to avoid generating redundant rules. The *CandidateGen* function generates candidate rule items that include the best split attribute (line 7). Strategy 2 is applied here for avoid conflicting rules so as to reduce the number of candidate rules. The generated fuzzy CARs of each pass, with rule redundancy and rule conflicts eliminated, construct the compact set of fuzzy CARs (line 12). Finally, all the rules in the compact set will be sorted with precedence (line 13). In addition, the algorithm will terminate when $F_{k-1} = \varnothing$ or otherwise, in at most $m$ passes, which should be less than the number of attributes in D.

## 8 Data Experiments

This section shows the experimental results for the proposed $GARC_{f2}$ algorithm, along with some comparisons with GARC. Briefly, the results show that $GARC_{f2}$ is at the same level of accuracy as GARC statistically, while the number of rules is significantly fewer than that of GARC.

The experimental environment is with Window XP, Intel Core Duo 3GHz, 2G RAM and MatLab 2009a. A commonly used benchmarking database in this field, namely the UCI Machine Learning Repository (Merz & Murphy, 1996) is used, including the 20 datasets that GARC selected with continuous attributes. Table 4 provides the basic information about the datasets, each containing two parts: a training data part and a testing data part. The former was used for generating the classifiers while the latter was used for testing purposes.

**Table 4** Basic Information of the UCI Datasets

| Dataset | Attributes | Number of attributes | Null Value (Y/N) | Number of Classes | Number of training data | Number of testing data |
|---|---|---|---|---|---|---|
| 1. Anneal | Discrete, continuous | 38 | Y | 6 | 598 | 300 |
| 2. Australian | Discrete, continuous | 14 | N | 2 | 460 | 230 |
| 3. Balance | Continuous | 4 | N | 3 | 416 | 209 |
| 4. Breast | Continuous | 10 | Y | 2 | 466 | 233 |
| 5. Cleve | Discrete, continuous | 13 | Y | 2 | 202 | 101 |
| 6. Crx | Discrete, continuous | 15 | Y | 2 | 460 | 230 |
| 7. Diabetes | Continuous | 8 | N | 2 | 512 | 256 |
| 8. Glass | Continuous | 9 | N | 7 | 142 | 72 |
| 9. Heart | Discrete, continuous | 13 | N | 2 | 180 | 90 |
| 10. Hepatitis | Discrete, continuous | 19 | Y | 2 | 103 | 52 |
| 11. Hypothyroid | Discrete, continuous | 25 | Y | 2 | 2108 | 1055 |
| 12. Ionosphere | Continuous | 34 | Y | 2 | 234 | 117 |
| 13. Iris | Continuous | 4 | N | 3 | 100 | 50 |
| 14. Labor | Discrete, continuous | 16 | Y | 2 | 38 | 19 |
| 15. Pima | Continuous | 8 | N | 2 | 512 | 256 |
| 16. Sick | Discrete, continuous | 29 | Y | 2 | 2515 | 1257 |
| 17. Sonar | Continuous | 60 | N | 2 | 138 | 70 |
| 18. Vehicle | Continuous | 18 | N | 4 | 564 | 282 |
| 19. Waveform | Continuous | 40 | N | 3 | 3333 | 1667 |
| 20. Wine | Continuous | 13 | N | 3 | 118 | 60 |

## 9 Accuracy

Accuracy, as one of the basic and important measures for classifiers, is the ratio of the number of cases truly predicted by the classifier over the total number of cases in the test dataset. In this experiment, $GARC_{f2}$ with different $f_2$ values is compared to GARC (at $\alpha = 0.01$ and $\beta = 0.7$ as in GARC (Chen et al., 2006)). The results are tabulated in Table 5.

**Table 5** Accuracies of GARC and $GARC_{f2}$ ($f_2$ = 0.1, 0.2, 0.3, 0.4, 0.5)

| | Dataset | GARC | $GARC_{f2}$ ($f_2$=0.1) | $GARC_{f2}$ ($f_2$=0.2) | $GARC_{f2}$ ($f_2$=0.3) | $GARC_{f2}$ ($f_2$=0.4) | $GARC_{f2}$ ($f_2$=0.5) |
|---|---|---|---|---|---|---|---|
| 1. | Anneal | 87.00 | 88.33 | 87.33 | 87.33 | 87.33 | 87.33 |
| 2. | Australian | 87.39 | 87.39 | 87.39 | 87.39 | 87.39 | 87.39 |
| 3. | Balance | 71.29 | 71.29 | 71.29 | 71.29 | 71.29 | 71.29 |
| 4. | Breast | 94.42 | 94.42 | 94.42 | 93.99 | 94.42 | 94.42 |
| 5. | Cleve | 66.34 | 69.31 | 69.31 | 69.31 | 69.31 | 69.31 |
| 6. | Crx | 85.65 | 85.65 | 85.65 | 85.22 | 85.22 | 85.22 |
| 7. | Diabetes | 67.97 | 68.36 | 68.75 | 67.58 | 63.67 | 64.06 |
| 8. | Glass | 62.50 | 62.50 | 62.50 | 61.11 | 51.39 | 55.56 |
| 9. | Heart | 78.89 | 77.78 | 77.78 | 77.78 | 77.78 | 77.78 |
| 10. | Hepatitis | 86.54 | 86.54 | 86.54 | 86.54 | 86.54 | 86.54 |
| 11. | Hypothyroid | 94.79 | 94.79 | 94.79 | 94.79 | 94.79 | 94.79 |
| 12. | Ionosphere | 91.45 | 93.16 | 93.16 | 92.31 | 92.31 | 90.60 |
| 13. | Iris | 94.00 | 94.00 | 94.00 | 94.00 | 94.00 | 94.00 |
| 14. | Labor | 84.21 | 84.21 | 84.21 | 84.21 | 84.21 | 84.21 |
| 15. | Pima | 76.17 | 75.39 | 75.39 | 74.22 | 74.61 | 75.39 |
| 16. | Sick | 93.79 | 93.79 | 93.79 | 93.79 | 93.79 | 93.79 |
| 17. | Sonar | 74.29 | 70.00 | 71.43 | 71.43 | 71.43 | 74.29 |
| 18. | Vehicle | 60.99 | 61.35 | 61.70 | 58.87 | 57.80 | 58.87 |
| 19. | Waveform | 73.73 | 73.67 | 71.51 | 71.51 | 71.63 | 71.57 |
| 20. | Wine | 83.33 | 91.67 | 93.33 | 93.33 | 90.00 | 91.67 |
| | Mean | 80.74 | 81.18 | 81.21 | 80.80 | 79.95 | 80.40 |
| | Std. Deviation | 11.01 | 11.37 | 11.39 | 11.80 | 12.99 | 12.29 |

For illustrative purposes, Figure 5 depicts the mean accuracy with respect to different $f_2$ values. It is worth noticing that, at $f_2 = 0$, $GARC_{f2}$ degenerates to GARC.

Furthermore, Table 6 lists the results of significance tests on pairwise mean difference for GARC and $GARC_{f2}$ ($f_2$ = 0.1, 0.2, 0.3, 0.4, 0.5) with t-test, which indicates that on average the accuracy of $GARC_{f2}$ (though varying with $f_2$) is not significantly different from that of GARC.

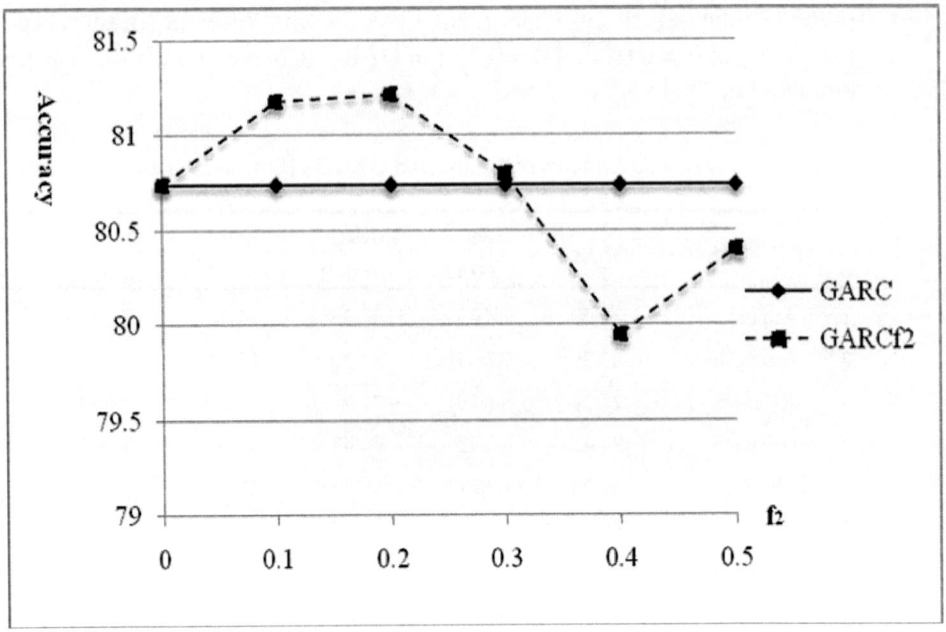


**Fig. 5** Mean Accuracy of GARC, $GARC_{f2}$ ($f_2$ = 0.1, 0.2 ,0.3, 0.4, 0.5)

**Table 6** Significance Tests on Pairwise Mean Difference for Accuracy of Classifiers

| | T test | |
|---|---|---|
| | 95% | 90% |
| $GARC_{f2}$ ($f_2$ = 0.1) – GARC | No | No |
| $GARC_{f2}$ ($f_2$ = 0.2) – GARC | No | No |
| $GARC_{f2}$ ($f_2$ = 0.3) – GARC | No | No |
| $GARC_{f2}$ ($f_2$ = 0.4) – GARC | No | No |
| $GARC_{f2}$ ($f_2$ = 0.5) – GARC | No | No |

## 10 Number of Rules

As stated in previous discussions, the number of rules reflects the understandability/ compactness of a classifier. Briefly, the smaller the number of rules is, the better understandability a classifier obtains (e.g., at the same accuracy level). The introduction of fuzzy partitioning could express appropriate semantics of data, which may lead to a decrease in the number of rules and therefore an increase in the understandability. The experiments on the number of rules as listed in Table 7 confirmed this statement.

**Table 7** Number of Rules of GARC and $GARC_{f2}$ ($f_2$ = 0.1, 0.2, 0.3, 0.4, 0.5)

| | Dataset | GARC | $GARC_{f2}$ ($f_2$=0.1) | $GARC_{f2}$ ($f_2$=0.2) | $GARC_{f2}$ ($f_2$=0.3) | $GARC_{f2}$ ($f_2$=0.4) | $GARC_{f2}$ ($f_2$=0.5) |
|---|---|---|---|---|---|---|---|
| 1. | Anneal | 51 | 48 | 47 | 47 | 48 | 48 |
| 2. | Australian | 17 | 17 | 17 | 17 | 17 | 17 |
| 3. | Balance | 4 | 5 | 5 | 5 | 5 | 5 |
| 4. | Breast | 22 | 21 | 21 | 21 | 21 | 21 |
| 5. | Cleve | 15 | 14 | 14 | 14 | 14 | 13 |
| 6. | Crx | 16 | 16 | 16 | 16 | 16 | 16 |
| 7. | Diabetes | 10 | 10 | 10 | 9 | 8 | 8 |
| 8. | Glass | 21 | 21 | 17 | 14 | 9 | 9 |
| 9. | Heart | 12 | 12 | 11 | 11 | 11 | 11 |
| 10. | Hepatitis | 22 | 22 | 22 | 22 | 22 | 22 |
| 11. | Hypothyroid | 43 | 44 | 44 | 44 | 44 | 44 |
| 12. | Ionosphere | 67 | 67 | 65 | 65 | 65 | 65 |
| 13. | Iris | 7 | 7 | 7 | 7 | 7 | 7 |
| 14. | Labor | 17 | 17 | 17 | 17 | 17 | 16 |
| 15. | Pima | 5 | 5 | 5 | 5 | 5 | 5 |
| 16. | Sick | 51 | 51 | 51 | 51 | 51 | 51 |
| 17. | Sonar | 16 | 15 | 13 | 11 | 11 | 16 |
| 18. | Vehicle | 122 | 102 | 86 | 82 | 68 | 78 |
| 19. | Waveform | 33 | 33 | 27 | 27 | 25 | 20 |
| 20. | Wine | 16 | 16 | 14 | 14 | 14 | 12 |
| | Mean | 28.35 | 27.15 | 25.45 | 24.95 | 23.90 | 24.20 |
| | Std. Deviation | 27.86 | 24.39 | 21.85 | 21.45 | 19.90 | 21.15 |

Similarly, Figure 6 represents the mean number of rules with respect to different $f_2$ values. Again, at $f_2 = 0$, $GARC_{f2}$ degenerates to GARC.

Next, statistical tests had been performed to examine the significance on the difference of number of rules of GARC and $GARC_{f2}$ ($f_2$ = 0.1, 0.2, 0.3, 0.4, 0.5) using the non-parametric Friedman test (Conover, 1999; Vapnik, 1998), with the results shown in Table 8, revealing the advantage of $GARC_{f2}$ over GARC in number of rules. That is, in general, the number of rules in the $GARC_{f2}$ classifier was fewer than that of GARC, which appeared to be significant in the case of $f_2 > 0.1$ (while insignificant at $f_2 = 0.1$).

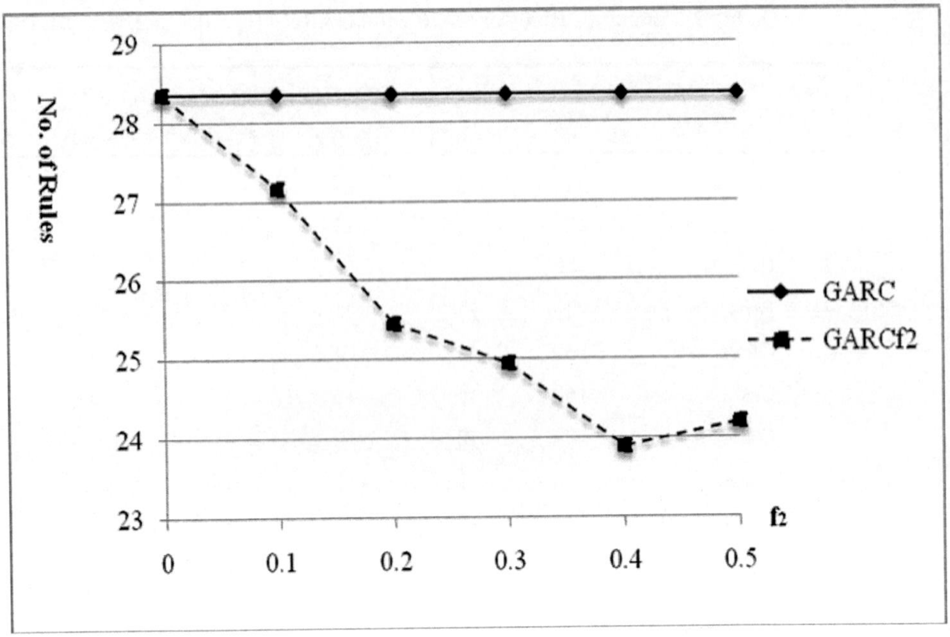


**Fig. 6** Mean Number of Rules of GARC, $GARC_f$ ($f$ = 0.1, 0.2, 0.3, 0.4, 0.5)

**Table 8** Significance Test on Pairwise Difference of Number of Rules of GARC and $GARC_{f2}$

| | Friedman Test | |
|---|---|---|
| | 95% | 90% |
| $GARC_{f2}$ ($f_2$ = 0.1) – GARC | No | No |
| $GARC_{f2}$ ($f_2$ = 0.2) – GARC | Yes (-)* | Yes (-) |
| $GARC_{f2}$ ($f_2$ = 0.3) – GARC | Yes (-) | Yes (-) |
| $GARC_{f2}$ ($f_2$ = 0.4) – GARC | Yes (-) | Yes (-) |
| $GARC_{f2}$ ($f_2$ = 0.5) – GARC | Yes (-) | Yes (-) |

**Note*. Yes (-) for A – B represents that number of rules of A is significantly smaller than that of B.

To summarize, from the data experiments, in a statistically significant sense, the accuracy performance of $GARC_{f2}$ is as good as that of GARC, and the compactness/understandability performance of $GARC_{f2}$ is generally better than that of GARC. Overall, these are deemed desirable.

## 11 A Case Study: Bank Credit Ratings

This section examines the effectiveness of $GARC_{f2}$ in a real-world classification application, namely, bank credit rating.

The data were collected from a rating institution in China, composed of two sets: one is 140 credit rating records of bank G in years 2000-2002 as the training sample, the other is 104 credit rating records of banks Z, J, X and R in years 1999-2002 as the testing sample. The ratings consist of three categories with a total of nine levels, i.e., AAA, AA, A, BBB, BB, B, CCC, CC, C. Further, in usual practice, they were grouped into two classes: Investment Grade (for the ratings of BBB or higher) and Speculative Grade (for the ratings of BB or lower). Tables 9 and 10 show the data and their attributes.

**Table 9** Class distribution in the credit rating dataset

| Class | No. of records in Training Dataset | No. of records in Testing Dataset |
|---|---|---|
| Investment Grade | 88 | 72 |
| Speculative Grade | 52 | 32 |
| Total | 140 | 104 |

**Table 10** Attributes of the credit rating dataset

| | Attribute | Crisp Boundaries* | Fuzzy Partitions |
|---|---|---|---|
| $A_1$: | Net asset to loan for the period ending at Dec 31 | 65.985 | Low, High |
| $A_2$: | Asset to liability ratio | 69.72 | Low, High |
| $A_3$: | Capital expenditure to equity ratio | 130.275, 203.85 | Low, Medium, High |
| $A_4$: | Current ratio | 95.965, 139.385 | Low, Medium, High |
| $A_5$: | Acid ratio | 62.845, 106.96 | Low, Medium, High |
| $A_6$: | Net non-financing cash inflow to liquidity liability ratio | -12.755, 9.785 | Low, Medium, High |
| $A_7$: | Operating cash inflow to liquidity liability ratio | -11.97, 7.305 | Low, Medium, High |
| $A_8$: | Account receivable turnover ratio | 5.03 | Low, High |
| $A_9$: | Inventory turnover ratio | 1.34, 3.555 | Low, Medium, High |
| $A_{10}$: | Fixed asset turnover ratio | 2.025 | Low, High |
| $A_{11}$: | Gross profit margin | 23.035 | Low, High |
| $A_{12}$: | Operating income margin | 0.305 | Low, High |
| $A_{13}$: | ROE | 0.745 | Low, High |
| $A_{14}$: | Return on total assets | 3.975 | Low, High |

**Note*. Crisp boundaries are the results of the discretization of continuous attributes, done by the Entropy method in (Fayyad & Irani, 1993), which is used in CBA/GARC. Crisp boundaries for each attribute correspond to a group of crisp partitions. For example, the crisp boundaries of attribute $A_4$ correspond to crisp partitions: $A_4<95.965$, $95.965< A_4<139.385$, $A_4>139.385$.

In Table 10, fuzzy partitions were defined upon the intervals with crisp boundaries determined in CBA/GARC, resulting in respective fuzzy sets such as high, medium or low using the trapezoidal membership functions discussed in previous sections. Subsequently, GARC and $GARC_{f2}$ were applied to the data with the settings of $\alpha$ and $\beta$ in order to get a classifier at the highest level of accuracy then smallest number of rules for each dataset (see Figure 7).

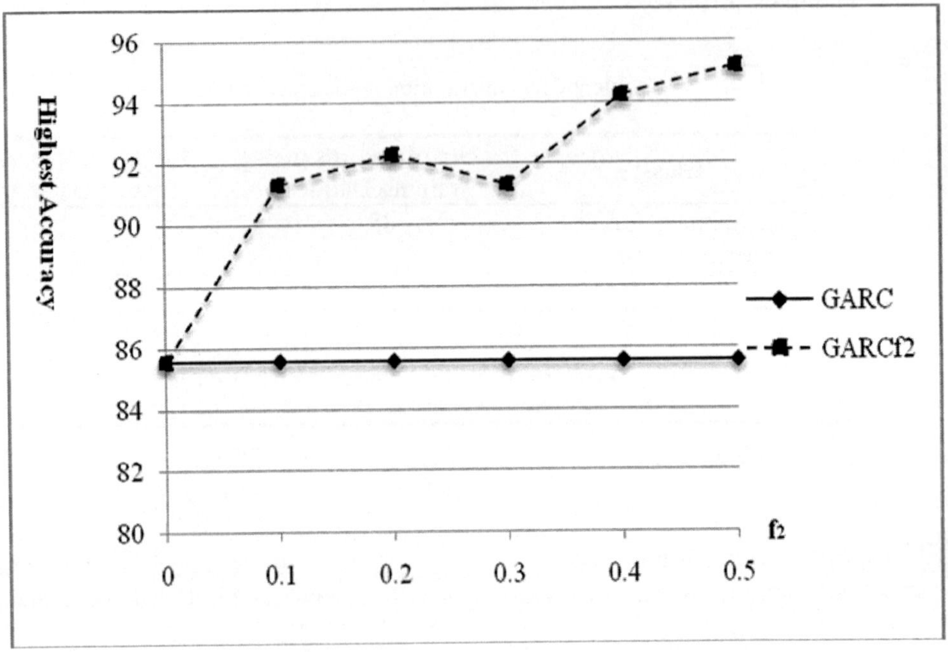


**Fig. 7** Highest Accuracy of GARC and $GARC_{f2}$ on the credit rating dataset

It can be seen from Figure 6 that at the best cases $GARC_{f2}$ was advantageous over GARC in accuracy. Furthermore, looking at the case with $f_2 = 0.5$, we could reach a situation where $GARC_{f2}$ performed better than GARC in accuracy and the number of rules (Table 11).

**Table 11** Performances of GARC and $GARC_{f2}$

| | Highest accuracy | No. of rules | Parameter Settings | |
|---|---|---|---|---|
| | | | $\alpha$ | $\beta$ |
| GARC | 85.58% | 15 | 0.06 | 0.95 |
| $GARC_{f2}$ ($f_2 = 0.5$) | 95.19% | 14 | 0.01 | 0.90 |

Finally, for illustrative purposes, the fuzzy CARs generated by $CARC_{f2}$ in this application are exemplified in Table 12 (with $f_2$=0.5).

**Table 12** Rules in the classifier built by $GARC_{f2}$ ($f_2 = 0.5$)

| | |
|---|---|
| Rule 1: | (Capital expenditure to equity ratio is low)∧(Account receivable turnover ratio is high) ⇒ Investment Grade |
| Rule 2: | (Current ratio is high) ⇒ Investment Grade |
| Rule 4: | (Acid ratio is high) ⇒ Investment Grade |
| Rule 6: | (Inventory turnover ratio is low) ⇒ Speculative Grade |

## 12 Conclusion

This chapter has presented a fuzzy extension to a compact and accurate associative classifier (namely GARC) by introducing fuzzy sets on the attribute domains in order to smooth the belongings of the data values around the boundaries of crisp intervals. This has been done via fuzzy partitioning using trapezoidal membership functions, designed with a single parameter $f_2$ in reflecting the degree of fuzziness around the boundaries, giving rise to the $GARC_{f2}$ approach. Moreover, the notions of support, confidence, information gain, rule redundancy and conflicts have also been extended in the fuzzy context, including the incorporation of pruning strategies in the rule generation process. This incorporation as well as the extended information gain have been proven to be important for compactness (i.e., with a reduced number of fuzzy CARs). Benchmarking datasets showed that on average $GARC_{f2}$ had accuracy similar to that of GARC, and was significantly advantageous over GARC in compactness. Finally, a real application on bank credit ratings also demonstrated the effectiveness of the extended approach.

**Acknowledgements.** The work was partly supported by the National Natural Science Foundation of China (70890080), the MOE Project of Key Research Institute of Humanities and Social Sciences at Universities of China (07JJD63005) and Tsinghua University's Research Center for Contemporary Management.

## References

Agrawal, R., Imielinski, T., Swami, A.: Mining association rules between sets of items in large databases. In: Proceedings of ACM-SIMOD international conference on management of data, Washington, DC, pp. 207–216 (1993)

Agrawal, R., Srikant, R.: Fast algorithm for mining association rules. In: Proceedings of the 20th VLDB Conference, pp. 487–499. Morgan Kaufmann, San Francisco (1994)

Ali, K., Manganaris, K., Srikant, R.: Partial Classification using Association Rules. In: Proceeding of the Third International Conference on Knowledge Discovery and Data Mining, Newport Beach, California, pp. 115–118 (1997)

Aumann, Y., Lindell, Y.: A statistical theory for quantitative association rules. Knowledge Discovery and Data Mining, 261–270 (1999)

Berzal, F., Cubero, J., Sanchez, D., et al.: ART: A Hybrid Classification Model. Machine Learning 54, 67–92 (2004)

Bezdek, J.C.: Pattern Recognition with Fuzzy Objective Function Algorithms. Plenum Press, New York (1981)

Bosc, P., Pivert, O.: On some Fuzzy Extensions of Association Rules. In: Proceeding of IFSA/NAFIPS 2001 Congress, Vancouver, Canada, pp. 1104–1109 (2001)

Breiman, L.: Classification and Regression trees. Wadsworth, Belmont (1984)

Chen, G.Q.: Fuzzy Logic in Data Modeling: semantics, constraints and database design. Kluwer Academic Publishers, Boston (1998)

Chen, G.Q., Liu, H.Y., Yu, L., Wei, Q., Zhang, X.: A New Approach to Classification Based on Association Rule Mining. Decision Support Systems 42, 674–689 (2006)

Chen, G.Q., Wei, Q.: Fuzzy Association Rules and the Extended Mining Algorithms. Information Sciences 147, 201–228 (2002)

Chen, G.Q., Wei, Q., Kerre, E.E.: Fuzzy logic in discovering association rules: an overview. In: Data Mining and Knowledge Discovery Approaches Based on Rule Induction Techniques. Massive Computing Series, pp. 474–480. Springer, Heidelberg (2006)

Chen, G.Q., Wei, Q., Liu, D., Wets, G.: Simple Association Rules (SAR) and the SAR-Based Rule Discovery. Computer & Industrial Engineering 43, 721–733 (2002)

Chen, G.Q., Yan, P., Kerre, E.E.: Computationally Efficient Mining for Fuzzy Implication-Based Association Rules in Quantitative Databases. International Journal of General Systems 33, 163–182 (2004)

Chen, G.Q., Yan, P., Wei, Q.: Discovering Associations with Uncertainty from Large Databases. In: Rakus-Andersson, E., Yager, R., Ichalkaranje, N., Jain, L. (eds.) Recent Advances in Decision Making, book series of Studies in Computational Intelligence, vol. 222, pp. 45–66 (2009)

Chen, G.Q., Zhang, X., Yu, L.: An improved GARC algorithm for building accurate and more understandable associative classifiers. Dynamics of Continuous, Discrete and Impulsive Systems – Series B – applications & algorithms 13, 161–165 (2006)

Chen, Z.L., Chen, G.Q.: Building an Associative Classifier Based on Fuzzy Association Rules. International Journal of Computational Intelligence Systems 1(3), 262–273 (2008)

Conover, W.J.: Practical Nonparametric Statistics, 3rd edn. Wiley, New York (1999)

De Luca, A., Termini, S.: A definition of Non-Probabilistic Entropy in the Setting of Fuzzy Sets. Information Control 20, 301–312 (1972)

Dong, G., Zhang, X., Wong, L., Li, J.: CAEP: Classification by aggregating emerging patterns. In: Proceedings of the Second International Conference on Discovery Science, Tokyo, Japan, pp. 30–42 (1999)

Dougherty, J., Kohavi, R., Sahami, M.: Supervised and unsupervised discretization of continuous features. In: The Twelfth International Conference on Machine Learning, pp. 194–202 (1995)

Dunn, J.C.: A Fuzzy Relative of the ISODATA Process and Its Use in Detecting Compact Well Separated Cluster. Cybernetics and Systems 3(3), 32–57 (1974)

Fayyad, U., Uthurusamy, R. (eds.): AAAI Workshop on Knowledge Discovery in Databases, Seattle, Washington, DC, USA (1994)

Fayyad, U.M., Irani, K.B.: Multi-interval discretization of continuous-valued attributes for classification learning. In: Proceedings of the 13th International Joint Conference on Artificial Intelligence, pp. 1022–1027 (1993)

Friedman, N., Geiger, D., Goldszmidt, M.: Bayesian network classifier. Machine Learning 29, 131–163 (1997)

Han, J.W., Kamber, M., Pei, J.: Data Mining: Concepts and Techniques, 2nd edn. Morgan Kaufmann Publishers, San Francisco (2005)

Klir, G.J., Yuan, B.: Fuzzy sets and fuzzy logic theories and applications. Prentice Hall PTR, Englewood Cliffs (1995)

Li, W.M., Han, J., Pei, J.: CMAR: Accurate and efficient classification based on multiple class-association rules. In: IEEE International Conference on Data Mining, San Jose, CA, pp. 69–376 (2001)

Liu, B., Hsu, W., Ma, Y.: Integrating Classification and Association Rule Mining. In: The 4th International Conference on Knowledge Discovery and Data Mining, New York, U.S. pp. 80–86 (1998)

Liu, B., Ma, Y.M., Wong, C.K.: Improving an association rule based classifier. Principles of Data Mining and Knowledge Discovery, 504–509 (2000)

Mannila, H., Toivonen, H., Verkamo, A.I.: Efficient Algorithms for Discovering Association Rules. In: AAAI Workshop on Knowledge Discovery in Databases, Seattle, Washington, pp. 181–192 (1994)

Merz, C.J., Murphy, P.: UCI repository of machine learning databases (1996) `http://www.cs.uci.edu/~mlearn/MLRepository.html`

Quinlan, J.R.: C4.5: Programs for Machine Learning, pp. 23–24. Morgan Kaufmann, San Francisco (1993)

Shafer, J., Agrawal, R., Mehta, M.: SPRINT: a scalable parallel classifier for data mining. In: Proceedings of the 22nd VLDB conference, India, pp. 544–555 (1996)

Sharpe, P.K., Glover, R.P.: Efficient GA based techniques for classification. Applied Intelligence 11, 277–284 (1999)

Shie, J.D., Chen, S.M.: Feature subset selection based on fuzzy entropy measures for handling classification problems. Applied Intelligence 28, 69–82 (2008)

Srikant, R., Agrawal, R.: Mining Quantitative Association Rules in Large Relational Tables. In: Proceedings of the ACM-SIGMOD Conference on Management of Data, Montreal, Canada, pp. 1–12 (1996)

Srikant, R., Vu, Q., Agrawal, R.: Mining Association Rules with Item Constraints. In: Proceedings of the 3rd Int'l Conference on Knowledge Discovery in Databases and Data Mining, Newport Beach, California, USA (1997)

Vapnik, V.N.: Statistical Learning Theory. John Wiley, Chichester (1998)

Wang, K., Zhou, S.Q., He, Y.: Growing decision tree on supportless association rules. In: KDD 2000, Boston, MA, USA (2000)

Weiss, S., Kulikowski, C.: Computer Systems that Learn: Classification and Prediction Methods from Statistics. In: Neural Nets, Machine Learning, and Expert Systems. Morgan Kaufma, San Francisco (1991)

Xiong, Y.X.: Fuzzy extensions to GARC associative classification (Master Thesis). Tsinghua University (2010)

Yin, X.X., Han, J.W.: CPAR: Classification based on predictive association rules. In: Proceedings of the third SIAM International Conference on Data Mining, pp. 331–334. SIAM, Philadelphia (2003)

Zadeh, L.A.: Fuzzy sets. Information Control 8, 338–353 (1965)

Zhang, X., Chen, G.Q., Wei, Q.: Building a Highly-Compact and Accurate Associative Classifier. Applied Intelligence (2009), doi:10.1007/s10489-009 -0181-z

# Part III
# Media Applications

# Fuzzy Techniques in Image Processing at Ghent University: Summary of a 12-Year Journey

Mike Nachtegael*, Tom Mélange, Stefan Schulte, Valérie De Witte, and Dietrich Van der Weken

*Dedicated to our promotor Etienne E. Kerre,*
*the captain during each of our journeys!*

**Abstract.** Fuzzy set theory and fuzzy logic provide a mathematical model for approximate reasoning, which mimics human reasoning and can therefore be very powerful. Also in the field of image processing many applications can be found. It was in the late nineties that Etienne's research group started to work on image processing, and took off on a journey that lasts until today. This chapter is a "crew member's log" of this journey, with Etienne as our captain, and provides an overview of our work.

## 1 Getting Ready to Sail Off

The first papers on fuzzy techniques in image processing, originating from our research group at Ghent University, were published in the period 1995-1997. They were co-authored by Bernard De Baets, Etienne Kerre, and foreign colleagues, and dealt with the construction of fuzzy mathematical morphology [1, 2, 3, 4, 5]. Inspired by the potential that fuzzy set theory had to offer in this field, Etienne submitted and obtained a large university research project. This marked the beginning of a 12-year period, starting in 1998 until today, in which fuzzy set theory and fuzzy logic were used for and applied in image processing problems.

Looking back at these years, one can distinguish 3 different research "axes": fuzzy mathematical morphology (which is mainly developed from a theoretical point of view), filters for noise reduction in images (which is very practical

Mike Nachtegael
Ghent University, Dept. of Applied Mathematics and Computer Science,
Krijgslaan 281 - S9, 9000 Gent, Belgium
e-mail: `Mike.Nachtegael@ugent.be`

* Corresponding author.

C. Cornelis et al. (Eds.): 35 Years of Fuzzy Set Theory, STUDFUZZ 261, pp. 197–217.
springerlink.com 

oriented), and similarity measures for image comparison (which holds the balance between theory and practical applicability). It is remarkable to see how each of these fields has evolved in this 12-year period.

We invite the reader to embark on the summary of this 12-year journey. It is our tribute to Etienne Kerre (who was promotor of our PhD's, which all dealt with image processing), and our way to say "thank you" for the wonderful years we had in his research group. As such, this chapter is intended to serve as a kind of overview chapter, in which we focus on the work w.r.t. image processing that has been done in Etienne's research group.

## 2 Crew Member's Logs of a 12-Year Journey

### 2.1 *Journey 1: Fuzzy Mathematical Morphology*

In the nineties several models for mathematical morphology based on fuzzy set theory were developed. In general, three types of models could be distinguished: models based on a fuzzification of the underlying logical operators (e.g., the model of Bloch & Maître and the model of De Baets), models based on a fuzzification of the concept of inclusion (e.g., the models of Zadeh, Sinha & Dougherty, Kitainik, Bandler & Kohout), and other models (e.g., the model based on the Minkowski addition, the model based on fuzzy integrals, and the model based on the erosions of Di Gesu). This wide variety of new models illustrated the fact that fuzzy set theory indeed had potential in this field, but on the other hand this diversity also made it difficult to keep the general overview and raised the question on how to choose which model is best fitted for a specific problem.

One of our first contributions to fuzzy mathematical morphology was an extensive study of all the different existing models. This was a very interesting work, as it turned out that several connections between the different models could be established [6, 7, 8]. The model developed by De Baets, based on morphological dilations and erosions that are defined using general conjunctors and implicators, is the most general one and includes every of the other mentioned models. The results of our study made it possible to gain insight in the variety of fuzzy morphological models, and also implied that future research on fuzzy mathematical morphology should focus on the most general model, since properties of other models could easily be derived from the properties of the general model.

Just as an example to familiarize the reader with some typical expressions, we recall the definition of the two basic binary morphological operations, namely the dilation and erosion:

$$\begin{aligned} D(A,B) &= \{y \in X | T_y(B) \cap A \neq \emptyset\}, \\ E(A,B) &= \{y \in X | T_y(B) \subseteq A\}, \end{aligned}$$

where $A, B$ are crisp subsets of a universe $X$ ($A$ represents the binary image and $B$ is the so-called structuring element which can be chosen by the user), and $T_y(B)$ is the translation of $B$ over the vector $y$. In the most general fuzzy morphological model, these definitions can be fuzzified as follows:

$$D_C(A,B)(y) = \sup_{x \in T_y(d_B) \cap d_A} C(B(x-y), A(x)),$$
$$E_I(A,B)(y) = \inf_{x \in T_y(d_B)} I(B(x-y), A(x)),$$

where $A,B$ are fuzzy sets in a universe $X$ ($A$ represents a grayscale image and $B$ a grayscale structuring element), $C$ is a conjunctor on $[0,1]$ (to model the intersection) and $I$ is an implicator on $[0,1]$ (to model the concept of subsetness). Conjunctors and implicators are fuzzifications of the Boolean conjunction and the binary implication, respectively, and are defined as follows: a binary operator $C$ on $[0,1]$ is a conjunctor if it is an increasing mapping (i.e. it has increasing partial mappings) that satisfies $C(0,0) = C(0,1) = C(1,0) = 0$ and $C(1,1) = 1$, and a binary operator $I$ on $[0,1]$ is an implicator on $[0,1]$ if it is a hybrid monotonous mapping (i.e. it has decreasing first and increasing second partial mappings) that satisfies $I(0,0) = I(0,1) = I(1,1) = 1$ and $I(1,0) = 0$. $d_A$ and $d_B$ in the above formulas represents the support of the fuzzy sets $A$ and $B$, respectively, e.g. $d_A = \{x \in X | A(x) > 0\}$.

Regarding the investigation of properties, we explicitly paid attention to the so-called decomposition and construction properties [9, 10]. Decomposition properties reveal the relation between fuzzy morphological operations and their $\alpha$-cuts, while construction properties show which fuzzy morphological operations can be constructed from their binary counterparts. For a brief period we also looked into the fuzzification of the concept of adjunction, which plays a crucial role in the lattice based approach to mathematical morphology [11]. We vividly remember our nice cooperation with Henk Heijmans (which included a short stay of Mike at Henk's institution), one of the most well-known researchers in this field. Unfortunately, Henk got very ill and until today is no longer able to perform academic work. Several years later, we published a second paper on fuzzy adjunctions, inspired by the work of our colleagues Yun Shi and Da Ruan on fuzzy implications [12].

In the same period that we were working on fuzzy mathematical morphology, our colleague Martine De Cock was working on linguistic modifiers and our Polish colleague Anna Radzikowska was working on fuzzy rough set theory. Looking at each other's work we noticed that there was a common ground: the operators that constitute the basis of each of the three different theories (mathematical morphology, linguistic modifiers, fuzzy rough sets) showed a lot of similarity. A deeper investigation of this similarity led to the conclusion that the three theories are actually special cases of a more general theoretical framework based on images under fuzzy relations. This theoretical generality was further investigated, which lead to several publications [13, 14, 15, 16, 17]. The practical result was that we could investigate the general framework, and translate the properties of this general framework to each specific application area.

A few years later a completely different contribution to the field of mathematical morphology was realized by constructing morphological models for colour images. Indeed, it must be noted that all of the above was developed in the context of grayscale images. The extension to colour images, whether this extension is realized for morphological operators such as in this case or for noise reduction filters (see further in this chapter), certainly is not straightforward. From an abstract point of view a colour image can be regarded as a triplet of grayscale images (e.g., the well-known RGB colour model), but it is not sufficient to treat these three components separately to obtain a good extension. On the contrary, such a simplistic approach does not take into account any correlations between the different components and will most likely introduce artefacts. Therefore, we looked at several alternative extensions to colour morphology [18, 19]. We also applied the developed colour morphology in the context of image magnification [20].

More recently, we shifted our attention to a third track in the world of (fuzzy) mathematical morphology. Until now, we only used fuzzy set theory and fuzzy logic as a tool. The application of fuzzy set theory in mathematical morphology was inspired by the fact that a grayscale image and a fuzzy set are mathematically modelled in exactly the same way, namely as mappings from some universe to the unit interval [0,1]. The interpretation of course might be completely different, but techniques from one field could easily be applied to the other field. In this case, this is best illustrated by the fact that fuzzy logical operators were tools to construct alternative morphological operators. The reader familiar with the field will know that this has nothing to do with modeling uncertainty or imprecision. However, since 2008 we established a very nice and fruitful cooperation with our colleague Peter Sussner. The shift in the work is quite clear: we do not only apply fuzzy techniques as a tool, but we also apply concepts related to fuzzy set theory to actually model uncertainty that comes along with image capture [21, 22, 23, 24]. In that way the picture gets more complete: fuzzy set theory not only serves as a tool, but also as a model in mathematical morphology.

The basic idea is to model grayscale images no longer as mappings from a universe to the unit interval [0,1], where the value between 0 and 1 represents the measured gray value, but as mappings from a universe to the set of closed subintervals of the unit interval. This means that a pixel is no longer mapped onto a single value, but to a closed interval of likely values. This actually corresponds to the real-life situation, in which one is not always sure of the measured gray value: such values are not only rounded up or down (for computational storage and processing), but might also slighty shift in value depending on the recording circumstances (e.g., a take of a scene with a sunny sky versus the take of the same scene with a cloudy sky). The uncertainty that comes along with this image capture can be modelled by closed intervals, and in a natural way coincides with the already developed framework of interval-valued fuzzy sets. An illustration of this interval-valued approach is shown in Figures 1 and 2. We can also make the link with intuitionistic fuzzy set theory (since interval-valued fuzzy set theory and intuitionistic fuzzy set theory are equivalent), and even attach a related interpretation to the corresponding image model [25, 26].

Once the new image model was established, the construction and deeper investigation of the corresponding morphological model was the next challenge. We constructed a general interval-valued framework [27], but also looked at very specific

**Fig. 1** Different captures of the cameraman image: top = take with cloudy sky, middle = take with sunny sky, bottom = take with distortion. This example illustrates that the capture circumstances can cause uncertainty regarding the real pixel values. Also, all recorded values and positions are approximations of the real situation due to technical limitations.

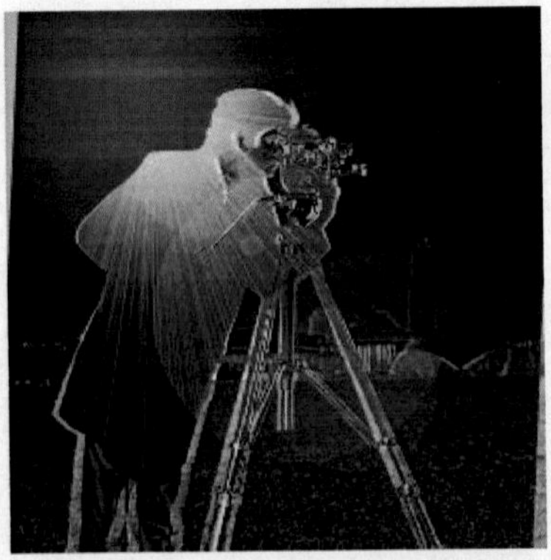

**Fig. 2** Representation of the interval width, resulting from the interval-valued representation of the cloudy/sunny/distorted take of the cameraman image. The more bright a pixel is, the more uncertain we are regarding its actual gray value.

models (such as in [28, 29]) or at further generalizations based on $L$-fuzzy set theory [30], where $L$ is a complete lattice. Just as we did for the "classical" fuzzy morphological operators, we also investigated decomposition and construction properties for interval-valued morphological operators [31, 32, 33].

The above overview of our activities in the field of mathematical morphology clearly illustrates the different "waves" that we experienced: starting from investigating the wide variety of fuzzy-inspired morphological models, we moved forward to colour extensions, and are currently on the track of interval-valued fuzzy morphology to also take into account the uncertainty that is involved with image capturing and modeling. Our future work will focus on theoretical items on the one hand (e.g., what is the relation between interval-valued morphological operators followed by defuzzification and defuzzification of the interval-valued image followed by classical morphological operators?), and on practical issues on the other hand (e.g., how can the interval-valued approach contribute to real-world edge detection applications?). One thing is sure: this journey is far from over . . .

## 2.2 *Journey 2: Similarity Measures*

There is a big need for objective similarity measures in image processing. Two different categories of applications demonstrate this. First, similarity measures are crucial in image retrieval applications: given a source image, similarity measures are required to retrieve those images from a database that are most similar to the source image. This also can have internet-based applications, where you feed a source image and try to retrieve similar images from the internet in order to obtain information regarding the content of the soure image. Second, similarity measures are also important for more theoretical purposes. For example, if one develops a filter for noise reduction (such as our research group did in the same period as we were looking at

similarity measures) it is very important that this filter is compared to other existing filters; otherwise one cannot demonstrate the usefulness and/or added value of that new filter. Popular similarity measures in that regard include the MSE (Mean Square Error) and the related PSNR (Peak-Signal-to-Noise-Ratio). However, experience learns that the numerical results of these measures do not always correspond to the visual evaluation of the images by humans. In other words: two images that have the lowest MSE (or highest PSNR) are not necessarily the two most similar images from a visual point of view. Of course, this observation is quite important when reading publications in which algorithms are compared to each other only using such measures. Either a visual evaluation should be included, or either new similarity measures that are more closely related to human evaluation have to be constructed.

It is the latter which inspired our research group to start working on similarity measures. We were familiar with similarity measures in the context of fuzzy set theory, which were developed to quantify the ressemblance between different fuzzy sets. Since fuzzy sets and grayscale images are modelled in the same way (namely as mappings from a universe to the unit interval $[0,1]$) the first step in our investigation was quite straightforward: to what extent can we apply existing similarity measures for fuzzy sets in image processing applications? For that purpose, more than 40 different measures were evaluated w.r.t. relevant image processing properties such as reflexivity (an image should be similar to itself to the degree 1), symmetry (the order in which the images are processed may not influence the result of the similarity measure), reaction to noise (a noisy image should have a large similarity w.r.t. the original image), reaction to enlightening or darkening (an enlightened or darkened image should have a large similarity w.r.t. the original image) and reaction to binary images (we expect similarity values between 0 and 1, and not only 0 or 1). From this investigation it followed that just a handful of the existing measures are appropriate for comparison of images [34, 35]. One of these measures is the following, which we refer to as $M_6$:

$$M_6(A,B) = \frac{|A \cap B|}{|A \cup B|} = \frac{\sum\limits_{(i,j)\in X} \min(A(i,j),B(i,j))}{\sum\limits_{(i,j)\in X} \max(A(i,j),B(i,j))},$$

where $A$ and $B$ are grayscale images (fuzzy sets in $X$) and $A(i,j)$ and $B(i,j)$ represent the gray value of the pixel at position $(i,j)$ in the respective images.

Because the perceptual behaviour of the remaining similarity measures was not always convincing (mainly due to the fact that the measures are pixel-based, and do not take into account any other relevant information), we constructed new similarity measures ourselves. First of all, we applied the similarity measures to partitioned images in order to construct neighbourhood-based similarity measures, with a more robust behaviour. However, simply applying the similarity measures to corresponding image parts did not yield satisfactory results either, so we needed to look further to other techniques in order to improve the behaviour of the similarity measures. In that way, we constructed neighbourhood-based similarity measures which also incorporate some characteristics of the human visual system, such as contrast sensitivity, by taking into account the homogenity in the considered neighbourhoods [36, 37, 38, 39, 40]. The results were much better compared to the results of the pixel-based similarity measures.

In a next step we investigated whether similarity measures from fuzzy set theory could be applied in a meaningful way to the histograms of images [41, 42, 43, 44]. It is indeed meaningful to compare two histograms in the framework of fuzzy set theory, because the histogram of an image can be transformed to a fuzzy set in the universe of gray levels by dividing the values of the histogram in every gray level by the maximum amount of pixels with the same gray value (resulting in so-called normalized histograms). In this way the most typical gray value gets membership degree 1 in the fuzzy set associated with the histogram and every other less typical gray value gets a smaller membership degree. Consequently, a normalized histogram is in accordance with the intuitive idea behind a fuzzy set: the most typical element in the universe gets membership degree 1 and all other less typical elements belong to the fuzzy set to a less extent. We found 15 similarity measures to be useful for direct application to (normalized) histograms. Furthermore, 22 similarity measures turned out to be appropriate for application to (normalized) ordered histograms; ordered histograms are obtained by placing the least occurring gray value in the first position of the histogram and by ordering the remaining frequencies in increasing order. Also in this approach, experimental results showed a better perceptual performance than the pixel-based similarity measures and the classical MSE or PSNR.

In order to confirm the applicability of neighbourhood-based and histogram-based similarity measures, a large psycho-visual experiment was conducted using a Multi-Dimensional Scaling approach (MDS) towards analyzing and modeling image quality variations [45]. This work was done in close cooperation with our colleagues from the Telecommunications and Information Department of the Faculty of Engineering Sciences. In total, 35 individuals evaluated a whole range of images (original images and slightly distorted ones) and these psycho-visual results were confronted with the output of our best neighbourhood- and histogram-based similarity measures. The conclusion of this large experiment was that the fuzzy similarity measures outperform the classical MSE and PSNR, i.e., they are better in accordance with the human visual evaluation. Another conclusion was that the neighborhood-based similarity measures perform better than the histogram-based similarity measures, which is probably due to the fact that histogram similarity measures do not incorporate the spatial properties of the different gray values, but only consider the frequency of occurence.

Since both neighbourhood-based and histogram-based similarity measures take into account different additional information from the images, we also constructed similarity measures which combine both approaches. This resulted in a new series of similarity measures with a strong global perceptual performance [46, 47].

Next, our research shifted to the construction of similarity measures for colour images. In a first approach, good similarity measures for grayscale images were applied to the different colour components of colour images (either in the RGB, HSV or Lab colour space) and then the results for each pixel were averaged. The results were promising, but we anticipated that a better approach would be to treat the colours as vectors, instead of just applying grayscale measures to each colour component. To achieve this, we introduced a new vector ordening in the RGB colour space which allowed us to extend the similarity measures that were constructed for grayscale images to similarity measures for colour images (such an ordering is required to be able to select minima and maxima). Experiments confirmed that the results were better than in the component-based approach [48, 49, 50, 51, 52].

Finally, we also constructed a colour image retrieval application. Colour image retrieval is becoming more and more important (also in the context of video), and so is the quest for automated and reliable retrieval systems. The novelty of our approach was the use of a fuzzy partition of the HSI colour space and the use of one of our similarity measures for histogram comparison (namely the one based on the above mentioned similarity measure $M_6$). The resulting retrieval system has the advantage that the images do not have to be characterized in advance using several features, and it is quite flexible since the database images are not required to have the same dimensions [53, 54]. We also situated our approach to image retrieval in a more larger framework in [55]. An example can be found in Figure 3, which shows an input image and the 9 most similar retrieved results from a large database of over 500 natural images of animals, flowers, buildings, cars, texture images, and so on. The results are quite good: the three most similar retrieved images are flowers in the same colour as the one in the query image; the other retrieved images do not contain flowers but have a very similar layout.

## 2.3 Journey 3: Fuzzy Filters for Noise Removal

Noise reduction filters are without doubt the most practical results from our research in image processing. During the past 12 years, a nice trajectory can be detected: starting with filters for grayscale images, the next step was filters for colour images, and finally filters for video sequences (both grayscale and colour). We focussed on two very common noise types: impulse noise (where a fraction of the pixel values is replaced by either fixed noise values or random noise values) and gaussian noise (addititive noise). Table 1 presents a bird's eye view on the filters that were developed in our group.

It all started with the so-called GOA-filter, named after the university project that enabled us to do this research and which resulted from a very nice cooperation with our colleagues from the Faculty of Engineering Sciences, in particular Dimitri Van De Ville, Wilfried Philips and Ignace Lemahieu [56, 57]. The filter is designed for the removal of gaussian noise in grayscale images, and uses fuzzy rules to detect the degree to which the gradient in a certain direction is small (the idea is that a small gradient is caused by noise, while a large gradient is caused by image structure). Fuzzy rules are also applied to calculate the correction term that is used for the denoising; the contribution of neighbouring pixels depends on their gradient values. The results of this filter were very good, and confirmed the usefulness of fuzzy logic for the construction of noise reduction filters. The main advantage of fuzzy filters is

**Fig. 3** Results of our colour image retrieval system. The source image is on top, and below are the most similar retrieved images. The value of the used similarity measure is shown below every retrieved image.

that they allow us to work and to reason with linguistic information, just as experts do (approximate reasoning); see the scheme in Figure 4. In order to confirm these good results, we carried out extensive comparative studies of existing classical and fuzzy filters, both for impulse noise and gaussian noise [58, 59, 60, 61, 62, 63].

A second filter for the reduction of gaussian noise from grayscale images was presented a few years later [64]. This FuzzyShrink-filter can be seen as a fuzzy variant of an existing probabilistic shrinkage method, and was developed in the wavelet domain. The filter outperformed fuzzy non-wavelet methods and was comparable with other recent but more complex wavelet methods.

After the succesfull GOA-filter for gaussian noise, we developed a filter for the removal of fixed impulse noise in grayscale images. This filter was called the Fuzzy Impulse noise Detection and Reduction Method, or FIDRM for short [65]. The filter

**Table 1** A summary of the different fuzzy filters for noise reduction that were developed in our research group.

| | Still gray | Still colour | Video gray | Video colour |
|---|---|---|---|---|
| fixed impulse | FIDRM | FIDRMC<br>HFMRC<br>HFC<br>OWA | | |
| random impulse | FRINR | HFC | FRINV-G | FRINV-C |
| gaussian | GOA<br>FuzzyShrink | FCG<br>OWA | FMDAF | FMDAF-RGB<br>FMDAF-CR<br>FMDAF-YUV |

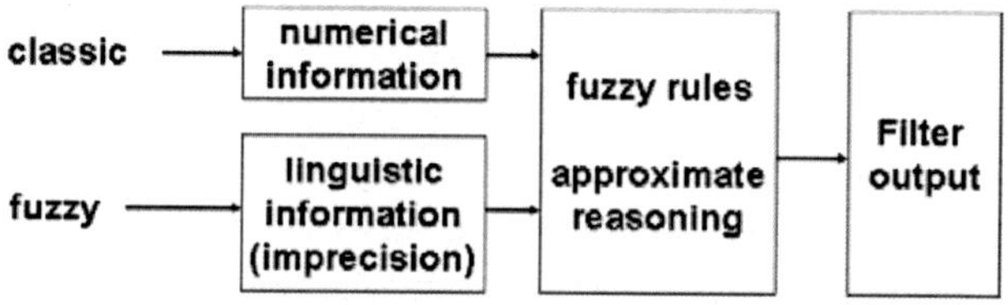


**Fig. 4** Fuzzy filters not only use numerical to filter out the noise in images, but can also work with linguistic information. Furthermore, fuzzy logic allows us to reason with this linguistic information and enables us to better approximate human reasoning.

followed a similar approach as the GOA filter, as it used gradient values to detect and remove the noise. Again, extensive experiments confirmed the state-of-the-art results of the filter [66, 67]. The filter could easily be extended to colour images by applying the filter on each of the corrupted colour bands separately. The results for colour images were relatively good [68], but the disadvantage of this approach is of course that correlations between colour band were neglected and small colour artefacts were introduced.

The latter observation inspired us to construct other filters, specifically to remove impulse noise from colour images, and led to the FIDRMC and HFMRC filters. The FIDRMC filter consists of two separated steps: the detection phase and the filtering phase. The detection phase is applied separately to each colour component, where fuzzy rules are used to determine whether a pixel pigment is corrupted with impulse noise or not. After the detection phase the filter only focuses on those pixel pigments which have a non-zero membership degree in the fuzzy set "impulse noise". In the filtering phase we also take into account the colour information of a certain neighbourhood around a given central pixel [69, 70]. The HFMRC filter follows a different approach and uses the histograms of the colour component differences to detect and filter the fixed impulse noise [71]. The HFMRC filter was later upgraded to the more complex HFC filter [72] that could also tackle randomly valued impulse noise in colour images. Previously, our FRINR filter already achieved the goal of removing randomly valued impulse noise in grayscale images [73, 74]. The detection phase of the FRINR filter consits out of two units that are both used to define corrupted impulse noise pixels. The first unit investigates the neighbourhood around a pixel to conclude whether the pixel can be considered as impulse noise or not, while the second unit uses fuzzy gradient values to determine the degree to which a pixel can be considered as impulse noise and the degree to which a pixel can be considered as noise free.

Regarding the removal of gaussian noise from colour images, we developed the FCG filter [75]. In contrast to most of the other existing methods, the first subfilter of the FCG filter distinguishes between local variations due to noise and local variations due to image structures (such as edges) by using the colour component distances instead of component differences. The second subfilter is used as a complementary filter which especially preserves differences between the colour components. A few years later, resulting from an idea of and our cooperation with Arya Basu, the OWA filter was developed for noise reduction in colour images corrupted by either impulse noise or a combination of salt & pepper noise and gaussian noise [76]. This OWA filter is an improvement of classical bilateral filtering, achieved by using Ordered Weighted Averaging operators.

With all of the above, we delivered fuzzy filters for fixed impulse noise, random impulse noise and gaussian noise, both for grayscale and colour images. The next step was the design of noise reduction filters for video. It is quite clear that video makes things much more complicated: the introduction of a temporal component implies that good filtering can only be achieved if the temporal aspect is taken well into account, i.e., if motion information is incorporated in the filter design. Filters for colour video are even more complex, because the temporal information has to be combined with the colour information. Nevertheless, a nice series of video filters were constructed.

The first achievement was the FMDAF or Fuzzy Motion and Detail Adaptive Filter, which was constructed in close cooperation with our colleagues from the Faculty of Engineering Sciences. We also developed a recursive version of the filter (RFMDAF), and a non-recursive and recursive version in the wavelet domain (WFDMAF and WRFMDAF). The method can be seen as a fuzzy variant of a multiple class video denoising method that automatically adapts to detail and motion. Experimental results show that the FMDAF filter efficiently removes gaussian noise from grayscale image sequences and outperforms other state-of-the-art filters of comparable complexity [77, 78].

Afterwards we extended this filter to colour video sequences in three different ways. First the FMDAF-RGB filter was an extension of the FMDAF filter in such a way that colours in the RGB colour space were treated as vectors; no separate filtering of colour bands occured [79]. Next we developed the FMDAF-CR filter [80]. Although the filtering here is performed in each colour band separately in the RGB colour space, the fuzzy rules also require information from the other colour bands such that correlations between colour bands are taken into account. We also constructed a third extension, called the FMDAF-YUV filter [80], in which the WFMDAF filter (in the wavelet domain) is applied to the Y-component in the YUV colour space and where the chrominance components U and V are averaged

**Fig. 5** 20-th frame of the "Deadline" video sequence: top = original frame, middle = noisy frame (25% random impulse noise), bottom = result of our FRINV-G filter.

over a small window. All these filters have a very competitive numerical and visual performance.

It is quite remarkable that most video filters that can be found in literature are designed for sequences corrupted by gaussian noise, and much less video filters exist for the removal of impulse noise. Existing 2D filters for impulse noise can of course be extended to video by applying the 2D filter to each frame of the video sequence, but temporal inconsistencies will arise when motion is neglected in the filtering process. Existing 3D filters on the other hand often suffer from detail loss because too many pixels are filtered or, vice versa, too many pixels are left unfiltered in order to preserve the image details. Quite recently, this observation encouraged us to develop a fuzzy filter for the removal of random impulse noise in both grayscale and colour video. The grayscale version of our filter (which we here refer to as the FRINV-G filter [81]) consists out of three different successive filtering steps and a fourth refinement step. In each filtering step, only the pixels that are detected as being noisy (a detection that is achieved by fuzzy rules) are filtered and to exploit the temporal information detected pixels are filtered in a motion compensated way. Experimental results clearly demonstrate that the FRINV-G filter outperforms other state-of-the-art filters, both numerically and visually. An illustrative example is shown in Figure 5. Finally, also a vector-based colour extension was constructed; here we refer to this colour video filter as the FRINV-C filter [82].

## 3 The Ship Has Sailed

Mathematical morphology, similarity measures, noise reduction filters: these were the three main tracks of the research on image processing in Etienne Kerre's research group. The short overviews above, together with the extensive literature list, show that each of these tracks was a very intensive and interesting journey. During all of these journeys we also had the opportunity to put some effort in the international promotion of soft computing in image processing in general. In 2002 we established the so-called "SCIP Working Group", where SCIP is an acronym for Soft Computing in Image Processing. The goal of this informal working group, of which the organisation is mainly internet-based, was and is to promote communication and cooperation between colleagues that are active in this exciting field. We firmly believe that it is very important that all efforts are undertaken to intensify this communication and cooperation: it leads to a cross-fertilization of ideas and new cooperations, and avoids that several groups are working on the same problems without knowing this from each other. The SCIP Working Group, which has around 160 members from over 40 countries, mainly tries to achieve these goals by regularly organizing special sessions on soft computing in image processing at international conferences. If the opportunity is there, we will also take the initiative to edit a book or journal issue that brings together interesting contributions from colleagues from all over the world. In that regard we certainly mention the three books in the series "Studies in Fuzziness and Soft Computing" [83, 84, 85] and the special issue of the international journal "Soft Computing - A Fusion of Foundations, Methodologies and Applications" [86] and of the "International Journal of Approximate Reasoning" [87].

Although each of the five authors of this chapter – which all five obtained their PhD under Etienne's guidance – were involved in one or more of the three discussed research tracks and in the SCIP Working Group, there always was one unifying factor: our promotor Etienne was there to support and guide us! Despite the fact that most of the authors have left the academic world, the work is still not done.

Currently, we are involved in a European Marie-Curie project on soft computing techniques in medical image processing. Our goal in this project is to use our expertise on similarity measures and noise reduction filters in order to apply these techniques on medical images. Two new Phd students are already working on the project! Etienne will not be around full-time, but we are all convinced that he will be part of this new journey ...

## References


1. De Baets, B., Kerre, E.E., Gupta, M.M.: The fundamentals of fuzzy mathematical morphology - part 1: basic concepts. International Journal of General Systems 23, 155–171 (1995)
2. De Baets, B., Kerre, E.E., Gupta, M.M.: The fundamentals of fuzzy mathematical morphology - part 2: idempotence, convexity and decomposition. International Journal of General Systems 23, 307–322 (1995)
3. De Baets, B.: Idempotent opening and closing operations in fuzzy mathematical morphology. In: Ayyub, B. (ed.) Proceedings of the joint 3rd International Symposium on Uncertainty Modelling and Analysis and NAFIPS 1995, pp. 228–233. IEEE Computer Society Press, Los Alamitos (1995)
4. De Baets, B.: Fuzzy morphology: a logical approach. In: Ayyub, B.M., Gupta, M.M. (eds.) Uncertainty Analysis in Engineering and Sciences: Fuzzy Logic, Statistics, and Neural Network Approach, pp. 53–67. Kluwer Academic Publishers, Boston (1997)
5. De Baets, B., Kwasnikowska, N., Kerre, E.E.: Fuzzy morphology based on conjunctive uninorms. In: Proceedings of IFSA 1997, pp. 215–220 (1997)
6. Nachtegael, M., Kerre, E.E.: Different Approaches towards Fuzzy Mathematical Morphology. In: Proceedings of EUFIT 1999 (7th European Conference on Intelligent Techniques and Soft Computing), Verlag Mainz GmbH, Aachen, pp. 303–304 (1999) (abstract); paper on CD
7. Nachtegael, M., Kerre, E.E.: Classical and Fuzzy Approaches towards Mathematical Morphology. In: Kerre, E.E., Nachtegael, M. (eds.) Fuzzy Techniques in Image Processing, pp. 3–57. Physica-Verlag, Heidelberg (2000)
8. Nachtegael, M., Kerre, E.E.: Connections Between Binary, Gray-scale and Fuzzy Mathematical Morphologies. Fuzzy Sets and Systems 124(1), 73–86 (2001)
9. Nachtegael, M., Kerre, E.E.: Fuzzy Mathematical Morphology: General Concepts and Decomposition Properties. In: Proceedings of KES 1999 (3rd International Conference on Knowledge-based Intelligent Information Engineering Systems), pp. 255–258. IEEE Inc., Los Alamitos (1999)
10. Nachtegael, M., Kerre, E.E.: Decomposing and Constructing of Fuzzy Morphological Operations over alpha-cuts: Continuous and Discrete Case. IEEE Transactions on Fuzzy Systems 8(5), 615–626 (2000)
11. Nachtegael, M., Heijmans, H., Van der Weken, D., Kerre, E.E.: Fuzzy Adjunctions in Mathematical Morphology. In: Proceedings of the 7th Joint Conference on Information Sciences (JCIS 2003), pp. 202–205 (2003) ISBN 0-970 890-2-5, ISBN: 0-9707890-2-5
12. Shi, Y., Nachtegael, M., Ruan, D., Kerre, E.E.: Fuzzy Adjunctions and Fuzzy Morphological Operations based on Fuzzy Implications. International Journal of Intelligent Systems 24(12), 1280–1296 (2009)

13. Nachtegael, M., Kerre, E.E., Radzikowska, A.M.: On Links between Fuzzy Morphology and Fuzzy Rough Sets. In: Proceedings of IPMU 2000 (8th International Conference on Information Processing and Management of Uncertainty in Knowledge-based Systems) Graficias, Spain, pp. 1381–1388 (2000)
14. De Cock, M., Nachtegael, M., Kerre, E.E.: Images under Fuzzy Relations: a Master-Key to Fuzzy Applications. In: Intelligent Techniques and Soft Computing in Nuclear Science and Engineering - Proceedings of FLINS 2000 (4th International Conference on Fuzzy Logic and Intelligent Technologies for Nuclear Science and Industry), pp. 47–54. World Scientific Publishing, Singapore (2000)
15. Nachtegael, M., De Cock, M., Van der Weken, D., Kerre, E.E.: Fuzzy Relational Images in Computer Science. In: Proceedings of RELMICS'6 (6th International Workshop on Relational Methods in Computer Science), Katholieke Universiteit Brabant, The Netherlands, pp. 156–170 (2001)
16. Nachtegael, M., De Cock, M., Van der Weken, D., Kerre, E.E.: Fuzzy Relational Images in Computer Science. In: de Swart, H. (ed.) RelMiCS 2001. LNCS, vol. 2561, pp. 134–151. Springer, Heidelberg (2002)
17. Kerre, E.E., Nachtegael, M.: Fuzzy Relational Calculus and its Application to Image Processing. In: Di Gesù, V., Pal, S.K., Petrosino, A. (eds.) Fuzzy Logic and Applications. LNCS, vol. 5571, pp. 179–188. Springer, Heidelberg (2009)
18. De Witte, V., Schulte, S., Nachtegael, M., Van der Weken, D., Kerre, E.E.: Vector Morphological Operators for Colour Images. In: Kamel, M.S., Campilho, A.C. (eds.) ICIAR 2005. LNCS, vol. 3656, pp. 667–675. Springer, Heidelberg (2005)
19. De Witte, V., Schulte, S., Nachtegael, M., Mélange, T., Kerre, E.E.: A Lattice-Based Approach to Mathematical Morphology for Greyscale and Colour Images. In: Kaburlasos, V.G., Ritter, G.X. (eds.) Computational Intelligence Based on Lattice Theory. de reeks Studies in Computational Intelligence, vol. 67, pp. 129–148. Springer, Heidelberg (2007) ISBN 978-3-540-72686-9
20. De Witte, V., Schulte, S., Kerre, E.E., Ledda, A., Philips, W.: Morphological Image Interpolation to Magnify Images with Sharp Edges. In: Campilho, A., Kamel, M.S. (eds.) ICIAR 2006. LNCS, vol. 4141, pp. 381–393. Springer, Heidelberg (2006)
21. Nachtegael, M., Sussner, P., Mélange, T., Kerre, E.E.: Modelling Numerical and Spatial Uncertainty in Grayscale Image Capture Using Extended Fuzzy Set Theory. In: Proceedings of NASTEC 2008 (1st International North American Simulation Technology Conference), pp. 15–22. Published by Eurosis (2008)
22. Nachtegael, M., Sussner, P., Mélange, T., Kerre, E.E.: Modelling Numerical and Spatial Uncertainty in Grayscale Image Capture Using Extended Fuzzy Set Theory. Journal of Advanced Computational Intelligence and Intelligent Informatics 13(5), 529–536 (2009) ISSN: 1343-0130
23. Nachtegael, M., Sussner, P., Mélange, T., Kerre, E.E.: Fuzzy Set Theory in Mathematical Morphology: from Tool to Uncertainty Model. In: Proceedings of WMSCI 2009 (13th World Multi-Conference on Systemics, Cybernetics and Informatics), vol. IV, pp. 149–155 (2009) ISBN-10: 1-934272-62-0, ISBN-13: 978-1-934272-62-6
24. Nachtegael, M., Sussner, P., Mélange, T., Kerre, E.E.: Fuzzy Set Theory in Mathematical Morphology: from Tool to Uncertainty Model, Accepted for: Information Sciences (special issue, edited by Vassilis Kaburlasos)
25. Nachtegael, M., Sussner, P., Mélange, T., Kerre, E.E.: Some Aspects of Interval-Valued and Intuitionistic Fuzzy Mathematical Morphology. In: Proceedings of IPCV 2008 (3rd International Conference on Image Processing, Computer Vision and Pattern Recognition), pp. 538–543 (2008) ISBN 1-60132-076-0 en 1-60132-077-9

26. Nachtegael, M., Sussner, P., Mélange, T., Kerre, E.E.: Foundations of the Interval-Valued Image Model to Model and Process Uncertainty in Image Capture. In: Proceedings of IWSSIP 2010 (17th International Conference on Systems, Signals and Image Processing), Eduff - Editora da Universidade Federal Fluminense, pp. 235–238 (2010), `http://www.creacteve.com.br/iwssip/index.htm`, ISBN 978-85-228-0565-5
27. Mélange, T., Nachtegael, M., Sussner, P., Kerre, E.E.: Basic Properties of Interval-Valued Fuzzy Morphological Operators. Accepted for: WCCI 2010 (2010)
28. Nachtegael, M., Sussner, P., Mélange, T., Kerre, E.E.: An Interval-Valued Fuzzy Morphological Model based on Lukasiewicz-Operators. In: Blanc-Talon, J., Bourennane, S., Philips, W., Popescu, D., Scheunders, P. (eds.) ACIVS 2008. LNCS, vol. 5259, pp. 601–612. Springer, Heidelberg (2008) ISBN 978-3-540-88457-6, ISSN 0302-9743
29. Nachtegael, M., Sussner, P., Mélange, T., Kerre, E.E.: A Study of Interval-Valued Morphology based on the Minimum-Operator. In: Proceedings of SPIE - 2nd International Conference on Digital Image Processing (ICDIP 2010), vol. 7546, pp. 75463H-1–75463H-7 (2010) ISBN 978-0-8194-7942-6
30. Sussner, P., Nachtegael, M., Mélange, T.: L-Fuzzy Mathematical Morphology: an Extension of Interval-Valued and Intuitionistic Fuzzy Mathematical Morphology. In: Proceedings of NAFIPS 2009 (28th North American Fuzzy Information Processing Society Annual Conference), paper on CD, 6 pages (2009) ISBN 978-1-4244-4577-6
31. Mélange, T., Nachtegael, M., Kerre, E.E., Sussner, P.: Decomposition of Interval-Valued Fuzzy Morphological Operations by Weak [a1, a2]-cuts. In: Proceedings of IFSA-EUSFLAT 2009 (2009 International Fuzzy Systems Association World Congress), pp. 167–172 (2009), ISBN 978-989-95079-6-8
32. Mélange, T., Nachtegael, M., Sussner, P., Kerre, E.E.: On the Decomposition of Interval-Valued Fuzzy Morphological Operations. Journal of Mathematical Imaging and Vision 36(3), 270–290 (2010)
33. Mélange, T., Nachtegael, M., Sussner, P., Kerre, E.E.: Construction of Interval-Valued Fuzzy Morphological Operators by Weak Cuts In: Proceedings of EUROFUSE 2009 (Eurofuse Workshop on Preference Modelling and Decision Analysis), pp. 227–232 (2009) (paper also on CD)
34. Van der Weken, D., Nachtegael, M., Kerre, E.E.: The Applicability of Similarity Measures in Image Processing. Intellectual Systems 6(1-4), 231–248 (2001) (paper translated in Russian)
35. Van der Weken, D., Nachtegael, M., Kerre, E.E.: An Overview of Similarity Measures for Images. In: Proceedings of ICASSP 2002 (IEEE International Conference on Acoustics, Speech and Signal Processing), vol. IV, pp. 3317–3320 (2002) ISBN: 0-7803-7403-7
36. Van der Weken, D., Nachtegael, M., Kerre, E.: Image Quality Evaluation. In: Knowledge-based Intelligent Information Engineering Systems and Allied Technologies. The series Frontiers in Artificial Intelligence and Applications - Proceedings of KES 2002 (6th International Conference on Knowledge-Based Intelligent Information and Engineering Systems), vol. 82, pp. 326–330. IOS Press, Amsterdam (2002)
37. Van der Weken, D., Nachtegael, M., Kerre, E.: Image Quality Evaluation. In: Proceedings of ICSP 2002 (6th International Conference on Signal Processing), pp. 711–714. IEEE Press, New York (2002) ISBN: 0-7803-7488-6
38. Van der Weken, D., Nachtegael, M., Kerre, E.E.: A New Similarity Measure for Image Processing. Journal of Computational Methods in Applied Sciences and Engineering 3(2), 209–222 (2003)

39. Van der Weken, D., Nachtegael, M., Kerre, E.E.: Constructing Neighbourhood-based Image Similarity Measures. In: Abraham, A., Koeppen, M., Franke, K. (eds.) Design and Application of Hybrid Intelligent Systems, Proceedings of HIS 2003 (3rd International Conference on Hybrid Intelligent Systems), pp. 878–886. IOS Press, Amsterdam (2003) ISBN: 1-58603-394-8
40. Van der Weken, D., Nachtegael, M., Kerre, E.E.: Using Similarity Measures and Homogeneity for the Comparison of Images. Image and Vision Computing 22(9), 695–702 (2004)
41. Van der Weken, D., Nachtegael, M., Kerre, E.E.: Using Similarity Measures for Histogram Comparison. In: De Baets, B., Kaynak, O., Bilgiç, T. (eds.) IFSA 2003. LNCS, vol. 2715, pp. 396–403. Springer, Heidelberg (2003)
42. Van der Weken, D., Nachtegael, M., Kerre, E.E.: Improved Image Quality Measures Using Ordered Histograms. In: Proceedings of the IEEE Signal Processing Society International Workshop on Multimedia Signal Processing (MMSP 2004), Siena (Italy), pp. 67–70 (2004) ISBN: 0-7803-8579-9, IEEE Catalog number: 04TH8761C
43. Van der Weken, D., Nachtegael, M., Kerre, E.E.: Some New Similarity Measures for Histograms In: Proceedings of 4th Indian Conference on Computer Vision, Graphics and Image Processing (ICVGIP 2004), Kolkata (India), pp. 441-446 (2004), ISBN: 81-7764-707-5
44. Van der Weken, D., Nachtegael, M., De Witte, V., Schulte, S., Kerre, E.E.: A Survey on the Use and the Construction of Fuzzy Similarity Measures in Image Processing. In: Proceedings of CIMSA 2005 (the 2005 IEEE International Conference on Computational Intelligence for Measurement Systems and Applications), Giardini Naxos (Italy), pp. 187 - 192 (2005) (paper on CD), ISBN 0-7803-9026-1, IEEE Catalog number: 05EX1037C
45. Van der Weken, D., Kerre, E.E., Vansteenkiste, E., Philips, W., Nachtegael, M., Zlokolica, V., Rombaut, J.: Evaluation of Fuzzy Image Quality Measures using a Multidimensional Scaling Framework. In: Proceedings of VPQM 2006 (2nd International Workshop on Video Processing and Quality Metrics for Consumer Electronics), Scottsdale (Arizona, United States), published on CD, ISBN 09774739
46. Van der Weken, D., Nachtegael, M., Kerre, E.: Combining Neighbourhood-Based and Histogram Similarity Measures for the Design of Image Quality Measures. In: Di Gesú, V., Masulli, F., Petrosino, A. (eds.) WILF 2003. LNCS (LNAI), vol. 2955, pp. 324–331. Springer, Heidelberg (2006)
47. Van der Weken, D., Nachtegael, M., Schulte, S., De Witte, V., Kerre, E.E.: Combining Neighbourhood-based and Histogram Similarity Measures for the Design of Image Quality Measures. Image and Vision Computing 25(2), 184–195 (2007)
48. Van der Weken, D., Nachtegael, M., Schulte, S., De Witte, V., Kerre, E.E.: Constructing Similarity Measures for Colour Images. In: Fuzzy Logic, Soft Computing and Computational Intelligence - Proceedings of IFSA 2005 (11th International Fuzzy Systems Association World Congress), Beijing (China), Vol. II, , pp. 1020 –1025 (2005) ISBN 7-302-11377-7
49. Van der Weken, D., De Witte, V., Nachtegael, M., Schulte, S., Kerre, E.E.: Colour Image Comparison Using Vector Operators. In: Proceedings of CIMCA 2005 (International Conference on Computational Intelligence for Modelling, Control and Automation), Vienna (Austria), pp. 295–300 (2005) ISBN 0-7695-2504-0
50. Van der Weken, D., De Witte, V., Nachtegael, M., Schulte, S., Kerre, E.E.: Fuzzy Similarity Measures for Colour Images. In: Proceedings of CIS-RAM 2006 (IEEE International Conferences on Cybernetics & Intelligent Systems and Robotics, Automation & Mechatronics), Bangkok (Thailand), pp. 806–810 (2006) ISBN 1-4244-0023-6, IEEE Catalog number 06EX1248C

51. Van der Weken, D., De Witte, V., Nachtegael, M., Schulte, S., Kerre, E.E.: A Component-based and Vector-based Approach for the Construction of Quality Measures for Colour Images In: Proceedings of IPMU 2006 (International Conference on Information Processing and Management of Uncertainty in Knowledge-Based Systems), Paris (France), pp. 1548–1555 (2006)
52. Van der Weken, D., De Witte, V., Nachtegael, M., Schulte, S., Kerre, E.E.: Colour Image Comparison Using Vector Operators. In: Bustince, H., Herrera, F., Montero, J. (eds.) Fuzzy Sets and Their Extensions: Representation, Aggregation and Models. Intelligent Systems from Decision Making to Data Mining, Web Intelligence and Computer Vision, pp. 645–662. Springer, Heidelberg (2007) ISBN 978-3-540-73722-3
53. Nachtegael, M., Schulte, S., De Witte, V., Mélange, T., Kerre, E.E.: Color Image Retrieval Using Fuzzy Similarity Measures and Fuzzy Partitions. In: Proceedings of ICIP 2007 (14th International Conference on Image Processing, San Antonio (United States), paper on CD, ISBN 1-4244-1437-7, ISSN 1522-4880
54. Nachtegael, M., Schulte, S., De Witte, V., Mélange, T., Kerre, E.E.: Image Similarity: from Fuzzy Sets to Color Image Applications. In: Qiu, G., Leung, C., Xue, X.-Y., Laurini, R. (eds.) VISUAL 2007. LNCS, vol. 4781, pp. 26–37. Springer, Heidelberg (2007)
55. Nachtegael, M., Schaefer, G., Mélange, T., Kerre, E.E.: Fuzzy Image Retrieval: a Selection of Recent Advances. In: Proceedings of IPCV 2009 (4th International Conference on Image Processing, Computer Vision and Pattern Recognition), vol. II, pp. 864–869. CSREA Press (2009) ISBN 1-60132-117-1 en 1-60132-118-X
56. Van De Ville, D., Nachtegael, M., Van der Weken, D., Philips, W., Lemahieu, I., Kerre, E.: A new Fuzzy Filter for Gaussian Noise Reduction. In: Visual Communications and Image Processing, Proceedings of SPIE, vol. 4310, pp. 1–9. SPIE - The international Society for Optical Engineering, United States (2001)
57. Van De Ville, D., Nachtegael, M., Van der Weken, D., Kerre, E.E., Philips, W., Lemahieu, I.: Noise Reduction by Fuzzy Image Filtering. IEEE Transactions on Fuzzy Systems 11(4), 429–436 (2003)
58. Nachtegael, M., Van der Weken, D., Van De Ville, D., Kerre, E., Philips, W., Lemahieu, I.: An Overview of Classical and Fuzzy-classical Filters for Noise Reduction In: Proceedings of FUZZ-IEEE 2001 (10th International IEEE Conference on Fuzzy Systems), The Institute of Electrical and Electronics Engineers, Melbourne, pp. 3–6 (2001) (CD) ISBN: 0-7803-7295-6.
59. Nachtegael, M., Van der Weken, D., Van De Ville, D., Kerre, E., Philips, W., Lemahieu, I.: An Overview of Fuzzy Filters for Noise Reduction. In: Proceedings of FUZZ-IEEE 2001 (10th International IEEE Conference on Fuzzy Systems), The Institute of Electrical and Electronics Engineers, Melbourne, pp. 7–10 (2001) (CD) ISBN: 0-7803-7295-6.
60. Nachtegael, M., Van der Weken, D., Van De Ville, D., Kerre, E., Philips, W., Lemahieu, I.: A Comparative Study of Classical and Fuzzy Filters for Noise Reduction. In: Proceedings of FUZZ-IEEE 2001 (10th International IEEE Conference on Fuzzy Systems), The Institute of Electrical and Electronics Engineers, Melbourne, pp. 11–14 (CD) (2001) ISBN: 0-7803-7295-6
61. Nachtegael, M., Van der Weken, D., Kerre, E., Van De Ville, D., Lemahieu, I., Philips, W.: On the Evaluation of Fuzzy Filters. In: Proceedings of SPS 2002 (3rd IEEE Benelux Signal Processing Symposium), Leuven, Belgium, pp. 97–100 (2002)
62. Nachtegael, M., Schulte, S., Van der Weken, D., De Witte, V., Kerre, E.E.: Fuzzy Filters for Noise Reduction: the Case of Gaussian Noise. In: Proceedings of FUZZ-IEEE 2005 (International IEEE Conference on Fuzzy Systems), p. 33. IEEE, Verenigde Staten (2005) (abstract booklet, ISBN: 0-7803-9158-6, ISSN: 1098-7584) + pp. 201–206 (paper on CD)

63. Nachtegael, M., Schulte, S., Van der Weken, D., De Witte, V., Kerre, E.E.: Gaussian Noise Reduction in Greyscale Images. International Journal of Intelligent Systems Technologies and Applications 1(3/4), 211–233 (2006)
64. Schulte, S., Huysmans, B., Pizurica, A., Kerre, E.E., Philips, W.: A New Fuzzy-Based Wavelet Shrinkage Image Denoising Technique. In: Blanc-Talon, J., Philips, W., Popescu, D., Scheunders, P. (eds.) ACIVS 2006. LNCS, vol. 4179, pp. 12–23. Springer, Heidelberg (2006)
65. Schulte, S., Nachtegael, M., De Witte, V., Van der Weken, D., Kerre, E.E.: A Fuzzy Impulse Noise Detection and Reduction Method. IEEE Transactions on Image Processing 15(5), 1153–1162 (2006)
66. Nachtegael, M., Schulte, S., Van der Weken, D., De Witte, V., Kerre, E.E.: Fuzzy Filters for Noise Reduction: the Case of Impulse Noise. In: Proceedings of SCIS & ISIS 2004 (Joint 2nd International Conference on Soft Computing and Intelligent Systems and 5th International Symposium on Advanced Intelligent Systems), Keio University, Japan, p. 89 (abstract booklet) + paper on CD (6 pages)
67. Nachtegael, M., Schulte, S., Van der Weken, D., De Witte, V., Kerre, E.E.: Do Fuzzy Techniques offer an Added Value for Noise Reduction in Images? In: Blanc-Talon, J., Philips, W., Popescu, D.C., Scheunders, P. (eds.) ACIVS 2005. LNCS, vol. 3708, pp. 658–665. Springer, Heidelberg (2005)
68. Schulte, S., Nachtegael, M., De Witte, V., Van der Weken, D., Kerre, E.E.: Fuzzy Impulse Noise Reduction Methods for Color Images. In: Proceedings of FUZZY DAYS 2006 (International Conference on Computational Intelligence), Dortmund (Germany), pp. 711–720 (2006)
69. Schulte, S., Nachtegael, M., De Witte, V., Van der Weken, D., Kerre, E.E.: A New Two Step Color Filter for Impulse Noise. In: Proceedings of the 11th Zittau Fuzzy Colloquium, Hochschule Zittau/Görlitz, Germany, pp. 185–192 (2004) ISBN: 3-9808089-5-5
70. Schulte, S., De Witte, V., Van der Weken, D., Nachtegael, M., Kerre, E.E.: Fuzzy Two-Step Filter for Impulse Noise Reduction from Color Images. IEEE Transactions on Image Processing 15(11), 3567–3578 (2006)
71. Schulte, S., De Witte, V., Nachtegael, M., Van der Weken, D., Kerre, E.E.: A Novel Histogram Based Fuzzy Restoration Method for Colour Images. In: Blanc-Talon, J., Philips, W., Popescu, D.C., Scheunders, P. (eds.) ACIVS 2005. LNCS, vol. 3708, pp. 626–633. Springer, Heidelberg (2005)
72. Schulte, S., De Witte, V., Nachtegael, M., Van der Weken, D., Kerre, E.E.: Histogram-Based Fuzzy Colour Filter for Image Restoration. Image and Vision Computing 25(9), 1377–1390 (2007)
73. Schulte, S., De Witte, V., Nachtegael, M., Van der Weken, D., Kerre, E.E.: A New Fuzzy Filter for the Reduction of Randomly Valued Impulse Noise. In: Proceedings of ICIP 2006 (13th International Conference on Image Processing), Atlanta (United States), pp. 1809–1812 (2006) Paper also on CD, ISBN 1-4244-0481-9, ISSN 1522-4880
74. Schulte, S., De Witte, V., Nachtegael, M., Van der Weken, D., Kerre, E.E.: Fuzzy Random Impulse Noise Reduction Method. Fuzzy Sets and Systems 158, 270–283 (2007)
75. Schulte, S., De Witte, V., Nachtegael, M., Mélange, T., Kerre, E.E.: A New Fuzzy Additive Noise Reduction Method. In: Kamel, M.S., Campilho, A. (eds.) ICIAR 2007. LNCS, vol. 4633, pp. 12–23. Springer, Heidelberg (2007)
76. Basu, A., Nachtegael, M.: OWA Filters: A Robust Filtering Method and its Application to Color Images. In: Proceedings of IEEE-CIIP 2009 (IEEE Symposium on Computational Intelligence for Image Processing), paper on CD, pp. 8–13 (2009) ISBN 978-1-4244-2760-4, part of the Proceedings of the 2009 IEEE Symposium Series on Computational Intelligence, ISBN 978-1-4244-2783-3

77. Mélange, T., Zlokolica, V., Schulte, S., De Witte, V., Nachtegael, M., Pizurica, A., Kerre, E.E., Philips, W.: A New Fuzzy Motion and Detail Adaptive Video Filter. In: Blanc-Talon, J., Philips, W., Popescu, D., Scheunders, P. (eds.) ACIVS 2007. LNCS, vol. 4678, pp. 640–651. Springer, Heidelberg (2007)
78. Mélange, T., Nachtegael, M., Kerre, E.E., Zlokolica, V., Schulte, S., De Witte, V., Pizurica, A., Philips, W.: Video Denoising by Fuzzy Motion and Detail Adaptive Averaging. Journal of Electronic Imaging 17(4), 43005-01–43005-19 (2008)
79. Mélange, T., Nachtegael, M., Kerre, E.E.: A Vector Based Fuzzy Filter for Colour Image Sequences. In: Proceedings of IPMU 2008 (12th International Conference on Information Processing and Management of Uncertainty in Knowledge-Based Systems), paper on CD (2008) ISBN 978-84-612-3061-7
80. Mélange, T., Nachtegael, M., Kerre, E.E.: A Fuzzy Filter for the Removal of Gaussian Noise in Colour Image Sequences. In: Proceedings of IFSA-EUSFLAT 2009 (2009 International Fuzzy Systems Association World Congress), pp. 1474–1479 (2009) ISBN 978-989-95079-6-8
81. Mélange, T., Nachtegael, M., Kerre, E.E.: A Fuzzy Filter for Random Impulse Noise Removal From Video. In: Proceedings of NAFIPS 2010 Annual Conference of the North-American Fuzzy Information Processing Society (2010)
82. Mélange, T., Nachtegael, M., Kerre, E.E.: A Fuzzy Filter for the Removal of Random Impulse Noise in Colour Video. In: Proceedings of WCCI 2010 IEEE World Conference on Computational Intelligence (2010)
83. Kerre, E.E., Nachtegael, M. (eds.): Fuzzy Techniques in Image Processing. Studies in Fuzziness and Soft Computing, vol. 52, 429 pagina's. Springer, Heidelberg (2000)
84. Nachtegael, M., Van der Weken, D., Van De Ville, D., Kerre, E.E. (eds.): Fuzzy Filters for Image Processing. Studies in Fuzziness and Soft Computing, vol. 122, 386 pagina's. Springer, Heidelberg (2003)
85. Nachtegael, M., Van der Weken, D., Kerre, E.E., Philips, W. (eds.): Soft Computing in Image Processing: Recent Advances. Studies in Fuzziness and Soft Computing, vol. 210, 521 pagina's. Springer, Heidelberg (2006)
86. Hassanien, A.E., Nachtegael, M., Van der Weken, D., Nobuhara, H., Kerre, E.E. (eds.): Soft Computing in Image Processing, Special issue of the international journal Soft Computing - A Fusion of Foundations, Methodologies and Application, vol. 10(12), pp. 1115–1201. Springer, Heidelberg (2006)
87. Nachtegael, M., Kerre, E., Damas, S., Van der Weken, D. (eds.): Recent Advances in Soft Computing in Image Processing. Special issue of the international journal "International Journal of Approximate Reasoning" 50(1), 1–61 (2009)

# The Need to Use Fuzzy Extensions in Fuzzy Thresholding Algorithms

Humberto Bustince, Miguel Pagola, Edurne Barrenechea, and Javier Fernández

**Abstract.** In this chapter we present some recent applications of fuzzy extensions in image segmentation. First we review some basic concepts of Interval-valued fuzzy sets, which is the extension that is mainly used. Next we present the fuzzy thresholding algorithm and we discuss its main problem that leads to use the extensions of fuzzy sets. In section 3 we review some methods recently published that use extensions of fuzzy sets in image thresholding. Finally we show some experimental results comparing the classical fuzzy thresholding algorithm against the algorithms based on extensions of fuzzy sets.

## 1 Introduction to Extensions of Fuzzy Sets: Interval Valued Fuzzy Sets

From the beginning it was clear that fuzzy set theory [29] was an extraordinary tool for representing human knowledge. Nevertheless, Zadeh himself established (see [30]) that sometimes, in decision-making processes, knowledge is better represented by means of some generalizations of fuzzy sets. A key problem of representing the knowledge by means of fuzzy sets is to choose the membership function which best represents such knowledge.

Sometimes, it is appropriate to represent the membership degree of each element to the fuzzy set by means of an interval. From these considerations arises the extension of fuzzy sets called *theory of interval-valued fuzzy sets* (IVFSs), that is, fuzzy sets such that the membership degree of each element of the fuzzy set is given by a closed subinterval of the interval $[0,1]$. Hence, not only vagueness (lack of sharp class boundaries), but also a feature of uncertainty (lack of information) can be addressed intuitively.

These sets were first introduced in the 1970s. In May 1975 Sambuc (see [24]) presented in his doctoral thesis the concept of an interval-valued fuzzy set named

Humberto Bustince · Miguel Pagola · Edurne Barrenechea · Javier Fernández
Universidad Pública de Navarra, Pamplona, Spain
e-mail: miguel.pagola@unavarra.es

C. Cornelis et al. (Eds.): 35 Years of Fuzzy Set Theory, STUDFUZZ 261, pp. 219–235.
springerlink.com 

a Φ-fuzzy set. That same year, Zadeh [30] discussed the representation of type 2 fuzzy sets and its potential in approximate reasoning.

The concept of a *type 2 fuzzy set* was introduced by Zadeh [30] as a generalization of an ordinary fuzzy set. Type 2 fuzzy sets are characterized by a fuzzy membership function, that is, the membership value for each element of the set is itself a fuzzy set in $[0,1]$.

Formally, given the referential set $U$, a type 2 fuzzy set is defined as an object $\overline{\overline{A}}$ which has the following form:

$$\overline{\overline{A}} = \{(u,x,\mu_u(x))|u \in U, x \in [0,1]\},$$

where $x \in [0,1]$ is the primary membership degree of $u$ and $\mu_u(x)$ is the secondary membership level, specific to a given pair $(u,x)$.

One year later, Grattan-Guinness [17] established a definition of an interval-valued membership function. In that decade interval-valued fuzzy sets appeared in the literature in various guises and it was not until the 1980s, that the importance of these sets, as well as their name, was definitely established.

A particular case of a type 2 fuzzy set is an *interval type 2 fuzzy set* (see [20]–[21]). An interval type 2 fuzzy set $\overline{\overline{A}}$ in $U$ is defined by

$$\overline{\overline{A}} = \{(u,A(u),\mu_u(x))|u \in U, A(u) \in L([0,1])\},$$

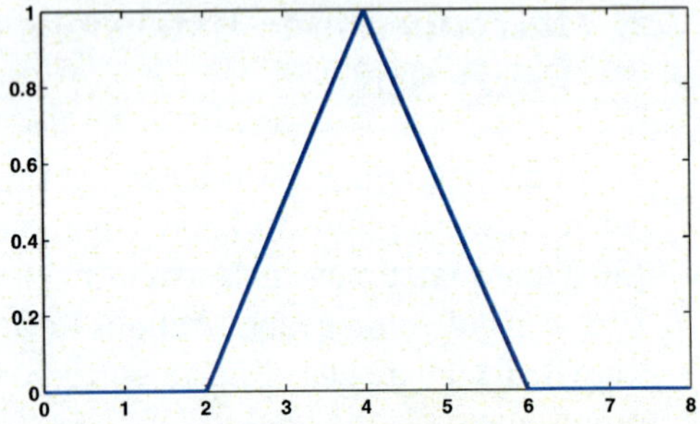


**Fig. 1** Fuzzy membership function.

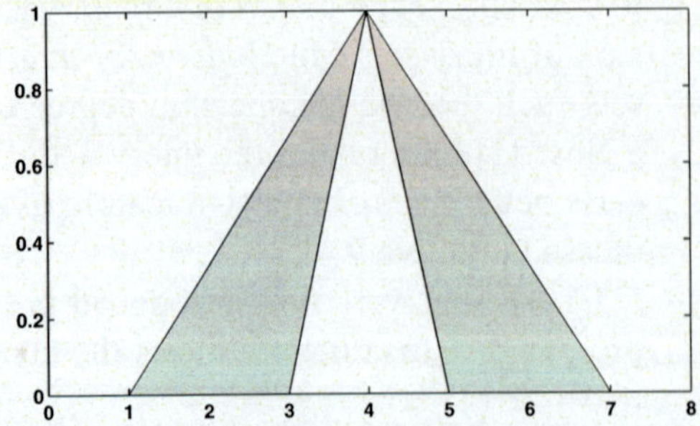


**Fig. 2** Interval valued fuzzy membership function. Lower bound and upper bound

where $A(u)$ is a closed subinterval of $[0,1]$, and the function $\mu_u(x)$ represents the fuzzy set associated with the element $u \in U$ obtained when $x$ covers the interval $[0,1]$; $\mu_u(x)$ is given in the following way:

$$\mu_u(x) = \begin{cases} a \text{ if } \underline{A}(u) \leq x \leq \overline{A}(u) \\ 0 \text{ otherwise} \end{cases},$$

where $0 \leq a \leq 1$. It turns out that an interval type 2 fuzzy set is the same as an IVFS if we take $a = 1$.

Another important extension of fuzzy set theory is the theory of Atanassov's intuitionistic fuzzy sets ([1], [2]). Atanassov's intuitionistic fuzzy sets (A-IFSs) assign to each element of the universe not only a membership degree, but also a nonmembership degree, which is less than or equal to 1 minus the membership degree.

An Atanassov's intuitionistic fuzzy set (A-IFS) on $U$ is a set

$$\hat{A} = \{(u, \mu_{\hat{A}}(u), \nu_{\hat{A}}(u)) | u \in U\},$$

where $\mu_{\hat{A}}(u) \in [0,1]$ denotes the membership degree and $\nu_{\hat{A}}(u) \in [0,1]$ the nonmembership degree of $u$ in $\hat{A}$ and where, for all $u \in U$, $\mu_{\hat{A}}(u) + \nu_{\hat{A}}(u) \leq 1$.

In [1] Atanassov established that every Atanassov intuitionistic fuzzy set $\hat{A}$ on $U$ can be represented by an interval-valued fuzzy set $A$ given by

$$\begin{aligned} A : U &\rightarrow L([0,1]) \\ u &\rightarrow [\mu_{\hat{A}}(u), 1 - \nu_{\hat{A}}(u)], \qquad \text{for all } u \in U. \end{aligned}$$

Using this representation, Atanassov proposed in 1983 that Atanassov's intuitionistic fuzzy set theory was equivalent to the theory of interval-valued fuzzy sets. This equivalence was proven in 2003 by Deschrijver and Kerre [12]. Therefore, from a mathematical point of view, the results that we obtain for IVFSs are easily adaptable to A-IFSs and vice versa. Nevertheless, we need to point out that, conceptually, the two types of sets are totally different. This is made clear when applications of these sets are constructed (see [28]).

In 1993, Gau and Buehrer introduced the concept of *vague sets* [16]. Later, in 1996, it was proven that vague sets are in fact A-IFSs [5].

A compilation of the sets that are equivalent (from a mathematical point of view) to interval-valued fuzzy sets can be found in [13]. Two conclusions are drawn from this study:

1.- Interval-valued fuzzy sets are equivalent to A-IFSs (and therefore vague sets), to grey sets (see [11]) and to $L$-fuzzy set in Goguen's sense with respect to a special lattice $L([0,1])$.
2.- IVFSs are a particular case of probabilistic sets (see [14]), of soft sets (see [3]), of Atanassov's interval-valued intuitionistic fuzzy sets and evidently of Type 2 fuzzy sets.

## 2 Interval Valued Fuzzy Sets and Image Segmentation

Image segmentation is the first step in image analysis and pattern recognition. It is a critical and essential component of image analysis and/or pattern recognition system and is one of the most difficult tasks in image processing, that can determine the quality of the final result of the system.

The goal of image segmentation is the partition of an image in different areas or regions.

**Definition 1.** *Segmentation is grouping pixels into regions such that*

1. $\cup_{i=1}^{k} P_i =$ *Entire image* ($\{P_i\}$ *is an exhaustive partitioning).*
2. $P_i \cap P_j = 0, i \neq j$ ($\{P_i\}$ *is an exclusive partitioning).*
3. *Each region* $P_i$ *satisfies a predicate; that is, all points of the partition have some common property.*
4. *Pixels belonging to adjacent regions, when taken jointly, do not satisfy the predicate.*

There exist three different approaches using fuzzy methods:

- Histogram thresholding.
- Feature space clustering.
- Rule based systems.

### 2.1 Thresholding

One of the earlier papers of fuzzy thresholding is [23] in 1983. The idea behind the fuzzy thresholding is to first transfer the selected image feature into a fuzzy subset by means of a proper membership function and then select and optimize a global or local fuzzy measure to attain the goal of image segmentation.

In [23] the authors used the S-function to fuzzify the image. They minimize the entropy in such a way that the final segmented image is the one which has less doubtful pixels.

Said membership function represents the brightness set within the image. The basic idea of using this membership function is that, if we take the value of a parameter as the threshold value, the dark pixels should have low membership degrees, and on the contrary, brighter pixels should have high membership degrees. The pixels with membership function near 0.5 should be the ones that are not clearly classified. Therefore the set with less entropy is the set with less amount of pixels with uncertain membership (around 0.5).

Since the first works that used S-functions up to now the problem is: *finding the membership function that will accurately represent the membership degree of each pixel to each region of the image* (in the case of two regions: the background and the object).

If the expert knows the exact membership degree of each pixel to each region of the image, the problem is solved. This does not normally happen, that is, there are pixels which the expert does not know whether they belong to the background or to the object.

Depending on the type of image we are working with, it is necessary to use one membership function or another. That is, in order to obtain a good segmentation the user or expert must choose the membership function that will best represent that image.

Therefore, in any segmentation algorithm there is always a degree of ignorance in the choice of the membership function made by the user. It is established (see [15, 18]) that the membership function must always try to represent:

- The similarity of the pixels within the regions.
- The connectivity of the pixels within the regions.
- The difference between the pixels of different regions.

We do not know how to correctly express these three properties in the membership function. For this reason there is always an uncertainty associated with thresholding algorithms. According to Klir and Wierman [19] "Uncertainty has a pivotal role in any efforts to maximize the usefulness of systems models ... uncertainty becomes very valuable when considered in connection to the other characteristics of systems models: a slight increase in uncertainty may often significantly reduce the complexity and, at the same time, increase the credibility of the model. Uncertainty is thus an important commodity in the modeling business, a commodity which can be traded for gains in the other essential characteristics of models".

In this sense Zadeh [30], Mendel [21], and other authors establish that the main characteristic of the extensions of fuzzy sets is their capacity to model uncertainty in the membership function selection.

In this work we are going to present some recent methods to add uncertainty by means of extensions of fuzzy sets into the thresholding process. All of these techniques are based on the classical fuzzy thresholding algorithm.

## 3 Fuzzy Algorithm of Image Thresholding and Its Extensions

The basic structure of the algorithms for the calculation of the threshold of an image that use fuzzy techniques (see [7, 15, 18, 22]) is composed of the following steps:

(a) Assign $L$ fuzzy sets $\tilde{Q}_t$ to each image $Q$. Each one associated to a level of intensity $t$, $(t = 0, 1, \cdots, L-1)$, of the grayscale $L$ used.

(b) Calculate the entropy of each one of the $L$ fuzzy sets $\tilde{Q}_t$ associated to $Q$.

(c) Take as *best threshold* $t$ the level of gray associated to the fuzzy set with lowest entropy.

Alg. Fuzzy

Membership functions are normally defined by the expert on the basis of his/her knowledge. The membership functions used in [15, 18] represent how similar the intensity of each pixel is to the mean of the intensities of the object or to the mean of the intensities of the background. This type of membership functions are the ones that provide the best results (see [22, 25]). By defining the membership functions in this way, the set with lowest entropy is the set that contains the greatest number of pixels around the mean of the intensities of the background and the mean of the intensities of the object.

In these conditions given an image $Q$ and an intensity threshold $t$ set, we have proposed a general method [7] of constructing the membership function of each intensity to the set $\tilde{Q}_t$ in the following way:

$$\mu_{\tilde{Q}_t}(q) = \begin{cases} F(REF(q, m_b(t)))\ if\ q \leq t \\ F(REF(q, m_o(t)))\ if\ q > t \end{cases} \quad (1)$$

Where $m_b(t)$ and $m_o(t)$ are given by the following expressions:

$$m_b(t) = \frac{\sum_{q=0}^{t} q h(q)}{\sum_{q=0}^{t} h(q)} \quad (2)$$

$$m_o(t) = \frac{\sum_{q=t+1}^{L-1} q h(q)}{\sum_{q=t+1}^{L-1} h(q)}. \quad (3)$$

$h(q)$ being the number of pixels of the image with intensity $q$.

Depending on the functions $F$ and $REF$ chosen we can construct most of the expressions of membership functions used in the fuzzy thresholding in the literature [18, 25].

Once the manner of constructing the membership function has been chosen, we calculate the $L$ fuzzy sets associated with each $t = 0, 1, \cdots, 255$. In the following step of the algorithm we calculate the entropy of each one of those sets, which is why we now present the definition of entropy and some expressions.

## 3.1 Examples of the Fuzzy Algorithm Using Different Membership Functions

Step (c) of the fuzzy algorithm establishes that once the entropy is calculated for all the fuzzy sets, we must take as threshold the value of $t$ associated with the fuzzy set $\tilde{Q}_t$ with lowest entropy. Taking into account that we have constructed the membership function using $REFs$, then the membership value equal to 1 indicates that the value of the pixel is very close to the value of the mean of the intensities of the background or to the value of the mean of the intensities of the object. As a result the expert is sure of the membership, either to the background or to the object, of most pixels.

In the following example we prove experimentally that the solution to the algorithm depends on the membership function selected.

**Table 1** Different membership functions

| | | | |
|---|---|---|---|
| 1) | $F(x) = 0.5(1+x)$<br>$REF(x,y) = 1 - \|x-y\|$ | 2) | $F(x) = 0.5(1+x)$<br>$REF(x,y) = (1 - \|x-y\|)^2$ |
| 3) | $F(x) = 0.5(1+x)$<br>$REF(x,y) = 1 - \|x-y\|^2$ | 4) | $F(x) = 0.5(1+x)$<br>$REF(x,y) = (1 - \|x-y\|)^{0.5}$ |
| 5) | $F(x) = 1/(2-x)$<br>$REF(x,y) = 1 - \|x-y\|$ | 6) | $F(x) = 1/(2-x)$<br>$REF(x,y) = (1 - \|x-y\|)^2$ |
| 7) | $F(x) = 1/(2-x)$<br>$REF(x,y) = 1 - \|x-y\|^2$ | 8) | $F(x) = 1/(2-x)$<br>$REF(x,y) = (1 - \|x-y\|)^{0.5}$ |

*Example 1.* The membership functions are constructed from the eq. (1). On Table 1 we show the different functions $F$ and $REF$ taken. To calculate the entropy of fuzzy sets we use the eq. (4) with

$$E(\tilde{A}) = \sum_{x \in X} 1 - |1 - 2\mu_{\tilde{A}}(x)|^2 \tag{4}$$

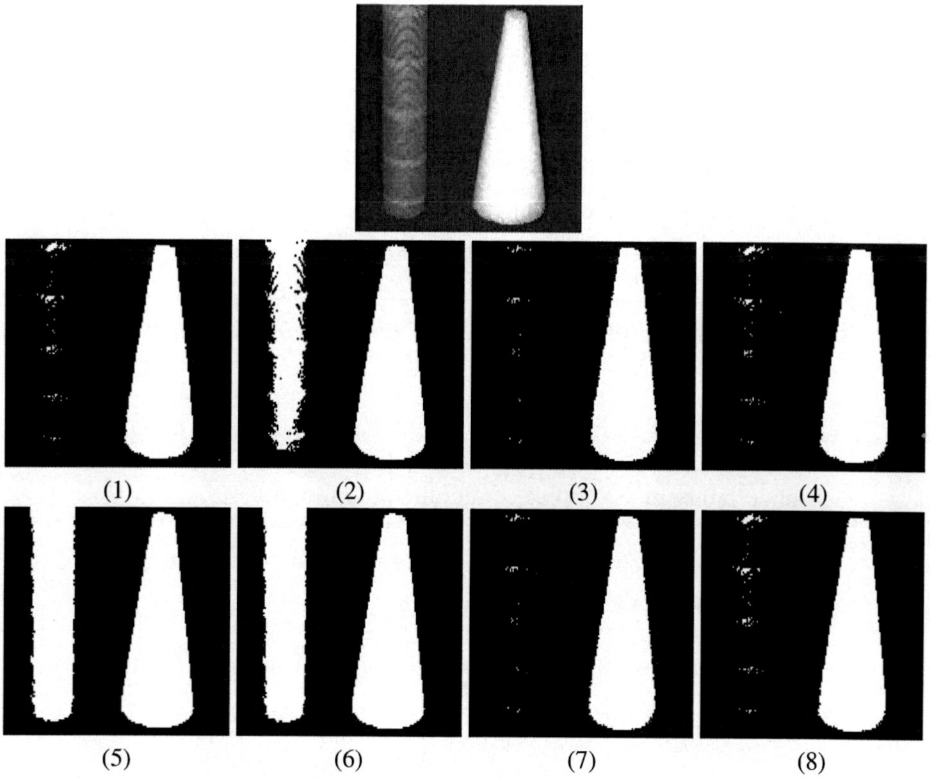

**Fig. 3** Different solutions to the fuzzy algorithm with different membership functions

In figure 3 we show the binary images obtained. From the visual analysis of these images we deduce that the results of the fuzzy algorithm depend on the membership function chosen (in this case functions (5) and (6) would be the most suitable solutions).

## 3.2 *Generalization of the Fuzzy Algorithm Using IVFS*

In 2005 Tizhoosh [26] presented a paper that uses Interval-valued fuzzy sets in image thresholding (we must point out that he tries to use type 2 fuzzy sets, however in the paper he only uses Interval-valued fuzzy sets [9]). His study is based on the modification of the classical fuzzy algorithm of Huang and Wang [18], so that he applies an $\alpha$ factor as interval generator to the membership function. Starting from a membership function, Tizhoosh obtains an interval valued fuzzy set that "contains" different membership functions and is useful for finding the threshold of an image. In this manner we can handle the inaccuracy of the selection of the membership function and obtain good results even though the best selection has not been made. That is why to each pixel we are going to assign two values:

- We will obtain the first one by changing the membership function chosen (constructed from the restricted equivalence functions), which we interpret as a pessimistic measure of the membership degree of said pixel to the set that represents the image.
- The second value, also obtained from the membership function chosen, we interpret as a optimistic measure of the membership degree of said pixel to the set that represents the image.

The structure of the algorithm is the following:

(a) Assign $L$ fuzzy sets $\tilde{Q}_t$ to each image $Q$. Each one associated to a level of intensity $t$, $(t = 0, 1, \cdots, L-1)$, of the grayscale $L$ used.

(b) Construct an Interval valued fuzzy set $Q_t$ with the parameters $\alpha$ and $\beta$ from each one of the $L$ fuzzy sets $\tilde{Q}_t$ associated to $Q$. It must reflect the uncertainty in the choice of membership function..

(c) Calculate the Interval valued entropy of each one of the $L$ interval valued fuzzy sets $Q_t$ associated to $Q$.

(d) Take as *best threshold* $t$ the level of gray associated to the fuzzy set with lowest IV entropy.

Alg. IVFS

So in this version the main fact is to construct the IVFSs in such a way that it represents the original shape of the membership function given by the expert and also brings the uncertainty of the election. The other different aspect is that we choose as threshold the $t$ associated with the set with minimum IV-entropy.

A generalization of the Tizhoosh algorithm can be made using two different parameters to estimate the uncertainty of the membership function, so we can construct an interval-valued fuzzy set in the following way:

$$A_{\alpha,\beta} = \{(x, [\mu_{\tilde{A}}^{\alpha}(x)\ \mu_{\tilde{A}}^{\frac{1}{\beta}}(x)] | x \in X\} \in IVFSs(X). \tag{5}$$

The verification that $A_{\alpha,\beta} \in IVFSs(X)$ is evident: $0 \leq \mu_{\tilde{A}}^{\alpha}(x) \leq \mu_{\tilde{A}}^{\frac{1}{\beta}}(x) \leq 1$. The parameters $\alpha$ and $\beta$ must be related with the ignorance of the expert in the membership function selection. Tizhoosh used only one parameter $\alpha$, so that the IVFS set he constructed is:

$$A_{\alpha} = \{(x, [\mu_{\tilde{A}}^{\alpha}(x)\ \mu_{\tilde{A}}^{\frac{1}{\alpha}}(x))] | x \in X\} \in IVFSs(X). \tag{6}$$

In [4] Bustince presents the general way of constructing $\mathscr{E}_F$. From all of the expressions presented we are going to use the simplest, Sambuc's indetermination index [24]. That is, for the set constructed in the eq. (5) the entropy is given by:

$$\mathscr{E}_F(Q_t) = \frac{1}{N \times M} \sum_{q=0}^{L-1} h(q)(\mu_{\tilde{Q}_t}^{1/\beta}(q) - \mu_{\tilde{Q}_t}^{\alpha}(q)). \tag{7}$$

The justification for choosing the set with lowest $\mathscr{E}_F$ is the following:

if $\mathscr{E}_F$ is close to zero, then we can consider that generally, for all $q$ this holds: $\mu_{\tilde{Q}_t}^{1/\beta}(q) \approx \mu_{\tilde{Q}_t}^{\alpha}(q)$.

By the way of constructing the membership functions we know that:

$$0.5 \leq \mu_{\tilde{Q}_t}(q) \leq 1 \text{for each } q \in \{0, \cdots, L-1\};$$

therefore $\mu_{\tilde{Q}_t}^{1/\beta}(q) \approx 1$, (as long as we take $\alpha \neq 1$ and $\beta \neq 1$). In these conditions, by the construction of $\tilde{Q}_t$ we have

$$\mu_{\tilde{Q}_t}(q) = \begin{cases} F(REF(q, m_b(t))) \text{ if } q \leq t \\ F(REF(q, m_o(t))) \text{ if } q > t \end{cases} \approx 1$$

By the condition: $F(x) = 1$ if and only if $x = 1$, we have that $\mu_{\tilde{Q}_t}(q)$ is close to one if $REF(q, m_b(t))$ when $q \leq t$ (or $REF(q, m_o(t))$ when $q > t$) is close to one; Bearing in mind the definition of function of $REF$ we have that this happens when q is very close to $m_b(t)$ or to $m_o(t)$ depending on the case. Furthermore we know that (see [15, 18]) that if $q$ is very close to $m_b(t)$ or to $m_o(t)$, then we are in the best situation for taking the threshold. Therefore if we take the set with lowest value of $\mathscr{E}_F$, by the reasoning above, then that IVFS is the one that best represents the image and the one with the minimum in that representation [9].

## 3.3 Modification of the Fuzzy Algorithm by Means of A-IFSs

Vlachos and Sergiaidis [28] also propose a modification of Tizhoosh's algorithm, using Atanassov's intuitionistic fuzzy sets (see [1]). Their basis are membership functions similar to Huang's but, instead of minimizing the entropy, the algorithm minimizes the divergence with set $\hat{\mathbf{1}}$ (see [10]). The structure of the intuitionistic algorithm is the same as Tizhoosh's. The construction of the intuitionistic fuzzy sets is done in the following way:

$$\mu_{\hat{A}}(g,t) = \lambda \mu_A(g,t) \tag{8}$$
$$\nu_{\hat{A}}(g,t) = (\hat{1} - \lambda \mu_A(g,t))^{\lambda}$$

With $\lambda \in [0,1]$, being $\hat{A}$ an intuitionistic fuzzy set, and the divergence:

$$D_{IFS}(\hat{A},\hat{1},t) = \sum_{g=0}^{L-1} h_A(g) \left( \mu_{\hat{A}(g,t)} ln \frac{2\mu_{\hat{A}}(g,t)}{1+\mu_{\hat{A}}(g,t)} + \nu_{\hat{A}}(g,t) ln2 + ln \frac{2}{1+\mu_{\hat{A}}(g,t)} \right) \tag{9}$$

In this case the uncertainty is represented by the hesitation index:

$$\pi_{\hat{A}}(g,t) = 1 - \mu_{\hat{A}}(g,t) - \nu_{\hat{A}}(g,t)$$

## 3.4 Construction of the IVFSs Using Ignorance Functions

In this case the classical fuzzy thresholding algorithm is modified due to the user should pick two functions, one to represent the background and another one to represent the object. We have chosen this representation since, in this way, the expert is able to get a better representation of the pixels for which he is not sure of their membership to the object or the background. In figure 4 we show two membership functions, one to represent the background and the other to represent the object.

As we have already said in the previous paragraph, we are going to represent the images by means of two different fuzzy sets. For this reason, in our proposed algorithm we introduce the concept of *ignorance function* $G_u$. Such functions are a way to represent the user's ignorance for choosing the two membership functions

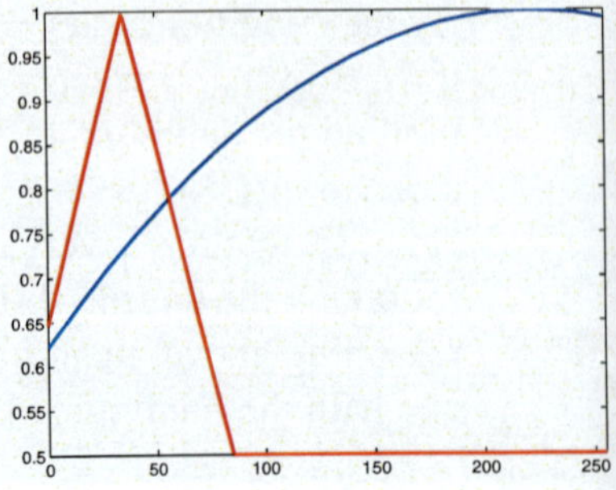


**Fig. 4** Two different membership functions to represent the background and the object with t=150

used to represent the image (object and background). Therefore, in our algorithm we will associate to each pixel three numerical values:

- A value for representing its membership to the background, which we will interpret as the expert's knowledge of the membership of the pixel to the background.
- A value for representing its belongingness to the object, which we will interpret as the expert's knowledge of the membership of the pixel to the object.
- A value for representing the expert's ignorance of the membership of the pixels to the background or to the object. This ignorance hinders the expert from making an exact construction of the membership functions described in the first two items and therefore it also hinders the proper construction of step $(a)$ of the fuzzy algorithm. The lower the value of ignorance is, the better the membership function chosen to represent the membership of that pixel to the background and the one chosen to represent the membership to the object will be. Evidently, there will be pixels of the image for which the expert will know exactly their membership to the background or to the object but there will also be pixels for which the expert is not able to determine if they belong to the background or to the object.

Under these conditions, if the value of the function of ignorance ($G_u$) for a certain pixel is zero, it means that the expert is positively sure about the belongingness of the pixel to the background or to the object. However, if the expert does not know at all whether the pixel belongs to the background or to the object he must represent its membership to both with the value 0.5, and under these conditions we can say that the expert has *total* ignorance regarding the membership of the pixel to the background and the membership of the same pixel to the object.

In [8] a methodology is proposed to construct ignorance functions. In the following example we show an ignorance function that can be constructed by means of the t-norm minimum.

*Example 2.* Using the t-norm $T = T_M$

$$G_u(x,y) = \begin{cases} 2 \cdot min(1-x, 1-y) \text{ if } min(1-x,1-y) \leq 0.5 \\ \frac{1}{2 \cdot min(1-x,1-y)} \text{ otherwise} \end{cases}$$

is a continuous ignorance function.

Also Ignorance functions can be constructed from functions like the geometric mean that are not t-norms (please see [8]).

*Example 3.* If we take $\varphi(x) = \sqrt{x}$ for all $x \in [0,1]$ we recover the following ignorance function:

$$G_u(x,y) = \begin{cases} 2\sqrt{(1-x)\cdot(1-y)} \text{ if } (1-x)\cdot(1-y) \leq 0.25 \\ \frac{1}{2\sqrt{(1-x)\cdot(1-y)}} \text{ otherwise} \end{cases}$$

In the following proposition we can see how we can construct IVFS by means of the ignorance function.

**Proposition 1.** *Let $\tilde{Q}_B$ and $\tilde{Q}_O$ be the fuzzy sets associated to the background and the object of the image built by an expert. Let $G_u$ be the ignorance function associated to the construction of the said fuzzy sets. Under these conditions, if we define*

$$\Phi : \mathscr{FS}s(U) \times \mathscr{FS}s(U) \longrightarrow \mathscr{IVFS}s(U) \text{ given by}$$
$$\Phi(\tilde{Q}_B, \tilde{Q}_O) = \{(u, [M_L(u), M_U(u)]) | u \in U\} \text{ such that}$$
$$[M_L(u), M_U(u)] = [G_u(0.5, 0.5) - G_u(\mu_{\tilde{Q}_B}(u), \mu_{\tilde{Q}_O}(u)), G_u(0.5, 0.5)],$$

*then for each membership $[M_L(u), M_U(u)]$ defined by $\Phi$ the following equation holds:*

$$W([M_L(u), M_U(u)]) = G_u(\mu_{\tilde{Q}_B}(u), \mu_{\tilde{Q}_O}(u))$$

If the IV entropy $\mathscr{E}_F(Q)$ that represents the total influence of the ignorance in the construction of fuzzy sets $Q_{Bt}$ and $Q_{Ot}$ is such that it verifies:

$$\mathscr{E}_F \rightarrow 0$$

then bearing in mind the properties required to $\mathscr{M}$ we have

$$G_u(\mu_{Q_{Bt}}(q), \mu_{Q_{Ot}}(q)) \rightarrow 0 \text{ for all } q \in \{0, \cdots, L-1\}.$$

By the property "$G_u(x,y) = 0$ if and only if $x = 1$ or $y = 1$" that is required to the functions $G_u$, we have $\mu_{Q_{Bt}}(q) \rightarrow 1$ or $\mu_{Q_{Ot}}(q) \rightarrow 1$, therefore $q$ is very close to $m_B(t)$ ($m_O(t)$), that is, we are certain that the pixels with intensity $q$ belong to the background (object). This is due to the following reasoning:

1. If $\mu_{Q_{Bt}}(q) \rightarrow 1$, then $REF(\frac{q}{L-1}, \frac{m_B(t)}{L-1}) \rightarrow 1$, therefore $q \approx m_B(t)$. In this case the pixels with intensity $q$ are such that their intensity is very close to the average intensity of the pixels that represent the background. This fact enables us to assure that the pixel in question belongs to the background.
2. If $\mu_{Q_{Ot}}(q) \rightarrow 1$, then $REF(\frac{q}{L-1}, \frac{m_O(t)}{L-1}) \rightarrow 1$, therefore $q \approx m_O(t)$. In this case the pixels with intensity $q$ are such that their intensity is very close to the average intensity of the pixels that represent the object. This fact enables us to assure that the pixel in question belongs to the object.

## 4 Examples and Experimental Results

In this section we are going to present some results obtained for the different algorithms presented before. Recalling the objective of these algorithms is that he IVFS algorithm produces solutions comparable in quality to those of the fuzzy algorithm when an adequate membership function has been used and surpasses the fuzzy algorithm when the membership function chosen does not represent the image properly. In this last case we deduce that there exists great uncertainty in the election of the membership function. In the first experiment we calculate the threshold using the fuzzy algorithm using 8 different membership functions (table 1) and then we calculate the threshold with the IVFS algorithm with $\alpha = 2$ and $\beta = 2$ for the set

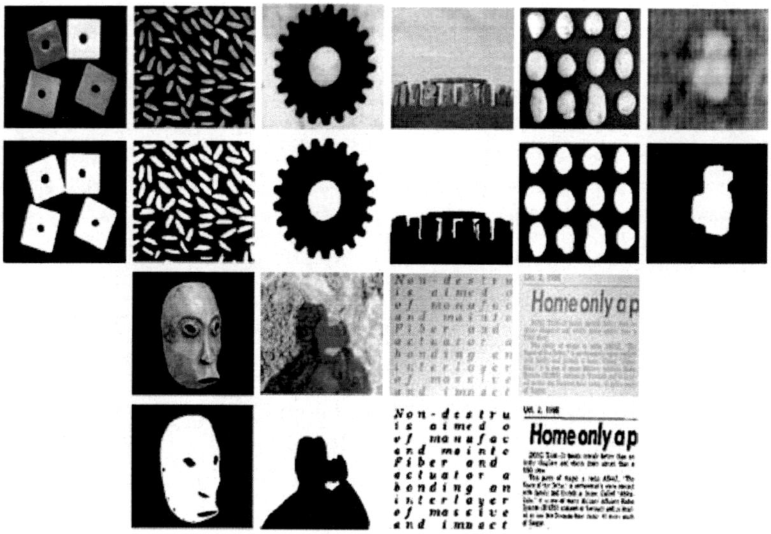

**Fig. 5** Natural images set.

of natural images depicted in figure 5. On Table 2 we present the percentage of wrongly classified pixels in each case. We observe again that for a single image the fuzzy algorithm returns different solutions depending on the membership function we select. The lower the number of wrongly classified pixels, the better the solution and therefore the better is the membership function for that image. In the last row the mean of the values of the ten images is shown, this way we can choose the best membership function for this set of images. In this case the best membership function is $\mu_7$.

On the next table, Table 3 we show the values obtained with the IVFS algorithm using in every case the same 8 membership functions and the parameters $\alpha = 2$ and $\beta = 2$ in order to construct IVFSs.

The values in which the mean of error of the IVFS algorithm is less than the fuzzy case are highlighted in bold type. We observe in this experiment that the IVFS algorithm, with the exception of a membership function $\mu_7$ (which is the best), is

**Table 2** Results of the Fuzzy algorithm

| Image | $\mu_1$ | $\mu_2$ | $\mu_3$ | $\mu_4$ | $\mu_5$ | $\mu_6$ | $\mu_7$ | $\mu_8$ |
|---|---|---|---|---|---|---|---|---|
| 0 | 2.5763 | 2.8441 | 5.7520 | 2.6247 | 2.8441 | 3.3262 | 3.9639 | 2.5916 |
| 1 | 7.2416 | 7.2135 | 7.7773 | 7.2569 | 7.2365 | 7.5502 | 7.7543 | 7.2238 |
| 2 | 1.6334 | 1.6234 | 1.9852 | 1.6611 | 1.6334 | 1.7314 | 2.0028 | 1.6234 |
| 3 | 4.2722 | 4.4088 | 4.1717 | 4.2722 | 4.4088 | 4.2722 | 4.1717 | 4.4088 |
| 4 | 4.2776 | 4.8056 | 4.1526 | 4.0379 | 4.8056 | 6.3004 | 4.1960 | 4.3388 |
| 5 | 3.4359 | 3.1119 | 4.3592 | 3.5915 | 3.1119 | 2.9538 | 4.3592 | 3.3211 |
| 6 | 3.2792 | 3.1740 | 3.9868 | 3.2792 | 3.1740 | 3.0458 | 3.8842 | 3.2792 |
| 7 | 10.4274 | 10.4274 | 10.9065 | 10.4274 | 10.4274 | 10.4274 | 10.9065 | 10.4274 |
| 8 | 39.8728 | 42.2905 | 30.8456 | 38.6832 | 42.2905 | 44.4647 | 30.8456 | 41.0650 |
| 9 | 26.5357 | 28.1151 | 20.3671 | 25.9384 | 28.9945 | 29.9303 | 20.7492 | 27.3382 |
| mean | 10.3552 | 10.8014 | 9.4304 | 10.1772 | 10.8927 | 11.4002 | **9.2833** | 10.5617 |

**Table 3** Results of the IVFS algorithm (a)

| Image | $\mu_1$ | $\mu_2$ | $\mu_3$ | $\mu_4$ | $\mu_5$ | $\mu_6$ | $\mu_7$ | $\mu_8$ |
|---|---|---|---|---|---|---|---|---|
| 0 | 2.5661 | 2.7217 | 5.7520 | 2.6247 | 2.7625 | 3.1961 | 3.9639 | 2.5763 |
| 1 | 7.2391 | 7.2340 | 7.7773 | 7.2620 | 7.1906 | 7.3666 | 7.7543 | 7.2416 |
| 2 | 1.6284 | 1.6108 | 1.9852 | 1.6962 | 1.6058 | 1.6636 | 2.0028 | 1.6334 |
| 3 | 4.2722 | 4.4088 | 4.1717 | 4.2722 | 4.4088 | 4.4088 | 4.1717 | 4.2722 |
| 4 | 4.2343 | 4.7521 | 4.1526 | 3.9894 | 4.7521 | 6.1830 | 4.1960 | 4.2776 |
| 5 | 3.4359 | 3.2165 | 4.3592 | 3.5915 | 3.1119 | 2.9767 | 4.3592 | 3.4359 |
| 6 | 3.2792 | 3.2176 | 3.9868 | 3.2792 | 3.1971 | 3.0817 | 3.9868 | 3.2792 |
| 7 | 10.4274 | 10.4274 | 10.9065 | 10.4274 | 10.4274 | 10.4274 | 10.9065 | 10.4274 |
| 8 | 39.8728 | 42.2905 | 29.9969 | 38.6832 | 42.2905 | 43.4468 | 30.8456 | 39.8728 |
| 9 | 26.5357 | 27.3382 | 20.3671 | 25.9384 | 28.1151 | 29.9303 | 20.7492 | 26.5357 |
| mean | **10.3491** | **10.7217** | **9.3455** | **10.1764** | **10.7862** | **11.2681** | 9.2936 | **10.3552** |

better than the fuzzy case. That is, if we choose a wrong membership function the IVFS algorithm provides a better result.

Similar results can be obtained for the A-IFS algorithm. Please see [28] for other results in pattern recognition.

**Case of IVFS constructed by ignorance functions**

Finally we evaluate the performance of the algorithm that uses ignorance functions in prostate ultrasound images. For each image the expert has provided a point at the center of the prostate making object detection/extraction easier. The image is then filtered via median filter in $5 \times 5$ neighborhoods. Further, selective contrast enhancement as described in [27] has been applied to increase the image quality.

We must take into account that, since ultrasound images depend on the particular settings of the machine is very important that our algorithm gives good solutions even if some membership functions that do not represent accurately the background and the prostate are chosen.

In this experiment we take 8 different membership functions. $\mu(q) = 1 - |q - m_z(t)|^{\lambda}$ with $\lambda = 0.1, 0.3, 0.8, 1, 1.3, 2, 6, 15$, where $z$ could denote the object or the background. For all the images of each set, we execute the Generalized fuzzy

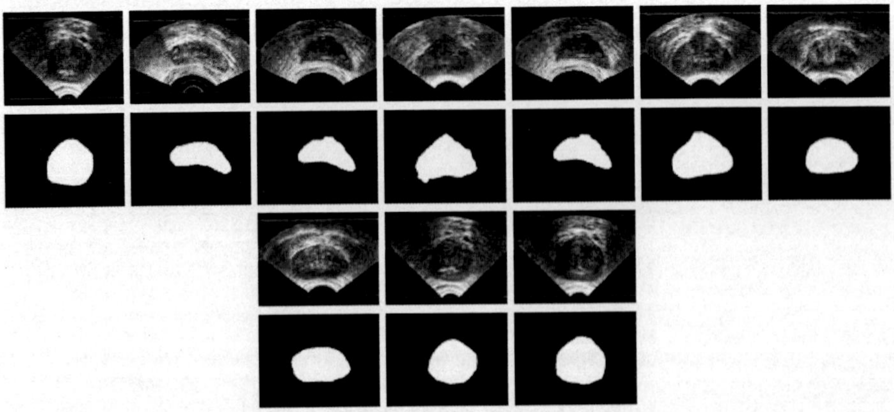

**Fig. 6** Set of Ultrasound prostate images with their ideal segmentation made by an expert radiologist.

algorithm with two membership functions using the classical fuzzy algorithm with all the possible combinations of the membership functions; i.e., $8^2 = 64$. Then we execute the ignorance based algorithm using the geometric mean based ignorance function for all the cases, so we'll obtain 64 different solutions.

To interpret the results of this experiment we study the graphic in Fig. 7. This graphic is obtained in the following way:

1. We arrange all the cases from the smallest to the biggest percentage of badly classified pixels (error) in the solution of the fuzzy algorithm.
2. For each pair of membership functions, we calculate the error obtained with the ignorance based algorithm.

The crosses represents the error we get with the ignorance based algorithm, and the dotted line, the error we get with the fuzzy algorithm. (If the dots are under the cross, it means that for that pair of membership functions the error of the ignorance based algorithm is smaller than the error of the fuzzy algorithm). Observe that:

1. For the pairs of membership functions such that the fuzzy algorithm solution is good (small error), the ignorance based algorithm does not provide better results.
2. If the error we get with the fuzzy algorithm begins to be high (i.e., if we have used bad-chosen membership functions), then the result of the ignorance based algorithm improves the other algorithm's result.

We observe that in the fuzzy algorithm case there are around 40% of the cases on which the membership functions do not represent correctly areas of the image and a very high error is obtained. However, the IVFS algorithm gets a good solution for almost all the cases.

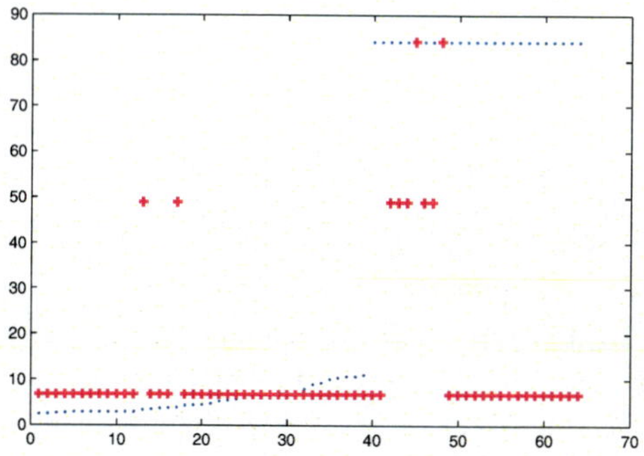


**Fig. 7** Percentage of error in all cases of first prostate ultrasound image in Fig. 7 with ignorance function using the geometric mean.

If we analyze the table 4 of mean errors for all the cases, we see that in the case of real ultrasound images ignorance based algorithm has a total mean error less than the fuzzy algorithm.

**Table 4** Average mean of the error of all cases in any image of the set of ultrasound images

| Image | Fuzzy | Geometric mean |
|---|---|---|
| 1 | 36,3 | 13,9 |
| 2 | 70,1 | 84,4 |
| 3 | 75,6 | 62,2 |
| 4 | 29,3 | 22,8 |
| 5 | 74,2 | 61,0 |
| 6 | 41,2 | 44,7 |
| 7 | 42,9 | 43,3 |
| 8 | 52,6 | 46,3 |
| 9 | 33,1 | 12,4 |
| 10 | 37,4 | 12,0 |
| Mean | 49.3 | 40.3 |

## 5 Conclusions and Future Research

A key problem of the fuzzy thresholding algorithm is the accurate election of the membership function. In this chapter we have presented methods which use extensions fuzzy sets as a tool to represent the uncertainty presented in the election of the correct membership function. The experimental results shown in section 4 allow us to conclude that algorithms that adds the uncertainty presented in the problem by means of extensions of fuzzy sets provides similar or better results that the ones obtained by fuzzy classical techniques.

As a future research, form the extensions point of view, must focus on the use of general type 2 fuzzy sets, but the computation time of general type usually is not tractable. Also, finding applications or image representations in which membership and non-membership can be generated independently in order to use all of the power of A-IFSs, should be researched.

## References

1. Atanassov, K.: Intuitionistic fuzzy sets. Fuzzy Sets and Systems 20, 87–96 (1986)
2. Atanassov, K.: Intuitionistic Fuzzy Sets. Theory and Applications. Physica-Verlag, Heidelberg (1999)
3. Basu, K., Deb, R., Pattanaik, P.K.: Soft sets: an ordinal formulation of vagueness with some applications to the theory of choice. Fuzzy Sets and Systems 45, 45–58 (1992)
4. Burillo, P., Bustince, H.: Entropy on intuitionistic fuzzy sets and on interval-valued fuzzy sets. Fuzzy Sets and Systems 78, 305–3016 (1996)
5. Bustince, H., Burillo, P.: Vague sets are intuitionistic fuzzy sets. Fuzzy Sets and Systems 79, 403–405 (1996)
6. Bustince, H., Barrenechea, E., Pagola, M.: Restricted Equivalence Functions. Fuzzy Sets and Systems 157, 2333–2346 (2006)
7. Bustince, H., Barrenechea, E., Pagola, M.: Image thresholding using restricted equivalence functions and maximizing the measures of similarity. Fuzzy Sets and Systems 158, 496–516 (2007)
8. Bustince, H., Pagola, M., Barrenechea, E., Fernandez, J., Melo-Pinto, P., Couto, P., Tizhoosh, H.R., Montero, J.: Ignorance functions. An application to the calculation of the threshold in prostate ultrasound images, Fuzzy Sets and Systems 161(1), 20–36 (2010)

9. Bustince, H., Barrenechea, E., Pagola, M., Fernandez, J., Sanz, J.: Comment on: "Image thresholding using type II fuzzy sets". Importance of this method, Pattern Recognition 43(9), 3188–3192 (2010)
10. Chaira, T., Ray, A.K.: Segmentation using fuzzy divergence. Pattern Recognition Letters 24, 1837–1844 (2003)
11. Deng, J.L.: Introduction to grey system theory. Journal of Grey Systems 1, 1–24 (1989)
12. Deschrijver, G., Kerre, E.E.: On the relationship between some extensions of fuzzy set theory. Fuzzy Sets and Systems 133(2), 227–235 (2003)
13. Deschrijver, G., Kerre, E.E.: On the position of intuitionistic fuzzy set theory in the framework of theories modelling imprecision. Information Sciences 177, 1860–1866 (2007)
14. Hirota, K.: Concepts of probabilistic sets. Fuzzy Sets and Systems 5, 31–46 (1981)
15. Forero, M.G.: Fuzzy thresholding and histogram analysis. In: Nachtegael, M., Van der Weken, D., Van de Ville, D., Kerre, E.E. (eds.) Fuzzy Filters for Image Processing, pp. 129–152. Springer, Heidelberg (2003)
16. Gau, W.L., Buehrer, D.J.: Vague sets. IEEE Transactions on Systems, Man and Cybernetics 23(2), 751–759 (1993)
17. Grattan-Guinness, I.: Fuzzy membership mapped onto interval and many-valued quantities. Z. Math. Logik Grundlag. Mathe. 22, 149–160 (1976)
18. Huang, L.K., Wang, M.J.: Image thresholding by minimimizing the measure of fuzziness. Pattern recognition 28(1), 41–51 (1995)
19. Klir, G., Weirman, M.: Uncertainty-Based information: elements of generalized information theory, 2nd edn. Physica-Verlag, Heidelberg (1999)
20. Mendel, J.M., John, R.I.: Type-2 fuzzy sets made simple. IEEE Transactions on Fuzzy Systems 10(2), 117–127 (2002)
21. Mendel, J.M.: Uncertain Rule-Based Fuzzy Logic Systems. Prentice-Hall, Upper Saddle River (2001)
22. Pal, N.R., Pal, S.K.: A review of image segmentation techniques. Pattern recognition 26, 1277–1294 (1993)
23. Pal, S.K., King, R.A., Hashim, A.A.: Automatic grey level thresholding through index of fuzziness and entropy. Pattern Recognition Letters 1(3), 141–146 (1983)
24. Sambuc, R.: Function $\Phi$-Flous, Application a l'aide au Diagnostic en Pathologie Thyroidienne, These de Doctorat en Medicine, University of Marseille (1975)
25. Sezgin, M., Sankür, B.: Survey over image thresholding techniques and quantitative performance evaluation. J. Electronic Imaging 13, 146–165 (2004)
26. Tizhoosh, H.R.: Image thresholding using type-2 fuzzy sets. Pattern Recognition 38, 2363–2372 (2005)
27. Tizhoosh, H., Krel, G., Muchaelis, B.: Locally Adaptive Fuzzy Image Enhancement. In: Proceedings of 5th fuzzy days, Computational Intelligence, Theory and Applications, pp. 272–276 (1997)
28. Vlachos, I.K., Sergiadis, G.D.: Intuitionistic fuzzy information - Applications to pattern recognition. Pattern Recognition Letters 28, 197–206 (2007)
29. Zadeh, L.A.: Fuzzy sets. Information Control 8, 338–353 (1965)
30. Zadeh, L.A.: The concept of a linguistic variable and its application to approximate reasoning – I. Information Sciences 8, 199–249 (1975)

# Fuzzy Methods on the Web: A Critical Discussion

Steven Schockaert, Nataliya Makarytska, and Martine De Cock

**Abstract.** Gradual concepts abound in many web-related domains, ranging from the notion of relevance in information retrieval, to the strength of connection in online social networks. As a result, fuzzy set theory is often a natural choice in implementing web systems. In this chapter, we give an overview of applications of fuzzy set theory in this area, focusing in particular on information retrieval, the semantic web, and recommender systems. In each case, we contrast fuzzy methods with other approaches, analyzing why and how the ideas of fuzzy set theory may be beneficial.

## 1 Introduction

The world wide web has often been promoted as a key application domain for fuzzy set theory [23, 86, 113]. Indeed, it is clear that to cope with the overwhelming amount of information on the web, intelligent techniques are needed to appropriately filter and preprocess the content of web pages. In traditional search engines, users convey their information need using a textual query, which is used to rank documents according to relevance. This ranked list is then presented to the user using well-chosen snippets from each of the documents. Modern information retrieval research attempts to replace the traditional keyword-based queries by more informative information requests, such as natural language questions, and to develop more advanced ways to present search results, typically by inducing some kind of structure from the set of relevant documents using clustering techniques. It appears that fuzzy methods have a natural role to play in this process. After all, the relevance of a

Steven Schockaert · Martine De Cock
Ghent University, Department of Applied Mathematics and Computer Science, Belgium
e-mail: {steven.schockaert,martine.decock}@ugent.be

Nataliya Makarytska
University of Washington, Institute of Technology, Tacoma, WA, USA
e-mail: natashka@u.washington.edu

C. Cornelis et al. (Eds.): 35 Years of Fuzzy Set Theory, STUDFUZZ 261, pp. 237–266.
springerlink.com 

text document to a topic clearly is a matter of degree, as is the interest of the user in a given topic. Already in the 1980s, before the advent of the web, the importance of flexibility in querying information systems was understood, leading to a variety of information retrieval methods based on fuzzy set theory [46]. Similarly, it has long been recognized that fuzzy clustering techniques, in which the membership of an object to a cluster centre is graded, are often more appropriate than their classical counterparts [43, 84], and clustering documents is no exception to this [48, 114]. In addition to structuring document collections or lists of search results, fuzzy clustering methods have also been used to build user profiles, based on past behavior or explicit user input [3, 57]. Such user profiles are useful to help the system personalize its output according to the desires and interests of each user.

Information retrieval is not the only domain in which fuzzy methods have been proposed. The semantic web [11] — a vision of a web of interlinked machine-readable information sources — has attracted considerable attention in the last decade. At the core of the semantic web vision is the notion of ontologies, which are shared formalizations of the concepts that appear in a certain domain. Ontologies are usually encoded in a standardized language, such as OWL, which can be modeled using a particular description logic. Dedicated description logic reasoners are then used to draw conclusions. Given appropriate ontologies, the content of a web resource or a web service can be described in a machine-readable way. This makes it possible to use formal reasoning to prove that some resource is relevant to a user (semantic search), or to automatically derive what behavior results from combining certain web services. In addition to description logics, also rule-based formalisms play a central role on the semantic web. The interest in fuzzy methods for the semantic web has mainly manifested itself at the foundational level; it has led to the development of fuzzy description logics [37, 98, 100], has further stimulated the development of fuzzy logic programming [44, 104], and its integration with (extensions of) description logics [52, 55].

A third area of the web where fuzzy methods are studied are recommender systems. Recommender systems provide users with recommendations (e.g. products the user may want to buy, movies that she wants to see, reviews that she wants to read, etc.) based on information about the users' preferences and about the items (the products, the movies, the reviews, etc.). Good and accurate recommender applications that guide users through the vast amounts of online information are gaining tremendous importance, as the wealth of information makes it increasingly difficult to find exactly what you want or need; all the more because every person has her own preferences. *Content-based systems* generate recommendations based on item similarity and, as such, tend to have their recommendation scope limited to the immediate neighbourhood of a user's past purchase or rating record. The performance of these systems can be improved significantly by (additionally) using *collaborative filtering*, which typically identifies users whose tastes are similar to yours and recommends items that these so-called neighbor users have liked. A more recent addition to the family of recommendation paradigms are the *social recommender systems*; these systems make suggestions based on likes and dislikes of users in your online social network. As similarity between items and users, and strength of connections in social networks, are graded concepts, it does not come as a surprise that a variety of interesting ideas has been proposed for the use of fuzzy set theory in content-based systems [110], colloborative filtering recommender systems [69, 79], as well as social recommenders [106].

However, despite the wide enthusiasm by the fuzzy set community, it is not clear what the real impact is of fuzzy methods on the web. Often, fuzzy methods are proposed to address problems that have not previously been considered. The lack of standardized benchmark data and strong baseline techniques then makes it difficult to provide a credible evaluation, and assessing the usefulness of the proposed solutions. In more classical domains, such as document retrieval, experimental results have been mixed. Due to the lack of clear experimental evidence for the success of fuzzy methods on the web, their impact outside the fuzzy set community remains limited. On the other hand, many popular approaches that do not refer to the term "fuzzy" are strongly related to the ideas of fuzzy set theory. Tag clouds [94], for instance, are little more than a fuzzy set of keywords. Conversely, techniques such as fuzzy clustering have little in common with the core ideas of fuzzy set theory and approximate reasoning, despite their reference to the term "fuzzy". In this sense, the question of whether or not fuzzy methods currently play an important role in web research is inherently vague, and is therefore best answered in linguistic terms: to some extent.

In this chapter, we provide a personal view on the benefits of fuzzy methods in web-related applications, as well as on the challenges that arise. It is not intended as a complete survey, but rather focuses on the three key domains that were sketched above: information retrieval, semantic web, and recommendation.

The chapter is structured as follows. In the next section, we provide a general introduction to fuzzy set theory, possibility theory and multi-valued logics, focusing especially on the different intuitions underlying these frameworks. Next, we discuss the use of fuzzy methods in each of the three aforementioned application domains: information retrieval in Section 3, the semantic web in Section 4 and recommender systems in Section 5. We end the chapter with a general conclusion on the role of fuzzy methods on the web.

## 2 Background

This section provides a non-technical introduction to fuzzy set theory and two related frameworks: possibility theory and multi-valued (or graded) logics. With the aim of clarifying the motivation of using fuzzy methods on the web in the following sections, our focus in this section is on the different intuitions that are behind these theories.

### 2.1 *Fuzzy Sets*

Fuzzy sets were introduced by Zadeh [111] with the aim of modeling human concepts. It is well-known that such concepts tend to be vague, in the sense that for some objects it is not clear whether or not they belong to the concept [62]: is architecture a science? Is food poisoning a disease? Are bookends furniture? Zadeh noted that there usually exists a continuous transition between those objects that clearly belong to some concept, and those that clearly do not. Taking this observation into account, the fuzzy set representation of a concept associates to each object a degree of membership, chosen from the unit interval [0,1]. Formally, a fuzzy set $A$ in a universe $X$ is any mapping from $X$ to $[0,1]$, where $A(x) = 1$ means that $x$ fully belongs to the concept and $A(x) = 0$ means that $x$ does not at all belong to the concept. In addition to modeling vague concepts, fuzzy sets are also used to model the intensity to which different objects satisfy some (well-defined) property. We may define, for instance, a fuzzy set of patients with fever, such that somebody with 37.5° C receives a membership degree of, say, 0.8, even though we may consider that fever by itself is a well-defined property. The concept of fuzziness, which is related to graded membership, should thus be distinguished from vagueness, which is related to the lack of precise meaning. A fuzzy relation from a universe $X$ to a universe $Y$ is a fuzzy set in the cartesian product $X \times Y$. A fuzzy relation from $X$ to $X$ is simply called a fuzzy relation in $X$. Fuzzy relations are typically used to model the strength of a certain relationship between objects of $X$ and objects of $Y$.

Set operations are generalized to fuzzy sets in an indirect way, by generalizing logical conjunction and disjunction to graded truth values, and relying on the intuition that e.g. an element belongs to the intersection $A \cap B$ if it belongs to $A$ *and* it belongs to $B$. Conjunction is usually generalized using t-norms, which are mappings $T$ from $[0,1]^2$ to $[0,1]$ that are symmetric, associative, increasing, and satisfy the boundary condition $T(1,a) = 1$ for all $a \in [0,1]$. Given a t-norm $T$, the intersection of two fuzzy sets $A$ and $B$ in the same universe $X$ is defined as $(A \cap B)(x) = T(A(x), B(x))$ for all $x \in X$. Similarly as for intersection, union is defined by generalizing disjunction. Typically, disjunction is generalized using a t-conorm $S$, which is a symmetric, associative, increasing $[0,1]^2 - [0,1]$ mapping that satisfies the boundary condition $S(0,a) = a$ for all $a \in [0,1]$. Given a t-conorm $S$, it is natural to define a generalized notion of implication, by $I(a,b) = S(1-a,b)$ for all $a,b \in [0,1]^2$, thus taking advantage of the classical equivalence $p \rightarrow q$ iff $\neg p \vee q$. Such generalized implications are called S-implicators. Another way to define generalized implications starts from a left-continuous t-norm $T$:

$$I(a,b) = \sup\{\lambda | \lambda \in [0,1] \text{ and } T(a,\lambda) \leq b\}$$

Such operators $I$ are called residual implicators. While their definition is less intuitive than that of S-implicators, they often turn out to be particularly useful, as they tend to preserve more properties from classical logic.

The membership degrees of a fuzzy set can essentially be interpreted in three different ways, which relate to measurement of cost, distance, and frequency [31]. When fuzzy sets are used to add flexibility to a query (e.g. give me a list of cheap hotels in Ghent), membership degrees are used to encode *preference*, in the sense that objects which satisfy the query to a larger extent are more preferred. Membership degrees are then related to utility or cost. When fuzzy sets are used to provide an interface between the numerical domain and linguistic terms, the membership degrees reflect the *similarity* of an object to prototypes of the concept being modeled. The fact that architecture is a science to degree 0.3 then intuitively means that there is an object which is a clear example of the concept 'science' (e.g. physics) and which is similar to degree 0.3 with 'architecture'. In this case, membership degrees are related to distance measures. Finally, membership degrees can be used to express *uncertainty*. When being told about some user that she is young, we may consider some ages to be more plausible than others. The membership degree of a certain age in a fuzzy set modeling the concept 'young' is then interpreted as a degree of plausibility. In this case, membership degrees are related to probability theory, although different interpretations may be given to the exact relationship. Among others, the membership degree of an age $\lambda$ in the fuzzy set *young* may be interpreted as the probability that somebody would assign the label young to the age $\lambda$. Fuzzy sets then correspond to likelihood functions [41]. Another way membership degrees can be related to probability is by interpreting fuzzy sets either as special cases or as approximations of random sets [30].

Regardless of the specific meaning that is given to membership degrees, it is important to note that taking the intersection of two fuzzy sets, for instance, is only meaningful if their membership degrees are *commensurable*. Let us take the example of querying a hotel reservation system. When we are interested in cheap hotels, interpreting the predicate 'cheap' is easy, as it can be done in a purely qualitative way (the cheaper the price, the more a hotel is cheap). However, when we rather ask for hotels that are at the same time 'cheap' and 'close to the city centre', the fuzzy sets modeling 'cheap' and 'close' should be such that the utility of being close to degree 0.7, for instance, is the same as the utility of being cheap to degree 0.7. Clearly, this puts strong constraints on how membership degrees should be obtained, which forms one of the most important practical difficulties in applying fuzzy set theory.

## 2.2 Possibility Theory and Approximate Reasoning

Possibility theory [29, 112] is an uncertainty calculus which is tightly related to fuzzy set theory. At its basis is the notion of a possibility distribution $\pi$, which is a mapping from a universe $X$ to the unit interval $[0,1]$, i.e. from a formal point of view, possibility distributions are fuzzy sets. A possibility distribution encodes for each $x \in X$ the degree of possibility that $x$ is the actual value of some variable. Each possibility distribution $\pi$ induces two uncertainty measures, called the possibility measure $\Pi$ and the necessity measure $N$; for a subset $A \subseteq X$, they are defined as

$$\Pi(A) = \sup_{x \in X} \pi(x) \qquad\qquad N(A) = 1 - \Pi(coA)$$

where $coA = X \setminus A$ denotes set complement. Note that, from these definitions it follows that $\Pi(A \cup B) = \max(\Pi(A), \Pi(B))$, which is why possibility theory is called a non-additive uncertainty theory. Although possibility theory has mainly been developed as an uncertainty calculus related to fuzzy set theory, its ideas go back to the work of Shackle [93], who advocated the need for a non-additive uncertainty theory based on degrees of surprise. In this view, the possibility degree $\pi(x)$ is interpreted as the degree to which one would be surprised to learn that $x$ is the actual value of the underlying variable. Note that possibility theory is not fully compositional. For instance, the value of $\pi(A \cap B)$ cannot be determined solely from the values of $\pi(A)$ and $\pi(B)$. This should come as no surprise as it is well-known that no uncertainty calculus can be completely compositional.

Possibility theory has played a central role in the theory of approximate reasoning. The basic idea is to interpret an assertion of the form "$V$ is $A$" by the inequality $\pi_V \leq A$ (i.e. $\pi_V(x) \leq A(x)$ for all $x \in X$), where $A$ is a fuzzy set, $V$ a variable, and $\pi_V$ a possibility distribution encoding knowledge about which values of $V$ are possible. Such a constraint is called a flexible restriction on $V$. Given a number of flexible restrictions $\pi_V \leq A_1,...,\pi_V \leq A_n$, our state of knowledge regarding the possible values of variable $V$ is encoded by the least specific solution to the constraints, i.e. $\pi_V(x) = \min(A_1(x),...,A_n(x))$.

A central issue in approximate reasoning is how we can derive a flexible restriction on the value of a variable $W$, given a possibility distribution which encodes the possible values of variable $V$ and an if–then rule of the form "**if** $V$ is $A$ **then** $W$ is $B$". If–then rules are useful to encode common-sense knowledge such as "**if** the weather is nice **then** driving time to the coast will be long". Zadeh's *compositional rule of inference* suggests to derive the following possibility distribution $\pi_W$ on $W$ from the possibility distribution $\pi_V$ on $V$ and an if–then rule $R$:

$$\pi_W(y) = \sup_{x \in X} T(\pi_V(x), R(x,y))$$

where $T$ is a t-norm and the if–then rule $R$ is interpreted as a fuzzy relation. The intuition is clear: a value $y$ for $W$ is possible to the extent that there is a value $x$ which is possible for $V$ and such that the combination $(x,y)$ does not violate the given if-then rule $R$. What remains to be decided is how to implement the if–then rule $R$ itself. The most natural choice is to take $R(x,y) = I(A(x),B(y))$ for some implicator $I$, although sometimes a t-norm $T'$ is used instead of an implicator. By extending this idea to multiple input variables, and multiple if–then rules, a powerful inference-scheme is obtained. Although it is computationally expensive in general, efficient methods can be obtained by assuming that the values of the input variables are precisely known and by taking fuzzy sets with membership functions that are easily processed (e.g. piecewise linear functions). Starting from the work by Mamdani [56], fuzzy reasoning with if–then rules has been widely used in industrial applications, as diverse as optimizing the program of washing machines and implementing auto-focusing techniques in digital cameras. It can be considered to be by far the most successful application of fuzzy set theory. In practice, if-then rules can be provided by an expert, or they can be derived automatically using appropriate training data. In domains where only limited training data is available, also a hybrid approach can be used: the expert provides a first version of the rules, which is subsequently refined using whatever training data that is available.

### 2.3 Multi-valued Logic

The term fuzzy logic is used in two different senses in the literature. The first sense, often called the broad sense, mainly refers to the idea of approximate reasoning with if–then rules, as described above. The second sense, called the narrow sense, refers to formal logics in which the notion of truth is graded. This idea of graded truth is not exclusively tied to the framework of fuzzy set theory. Indeed, the notion of truth degree is already present in the three-valued logics that were developed in the first half of the 20th century, by Łukasiewicz, Gödel, Post, and Kleene, among others [28]. In the case of infinite-valued logics, truth degrees are values from [0,1] and logical connectives are interpreted as suitable $[0,1]^2 - [0,1]$ (conjunction, disjunction, implication) and $[0,1] - [0,1]$ (negation) functions. Usually, (propositional) multi-valued logics use the same syntax as classical (propositional) logic, although truth constants other than 0 are sometimes allowed in the language (e.g. in rational Pavelka logic [76]). Sound and complete proof theories for propositional fuzzy logics have been established, based on modus ponens and particular subsets of the axioms of classical logic [36]. For instance, infinite-valued Gödel logic is characterized by the axioms of intuitionistic logic together with the axiom of prelinearity: $(x \rightarrow y) \vee (y \rightarrow x)$. The semantics of Gödel logic is obtained by interpreting conjunction and disjunction by the minimum and maximum operators, and implication as the residual implicator induced by the minimum. Negation is defined as $\neg a = a \rightarrow 0$. The most popular fuzzy logics are Gödel logic, product logic and Łukasiewicz logic. In each case, conjunction and implication are interpreted in terms of some continuous t-norm and its residual implicator. For a more complete introduction to fuzzy logic, both in the narrow and the broad sense, we refer to the chapter by V. Novák and I. Perfilieva in this volume.

Note that despite the use of fuzzy logic connectives, infinite-valued logics are based on a completely different intuition than approximate reasoning. In particular, formulas from an infinite-valued logic encode a precise relationship between variables of some continuous domain. As such, propositional fuzzy logics do not deal with uncertainty or vagueness.

## 3 Information Retrieval

### 3.1 Relevance Models

An abundance of techniques that are based on fuzzy set theory or possibility theory have been proposed to improve the effectiveness and flexibility of search engines. Although modern web search engines are considerably more sophisticated than traditional information retrieval (IR) systems (e.g. exploiting hyperlinks to obtain more accurate relevance estimates as well as indications of authoritativeness), they are still essentially based on the same ingredients: a boolean keyword-based formulation of queries, and a bag-of-words representation of documents. In particular, documents are represented as weighted collections of terms, thus ignoring the ordering of the terms in the document, as well as any structure the document may have. Because of this simplification, a document can formally be modeled as a vector in a multi-dimensional space, with one dimension for each term occurring in the document collection under consideration. The weight of a component of a document vector is calculated based on the number of times the corresponding term occurs in the document (term frequency), and on the number of documents of the collection in which this term appears (inverse document frequency). The intuition is that a given term should receive a high weight when it occurs a lot in the document, while being rare in the collection as a whole. The query of the user, which is provided as a list of keywords, can also be represented as a vector, by treating it as a (short) document. A common approach to estimate the relevance of a document to a query then consists of calculating the cosine of the angle between the corresponding vectors. This *vector-space model* of information retrieval [85] has traditionally been the most popular approach to information retrieval, and is still considered state-of-the-art. However, state-of-the art performance in the vector-space model is obtained for variants of the aforementioned cosine-similarity which are difficult to interpret intuitively and rely on careful tweaking of the parameters involved [116].

More recently, *probabilistic language models*, which were first developed in the area of speech recognition, have been successfully applied to the information retrieval problem [80], combining state-of-the-art performance with intuitively appealing probabilistic models. Documents are then formally represented as probability distributions, which are used to calculate the probability that a document is relevant to the user. Retrieval models in which documents are represented as fuzzy sets have also been proposed [14, 46]. Conceptually, *fuzzy IR models* are similar in spirit to the vector-space model, using the same formulas to weigh the importance of a term in a document. The key difference is in the way queries are formulated and evaluated. Rather than representing the query as a small document, the relevance of a document is calculated using fuzzy logic connectives, measuring the degree to which a document 'implies' a query term, and subsequently combining these degrees using flexible alternatives for the operations of boolean conjunction or disjunction. Fuzzy IR models typically allow the user to specify for each keyword to what extent it is important for the query in linguistic terms (e.g. *very important*, *rather important*, etc.), and how the keywords should be combined using linguistic quantifiers (e.g. *most* of the keywords should be present). The main advantage of fuzzy IR models is in the flexibility they give users to specify their queries. Recently, also possibilistic approaches have been proposed [17]. Similar to probabilistic models, possibilistic models attempt to estimate the likelihood that a document is relevant to a query. In contrast to probabilistic models, however, this leads to two scores: the necessity that a document is relevant and the possibility that it is relevant. Documents are then ranked primarily based on the necessity scores, using the possibility scores to break ties, and in particular to provide meaningful results in cases where the necessity of relevance is 0 for all documents.

### 3.2 Domain-Specific Retrieval

The traditional information retrieval models are very general. Due to the fact that they treat words as abstract entities, for instance, most models are language-independent (although effectiveness of IR models is often dependent on language-specific issues [38, 72]). By restricting attention to a narrower domain, however, additional resources may be available that can help the retrieval process. One example are thesauri, which encode semantic relationships between terms, indicating for example that two terms are related in meaning (e.g. synonyms), or that one term is a specialization of another term (e.g. 'mathematician' is a specialization of 'scientist'). It is natural to consider that relations such as 'related term' are graded, as e.g. football and FIFA are more strongly related than football and player, even though football and player are still somewhat related. Accordingly, approaches to information retrieval have been proposed which use *fuzzy thesauri* [66, 83]. While utilizing a thesaurus seems very natural, as it allows to retrieve documents that are relevant to a query without actually sharing any terms with it, experimental validations of thesaurus-based IR models have failed to show a consistent improvement over systems without thesauri [45]. Practical problems with the use of thesauri include the fact that many words have different senses, which may lead semantically unrelated documents to be considered relevant, and the difficulties and costs involved in manually building high-quality thesauri. Automatically generated thesauri, typically based on detecting co-occurrence of terms, may provide a solution to the latter problem, but such thesauri are of varying quality, and moreover, highly dependent on the collection from which they have been obtained. Going from classical thesauri to fuzzy thesauri makes some of the problems even worse: how should reliable and meaningful grades be obtained? A recent example of the use of fuzzy thesauri can be found in [96].

It is interesting to note that the idea of fuzzy thesauri is also considered in the probabilistic language modeling approach to IR, although the term 'fuzzy thesauri' is not used in this context. In particular, a document is represented as a probability distribution, which is initially obtained using maximum likelihood estimation, i.e. the probability $P(t|d)$ that a term $t$ is generated by the language model underlying document $d$ is estimated as $\frac{n_t}{\sum_{t'} n_{t'}}$, where $n_t$ is the number of occurrences of term $t$ in document $d$. As this leads to the undesired effect that terms which do not occur in the document receive a zero probability, different forms of *smoothing* are applied. One form of smoothing is to interpolate this initial document model with a corpus model (which models the probability that a given term appears in the collection as a whole), which has a similar effect as considering inverse document frequency in the vector-space model. Recently, however, an additional form of smoothing, called *semantic smoothing* has gained importance [115]. Essentially, semantic smoothing corresponds to using a fuzzy thesaurus to increase the probability of terms that do not occur in the document, but are related to terms that do occur.

Somewhat related to the use of thesauri is concept-based information retrieval, where documents are linked to concepts from an ontology. By abstracting away from the actual terms that appear in a document, it may be expected that documents and queries may be compared in a way which is semantically more meaningful. In [33], for instance, document are modeled as vectors of Wikipedia concepts, and experimental evidence is provided that the similarity between documents can thus be measured in a substantially more accurate way. Somewhat related, [7] proposes to represent documents and queries as subtrees of ontology concepts, and uses fuzzy logic connectives to compute relevance scores. Again, convincing experimental evidence is provided to demonstrate the usefulness of the approach. In [6] a variant based on possibilistic logic is proposed. The possibilistic view naturally allows to associate three different degrees with each pair of terms $(t_1, t_2)$: the possibility that $t_1$ and $t_2$ refer to the same thing, the necessity that $t_1$ is a specialization of $t_2$ and the necessity that $t_2$ is a specialization of $t_1$. An important advantage of this approach is that the degrees that are involved have a clear meaning.

In addition to retrieval of text documents, there is an increasing interest in retrieving other types of objects from the web [73], such as images[1], scientific papers[2], information about people[3], events[4], products[5], etc. Due to the fact that object-based retrieval is only applied in narrow domains, focusing on one particular type of objects, semantically richer, domain-dependent techniques may be applied, which are often of a very different nature than traditional text-based retrieval. For example, image retrieval systems often use a combination of textual evidence (e.g. the text surrounding the image on a web page) and visual features, and sometimes even focus exclusively on visual features [50, 95]. Fuzzy set-based approaches have been successfully applied to measure the similarity of visual features [12, 19, 49, 68]. Due to the use of richer semantics in object-based retrieval, often new types of opportunities arise for the application of fuzzy set theory. In [92], for instance, an approach is presented for retrieving events that satisfy given temporal restrictions, using a form of fuzzy temporal reasoning [89]. Here, the use of fuzzy set theory is directly motivated by the fact that many real-world events are of an inherently gradual nature, lacking a precise beginning and/or ending date. Similar considerations apply in the spatial domain, where the importance of vernacular geographic regions with imprecise boundaries is widely acknowledged [2, 67, 109], making approaches based on fuzzy set theory a natural choice [88, 91]. Finally, fuzzy methods have also been advocated in the area of music retrieval [15, 16].

## 3.3 Manipulation of Search Results

Fuzzy set based methods have been proposed for a variety of problems that in one way or another manipulate the results obtained from some standard information retrieval model. In [78], for instance, a fuzzy rule based system is presented to exploit the structure of HTML documents. While several retrieval models have

[1] `http://www.flickr.com`
[2] `http://scholar.google.com`
[3] `http://pipl.com`
[4] `http://upcoming.yahoo.com`
[5] `http://www.google.com/products`

already been proposed that can take advantage of the fact that e.g. terms in the title of an HTML document should be considered more important than terms in the body, most existing approaches are based on an assumption of statistical independence. The approach presented in [78] does not rely on such an assumption, as the overall importance of a term for an HTML document is determined by rules of the form

**if** 'Title' is *High* & 'Emphasis' is *Low* & 'Others' is *Low* **then** 'Result' is *Medium*

where 'Title' is the weight of the term in the title (i.e. treating the title as a small document), 'Emphasis' is the weight of the term when considering those occurrences that are otherwise emphasized in the document, and 'Others' is the weight of the remaining occurrences; *High*, *Low* and *Medium* refer to fuzzy sets with appropriately defined membership functions. The intuition of the rule above is that words which occur in the title, but not often in the remainder of the document may very well be irrelevant, e.g. because the title is metaphorical. Clearly, this intuition is very different from other approaches to retrieval of structured documents, which would boost the importance of terms that appear in the title, regardless of whether the term also appears elsewhere in the document. A total number of 9 such rules have been manually specified, and the resulting system is experimentally shown to outperform state-of-the-art methods such as BM25 and BM25F.

The motivation for using fuzzy set theory in this way is clearly not related to uncertainty or to the modeling of vagueness. Moreover, the same intuition can be implemented using other techniques, which after careful training and tweaking, might very well outperform such fuzzy rule based approaches. What fuzzy rules offer in this context, however, is the ability to easily implement an intuitive idea, using rules that are easy to understand. If the system does not perform as expected, it is straightforward to adapt the rules until the desired behavior is obtained, while many other methods crucially depend on the availability of good training data to arrive at 'black-box' models. Moreover, if such training data is actually available, the rules that have manually been constructed can be refined in an automated way [70]. A similar use of fuzzy rules is made in [90] with the aim of clustering web search results. There, fuzzy rules are used to implement the behavior of artificial agents, called ants, that move documents in a virtual environment and put them on heaps. Again, the use of if–then rules leads to a description which is easy to understand and to adapt to different intuitions about how documents should be clustered.

Other applications where fuzzy methods have been proposed to manipulate search results are: relevance feedback [21, 59, 108], meta-search [60], and query expansion [47, 58]. The techniques that are applied to this end are as diverse as fuzzy clustering [47], fuzzy association rules [58] and again fuzzy if-then rules [108]. This further illustrates the fact that fuzzy set theory can often provide a flexible vehicle for implementing advanced systems. In many cases, however, the authors provide very little experimental evidence to demonstrate the effectiveness of the proposed techniques over sufficiently strong baseline systems.

## 4 Semantic Web

The semantic web [11] is a vision of interlinked machine-readable resources that exist on top of the web of human-readable documents that we know today. The widespread availability of such machine-readable resources would allow for the development of a variety of intelligent systems, such as semantic search systems that can prove the relevance of an object to some query based on a semantic representation of both the resource and the query. Central in this view is the notion of an ontology, which, in this context, is essentially a formalization of a given domain, describing properties of the relevant concepts and relations. The realization of a semantic web requires that two important challenges are overcome. The first challenge is acquiring the machine-readable resources that constitute the semantic web, which could be achieved by human experts who manually build ontologies, by automated techniques based on natural language processing, or by a combination of both. The second challenge is to exploit available information in a scalable, robust and useful way. It is in addressing this second challenge that possibilistic and fuzzy methods have a key role to play.

Taken as a whole, the information that is asserted on the semantic web will inevitably be inconsistent. Uncertainty about the correctness of individual pieces of information is therefore a key issue, which could be tackled by either probabilistic or possibilistic methods. Fuzzy methods, on the other hand, serve a different, but arguably equally important purpose. In particular, when moving from classical retrieval to semantic search, we lose the idea of a ranking. Indeed, when both resources and queries are expressed using classical logic, then we cannot acquire a more refined conclusion than that a resource is relevant, or that it is not relevant. In practice, this is problematic, because it is important to discriminate between objects that best satisfy the user's information need and those that only satisfy it marginally. Moreover, when no resource completely satisfies a given query, it may still be of interest to identify resources that 'almost' satisfy it. Thus, concepts such as preference and similarity, which are at the heart of fuzzy set theory, are therefore of crucial importance.

### 4.1 Description Logics

Ontologies for the semantic web are usually modeled in description logics [4]. In such logics, knowledge is encoded in two separate knowledge bases, called the T-box and the A-box. The core idea is to describe properties of concepts and relations in the T-box and to describe in the A-box which objects are instances of which concepts, and which pairs of objects belong to which relations; usually relations are called roles in this context. Typically, atomic concepts are denoted by upper case letters $A$, $B$, etc. From such atomic concepts, complex concepts can be formed such as $A \sqcap B$, $A \sqcup B$, and $\neg A$, where e.g. $A \sqcap B$ is the concept whose instances are those objects that both belong to $A$ and to $B$. The formal semantics is defined in terms of interpretations $\mathscr{I}$ that map concepts to sets of objects from a given domain $\Delta^{\mathscr{I}}$, e.g. $(A \sqcap B)^{\mathscr{I}} = A^{\mathscr{I}} \cap B^{\mathscr{I}}$. Similarly, the interpretation of roles is as relations in

$\Delta^{\mathscr{I}} \times \Delta^{\mathscr{I}}$. In addition to the basic constructs, $\sqcap$, $\sqcup$, and $\neg$, also the concepts $\exists R.A$ and $\forall R.A$ are commonly used, where $A$ is a concept and $R$ a role; their semantics is as follows:

$$(\exists R.A)^{\mathscr{I}} = \{x \in \Delta^{\mathscr{I}} \mid \exists y \in \Delta^{\mathscr{I}}.(x,y) \in R^{\mathscr{I}} \wedge y \in A^{\mathscr{I}}\}$$
$$(\forall R.A)^{\mathscr{I}} = \{x \in \Delta^{\mathscr{I}} \mid \forall y \in \Delta^{\mathscr{I}}.(x,y) \in R^{\mathscr{I}} \rightarrow y \in A^{\mathscr{I}}\}$$

Thus, intuitively, $\exists R.A$ is the concept which contains all objects that are related (w.r.t. $R$) with some object in $A$, whereas $\forall R.A$ contains the objects that are only related to objects in $A$. Various variants of description logics can be defined, based on which type of constructs are allowed.

The T-box of a description logic theory encodes how different concepts relate to each other, using assertions of the form $A \sqsubseteq B$, where $A$ and $B$ are (not necessarily atomic) concepts, e.g.

$$\mathit{Professor} \sqsubseteq \mathit{FacultyMember} \qquad \exists \mathit{authorOf}.\mathit{ResearchPaper} \sqsubseteq \mathit{Researcher} \tag{1}$$

encoding that professors are faculty members, and that all individuals who have authored at least one research paper are researchers. The A-box of a description logic theory contains assertions about individual objects of the form $x : A$, where $x$ is an object and $A$ is a concept, as well as assertions of the form $(x_1, x_2) : r$, where $x_1$ and $x_2$ are objects and $r$ is a role; e.g.

$$\mathit{etienne} : \mathit{Professor} \qquad p_1 : \mathit{ResearchPaper} \qquad (\mathit{etienne}, p_1) : \mathit{authorOf}$$

Together (1) and (2) entail e.g. that $\mathit{etienne} : \mathit{FacultyMember} \sqcap \mathit{Researcher}$.

## 4.2 Fuzzy Description Logics

The main idea of fuzzy description logics [100] is to interpret concepts as fuzzy sets, acknowledging that many of the concepts that appear in real-world ontologies are vague. The most popular approach, initially proposed by Straccia [100], is based on a direct generalization of the semantics of classical description logics, e.g. the semantics of the concept $\exists R.A$ becomes for $x \in \Delta^{\mathscr{I}}$

$$(\exists R.A)(x) = \sup_{y \in \Delta^{\mathscr{I}}} T(R^{\mathscr{I}}(x,y), A^{\mathscr{I}}(y))$$

where $T$ is a t-norm, $A^{\mathscr{I}}$ is a fuzzy set in $\Delta^{\mathscr{I}}$ and $R$ is a fuzzy relation in $\Delta^{\mathscr{I}}$. Thus, each object $x$ belongs to a concept such as $\exists R.A$ to some degree in $[0,1]$. The T-box now contains assertions of the form $\langle A \sqsubseteq B \geq \lambda \rangle$, for $A$ and $B$ concepts or roles, and $\lambda \in [0,1]$. In the case where $A$ and $B$ are concepts, for instance, the semantics of this assertion is as follows:

$$\mathscr{I} \models \langle A \sqsubseteq B \geq \lambda \rangle \quad \text{iff} \quad \inf_{x \in \Delta^{\mathscr{I}}} I(A(x), B(y)) \geq \lambda \tag{2}$$

where $I$ is an implicator. Similarly, the A-box contains assertions of the form $\langle x : A \geq \lambda \rangle$, which, semantically, correspond to the condition that $A^{\mathscr{I}}(x) \geq \lambda$. Sound and complete reasoning procedures were introduced in [100] for a basic fuzzy description logic and a particular choice for the fuzzy logic connectives. More recently, among others, more expressive description logics have been considered [97, 99], larger classes of fuzzy logic connectives [37], and more complex reasoning tasks [54]. Of particular interest are fuzzy description logics with concrete domains [101], which allow to explicitly define fuzzy predicates which can then be used in the definition of concepts. For instance, in such logics, we could define a prolific researcher as a researcher who has published many papers as follows:

$$\mathit{Researcher} \sqcap \exists \mathit{numberOfPapers}.\mathit{Many} \sqsubseteq \mathit{ProlificResearcher}$$

together with an appropriate fuzzy set in $\mathbb{N}$ that encodes the predicate 'many', e.g.

$$\mathit{Many}(n) = \begin{cases} \frac{n-1}{n} & \text{if } n > 0 \\ 0 & \text{otherwise} \end{cases} \tag{3}$$

As we have already indicated, adding fuzziness to description logics serves two rather distinct purposes. First, the fact that concepts are fuzzy sets naturally leads to flexibility in the querying process. For instance, when a user indicates that he is interested in a list of prolific researchers, it suffices to rank all instances of the concept *Researcher* according to the degree to which they belong to $\exists \mathit{numberOfPapers}.\mathit{Many}$. The definition of *Many* which was chosen in (3) then essentially means that the ordering of researchers according to their membership degree in *ProlificResearcher* is identical to the ordering based on their number of publications. Thus, the use of fuzzy sets allows for flexibility, as it eliminates the need for a crisp threshold on the required number of publications, and it naturally allows the system to rank the objects that (partially) satisfy the query. Second, when specifying a knowledge base, the fuzziness naturally allows to encode the intensity by which certain properties are satisfied. For instance, rather than specifying that Etienne is a prolific researcher, we can specify to what extent he is a prolific researcher, e.g. by asserting

$$\langle \mathit{etienne} : \mathit{ProlificResearcher} \geq \frac{389}{390} \rangle \tag{4}$$

It is important to note, however, that fuzzy description logics are *not* suitable for modeling vague knowledge, despite a wide number of claims to the contrary in the literature. Indeed, asserting (4) is exactly the same as asserting that Etienne has published at least 390 papers, which is clearly not vague at all. Modeling vague knowledge, such as "Etienne has published many papers" requires a mechanism for dealing with uncertainty, which is not present in standard fuzzy description logics. In other words, fuzzy description logics are suitable to deal with information which is naturally graded, but which is precisely known.

Most work on fuzzy description logics has been theoretical, developing more expressive formalisms, or more scalable reasoning mechanisms [13, 75]. One notable application of fuzzy description logics is in multimedia retrieval [65], where fuzzy description logic are used to encode both semantic annotations of multimedia documents and the result of e.g. image processing analyses. When it comes to the semantic web, it is not clear which is the role to be played by fuzzy description logics. A core requirement on the semantic web is the ability to link two ontologies that have been developed independently from each other. This, however, introduces a problem of commensurability. How should we compare what is called a prolific researcher to degree 0.4 in one fuzzy description logic base to what is called a young researcher to degree 0.7 in another fuzzy description logic base? Which guarantees do we have that it actually makes sense to combine these two degrees, to answer the query $YoungResearcher \sqcap ProlificResearcher$? In such a case, it seems more reasonable to explicitly encode the number of publications and the age of the researcher (or the constraints on these values that are known), rather than to encode degrees of membership. Thus, in contexts where interoperability plays a role, it seems that the use of fuzzy description logics may be problematic. Along the same lines, what may be problematic for certain types of applications is that different users may have a different view on concepts such as 'young' or 'prolific'. In traditional approaches to flexible querying, it is indeed the user who (implicitly or explicitly) determines how such concepts should be understood and how the degrees to which somebody is 'young' or 'prolific' should influence the ranking of the results. In fuzzy description logics, such concepts have a fixed meaning, which is independent of the preferences of an individual user. To some extent, it seems that the need for flexible approaches to querying ontologies and web information systems has been confused with a need for fuzziness at the knowledge representation level.

## 4.3 Possibilistic Description Logics

Possibilistic description logics [42, 27, 82] keep the crisp representation of concepts from classical description logics, adding a mechanism for handling uncertainty instead. Essentially, a T-box axiom then takes the form $(A \sqsubseteq B, \lambda)$, meaning that it is certain to degree $\lambda$ that the concept inclusion $A \sqsubseteq B$ holds. Similarly, an A-box axiom takes the form $(x : A, \lambda)$, meaning that it is certain to degree $\lambda$ that $x$ is an instance of $A$. Semantically, possibilistic description logics generalize classical description logics, in a similar way as possibilistic logic generalizes propositional logic. Specifically, let $\mathscr{W}$ be the set of all description logic interpretations $\mathscr{I}$. An interpretation of a possibilistic description logic theory then is a possibility distribution $\pi$ on $\mathscr{W}$. For each classical interpretation $\mathscr{I}$, $\pi(\mathscr{I})$ expresses how plausible it is that $\mathscr{I}$ corresponds to the real world. Note that the notion of plausibility that is considered here is purely qualitative. Writing $[\![\alpha]\!] \subseteq \mathscr{W}$ for the set of models of a description logic formula $\alpha$ (i.e. either a concept inclusion axiom in the T-box or an A-box assertion), the possibilistic description logic formula $(\alpha, \lambda)$ is interpreted as $N(\alpha) \geq \lambda$, i.e. the possibilistic description logic models of $(\alpha, \lambda)$ are those possibility distributions $\pi$ whose induced necessity measure $N$ is such that $N(\alpha) \geq \lambda$.

In a possibilistic description logic, we may write, for instance, that

$$(etienne : Professor, 0.9) \qquad (etienne : AboveFourty, 0.6)$$

which means that we are quite certain that Etienne is a professor, and rather certain that his age is above 40. Note that by combining the ideas of fuzzy and possibilistic description logics, vague knowledge may be encoded, e.g. writing assertions as

$$(\langle etienne : ProlificResearcher \geq 0.7\rangle, 0.9)$$
$$(\langle etienne : ProlificResearcher \geq 0.8\rangle, 0.6)$$
$$(\langle etienne : ProlificResearcher \geq 0.9\rangle, 0.3)$$

When it comes to semantic web applications, possibilistic description logics share with their fuzzy counterparts the problem of commensurability. The certainty weights that appear in possibilistic description logic bases that have been developed independently cannot be compared. One solution would be to resort to possibilistic logic variants that can deal with partially ordered certainty weights [9]. Another avenue for applications is the combination of several classical description logic bases. Possibilistic certainty weights could then be added to the assertions that are made by each source, based on their reliability. In this way, when different sources are combined, the ones that are considered least reliable are discarded. Although this idea has not yet been considered for possibilistic description logics, similar ideas have been extensively studied for merging conflicting propositional knowledge bases [8, 10]. Especially when utilizing variants which do not suffer from the so-called 'drowning effect', such as the approach presented in [82], this seems to be a promising direction.

## 4.4 Logic Programming

Logic programming deals with inferring knowledge from rules of the form

$$c \leftarrow a_1, ..., a_n, not\, b_1, ..., not\, b_m \tag{5}$$

which encode the intuition that unless one of the terms $b_1, ..., b_m$ can be derived, it holds that $a_1 \wedge ... \wedge a_n$ implies $c$. In its simplest form, terms are restricted to atomic propositions and the semantics of logic programs may be given in a purely declarative way using the notion of stable models [34]; this approach is often referred to as *answer set programming*. The idea of logic programming in general, and answer set programming in particular, has been generalized to deal with graded properties. The intuition of (5) is then that the truth degree of $c$ is at least as high as the truth degree of $a_1 \wedge .... \wedge a_n$, unless one of the atoms $b_1, ..., b_m$ can be derived to a high degree. Note however, that there exist several ways to implement this intuition, leading to different semantics of fuzzy logic programming [44, 64, 102, 107]. In parallel, some possibilistic extensions to logic programming have been considered [5, 20, 26, 71, 74], in which it is possible to encode that a given rule or fact is more or less certain (or important, preferred, etc.). Although the idea of uncertainty or preference is clearly different from the idea of graded truth, at the formal level, extensions of answer set programming that deal with uncertainty are closely related to extensions dealing with graded truth [5, 24].

An interesting application of fuzzy logic programming for the semantic web is discussed in [63], where grades are used to encode similarity between terms. Among others, this is useful to deal with the fact that different resources may use a different terminology to refer to the same or similar concepts, for instance to tackle problems related to interoperability on the semantic web. In the proposed approach, a given set of logic programming rules is augmented with additional rules that encode which terms can be considered similar and to what degree. What is not entirely clear, from an application point of view, however, is whether these degrees relate to the certainty that two terms describe the same property/concept/object, or to the strength of the similarity between the two terms. Although the formal treatment may be analogous in both cases, in order to obtain meaningful results, a clear operational semantics of the grades is needed in applications, which may put constraints, for instance, on which fuzzy logic connectives can be used to combine the grades.

Motivated by the needs of semantic web applications, logic programming rules have also been combined with description logics, leading to *description logic programs* [32]. Accordingly, fuzzy description logic programs have been developed, which combine fuzzy description logics with fuzzy answer set programming [53, 103]. In [40], as an application of such fuzzy description logic programs, the problem of ranking web services according to the preferences of a given user is considered. Again, there is no clear distinction between uncertainty and graded truth, in the sense that the motivation of the paper is given in terms of graded truth, while the application example that is presented essentially deals with uncertainty.

## 5 Recommendation and Personalization

The wealth of information available on the web has made it increasingly difficult to find what one is really looking for. This is particularly true for exploratory queries where one is searching for opinions and views, not because it is difficult to look up this kind of information, but because there is simply so much of it that one does not know where to start consuming it. Hence, it comes at no surprise that personalization systems that guide the search process are gaining importance. On the popular consumer review site Epinions[6] for instance, the order in which reviews are presented to the user is personalized and depends on the user's previous ratings of other reviews (in terms of helpfulness) and the user's social network information. Another example is Google News[7], a computer-generated news site that aggregates headlines from news sources worldwide, groups similar stories together and displays them according to each reader's personalized interests. From an e-commerce perspective too, the value of a good recommender system cannot be underestimated: Cinematch, the recommender of the American online movie rental system Netflix[8],

[6] www.epinions.com

[7] news.google.com

[8] www.netflix.com

delivers two thirds of Netflix's rented movies, and Amazon.com claims that 35% of their sales result from recommendations [51]. In essence, the recommendation problem consists of predicting the extent to which a particular user, the so-called *target user*, will like a particular item, called the *target item*, which can be a review, a news article, a movie, a book, a song, a research paper, etc. The predicted degree is usually taken from a linear scale (for instance from 1 to 5 stars) which can, without loss of generality, be mapped to $[0,1]$. Hence the predicted degree is a fuzzy membership degree that encodes preference. This preference degree is however an outcome of (and not an input to) the recommendation process, and can be arrived at through various methods discussed below. Before we go on, note that a solution to the canonical recommendation problem also implies a solution to the problem of presenting the target user with a personalized list of items, as these can be ranked in order of preference degree.

## *5.1 Content-Based Recommendations*

The content-based approach to recommendation has its roots in information retrieval and employs many of the same techniques [77]. All content-based recommender systems take the *content* of items into account, which are usually described by vectors of attributes. In a movie recommender system, for example, a movie is typically represented by a vector that contains the title, the genre, the director, the lead actors, etc., while a personalized news website can use a term frequency–inverse document frequency (TF-IDF) representation of every news article. Furthermore, content-based recommenders rely on a *profile* of the target user, which can be either manually entered by the user or derived from past behavior, such as previous ratings or purchases. Some content-based recommender systems provide an interface that allows users to construct a representation of their own interests. In this case, the recommendation process, which compares the available items with the user profile, very much resembles information retrieval as discussed in Section 3, with the user profile playing the role of a query. This is especially so when the items contain textual information, such as news articles or research papers, and the user profile consists of keywords or topics that the target user is interested in. In addition, in the fuzzy research community, systems have been proposed in which users state their information need with linguistic labels, asserting for instance that *weight* is a *very important* consideration in a new laptop they want to buy, or which *research topics* are *more or less compatible* with their interests [18, 81, 110]. These linguistic labels are then mapped to fuzzy sets which are compared to a similar fuzzy set representation of the available items (consumer products, research funding opportunities,...). This approach's achilles heel for large scale deployment seems to be the need for domain experts to evaluate the features of every item and to establish item descriptions as vectors of linguistic labels (fuzzy sets).

Other content-based recommender systems learn the user profile automatically from past behavior and recommend items that are similar to items purchased or rated highly by the target user in the past. Implementing this requires a technique to compute the similarity between items, which varies with the domain. A common approach is to use the cosine similarity between the attribute vectors that describe the items, especially for textual items represented by TF-IDF vectors in term space. To this end, items are sometimes also represented as vectors in user space, with the $p$-th entry of the vector for an item containing the rating that the $p$-th user gave to the item, or, alternatively, a 1 if the $p$-th user purchased the item and a 0 otherwise. In this case, items are considered similar to the extent to which they have a common customer or fan base. Regardless of which of these techniques is used to compute it, let us denote the similarity of items $i$ and $j$ as $Sim(i,j)$, which, without loss of generality, can be thought of as a fuzzy relation in the set of items $\mathscr{I}$. The predicted rating $\widehat{P_{\text{CB}}}(u,i)$ for target user $u$ and target item $i$ can then be computed as the weighted mean [87]

$$\widehat{P_{\text{CB}}}(u,i) = \frac{1}{\sum\limits_{j \in \mathscr{I}_u} Sim(i,j)} \sum_{j \in \mathscr{I}_u} (Sim(i,j) \cdot P(u,j)) \tag{6}$$

in which $\mathscr{I}_u$ is the set of items previously rated by $u$, and $P(u,j)$ denotes the rating that user $u$ previously gave to item $j$. In the fuzzy set community, proposals have been made to generalize the product in (6) to an arbitrary t-norm, and to replace the weighted mean by the supremum as the aggregation operation, resulting in prediction formulas such as [79, 110]

$$\widehat{P_{\text{CB}}}(u,i) = \sup_{j \in \mathscr{I}_u} T(Sim(i,j), P(u,j)) \tag{7}$$

To the best of our knowledge, no experimental studies exist on which t-norm/aggregation combination performs best on benchmark datasets.

In content-based recommender systems, items for which no description is available can not be recommended, and the accuracy of the recommendations heavily relies on the quality of the representations. Furthermore, the technique to compute the similarities is domain dependent. For instance, a content-based system developed for recommendation of reviews or news articles in English requires adaption before it can be used for other languages as well. Another drawback of content-based systems is that they tend not to explore interests of the user besides those expressed in his rating record. In this sense, they can be improved significantly by (additionally) using collaborative methods, which do not require item descriptions.

## 5.2 *Collaborative Filtering*

While content-based methods depend on the computation of similarity between items, collaborative filtering relies on similarity between users. The main idea is to recommend items that have been rated highly by users similar to the target user. Similarity between users is typically assessed based on rating behavior, i.e. users are considered similar if they (dis)like the same items, and can be computed in the same way across different domains. The similarity between users $u$ and $v$ is commonly measured with Pearson's correlation coefficient (PCC) [39]

$$Sim(u,v) = \frac{\sum\limits_{j \in \mathscr{I}_u \cap \mathscr{I}_v} (P(u,j) - \overline{P}(u)) \cdot (P(v,j) - \overline{P}(v))}{\sqrt{\left(\sum\limits_{j \in \mathscr{I}_u \cap \mathscr{I}_v} (P(u,j) - \overline{P}(u))^2\right) \cdot \left(\sum\limits_{j \in \mathscr{I}_u \cap \mathscr{I}_v} (P(v,j) - \overline{P}(v))^2\right)}} \tag{8}$$

where the summations range over all items $j$ previously rated by both $u$ and $v$, and $\overline{P}(u)$ and $\overline{P}(v)$ are the average ratings given by $u$ and $v$ so far. The PCC ranges between $-1$ and 1. A positive PCC means that both users have similar taste in the sense that, when one of them rates an item above/below average, the other one does so too. The more negative the coefficient, the more the rating behaviors are opposites, and a correlation coefficient of 0 means that there is no relationship between the two sets of ratings. In practice, most often only users with a positive correlation with target user $u$ and who have rated target item $i$ are considered in the recommendation process. We denote this set by $\mathscr{U}_i$. The predicted rating $\widehat{P}_{\text{CF}}(u,i)$ for target user $u$ and target item $i$ can then be computed as the weighted mean [1]

$$\widehat{P}_{\text{CF}}(u,i) = \frac{1}{\sum\limits_{v \in \mathscr{U}_i} Sim(u,v)} \sum_{v \in \mathscr{U}_i} (Sim(u,v) \cdot P(v,i)) \tag{9}$$

Formula (9) does not take into account the fact that not every user exhibits the same rating behavior, in the sense that some users might be easy to please and regularly give high ratings, while others have a more pronounced taste and tend to give lower ratings more often. The classic collaborative filtering formula accounts for this [1]

$$\widehat{P}_{\text{CF}}(u,i) = \overline{P}(u) + \frac{1}{\sum\limits_{v \in \mathscr{U}_i} Sim(u,v)} \sum_{v \in \mathscr{U}_i} (Sim(u,v) \cdot (P(v,i) - \overline{P}(v))) \tag{10}$$

Note, however, that such an adaption still has its limitations. For instance, as there is no correction based on the standard deviation of the scores, users whose scores are almost always around 3 will influence recommendations to a lesser extent than users who make use of the entire range from 1 to 5 on a regular basis. More fundamentally, the theoretical justification of formulas such as (10) is very loose. In principle, user ratings can only be interpreted in an ordinal way, and imposing any kind of metric on these scores is always to some extent arbitrary. Approaches which are based on difference in ratings, average ratings, etc., should therefore be seen as heuristics rather than well-founded methods. One might imagine alternative, more principled techniques which are more in the spirit of qualitative decision making [25], e.g. ranking an item $i_1$ higher than an item $i_2$ iff the set of users who have rated $i_1$ higher than $i_2$ is more similar to the target user than the set of users who have rated $i_2$ higher than $i_1$. Such methods, however, would probably suffer from other issues, such as scalability. As in the domain of information retrieval, it thus seems that there is a trade-off between looking for techniques that make sense from a theoretical point of view, and exclusively relying on experimental studies to arrive at techniques that are efficient and effective in practice.

Similarly as with content-based recommendation, in the fuzzy set community proposals have been made to replace the product in (9) by an arbitrary t-norm and to use the supremum to aggregate over the users in the neighborhood of $u$, resulting in formulas such as [22, 69, 79]

$$\widehat{P}_{\mathrm{CF}}(u,i) = \sup_{v \in \mathscr{U}_i} T(Sim(u,v), P(v,i)) \tag{11}$$

Formulas (7) and (11) are very similar in structure. In (7) the supremum ranges over all items $j$ previously rated by target user $u$, and their similarity with target item $i$ is taken into account. In (11) on the other hand, the supremum ranges over all users $v$ who are already familiar with the target item $i$; in this case the similarity between users $u$ and $v$ is an important factor.

In [69], Formula (11) is used for a web page recommender system that dynamically appends a set of links to the contents of a web document returned in response to the most recent query of an ongoing user session. Recommendations are made based on access data instead of rating behavior. User sessions are represented as attribute vectors with the $p$-th attribute equal to 1 if the $p$-th url was accessed during the session, and 0 otherwise. To limit the number of user sessions over which (11) ranges, the set of user sessions from the access log files is replaced by a smaller set of prototypical user sessions, which represent clusters found in the original set. The attribute vector describing such a prototypical user session or cluster has values between 0 and 1; the $p$-th attribute indicates the relative frequency with which the $p$-th url was visited in all user sessions belonging to the clusters. $Sim(u,v)$ is computed as the cosine similarity of the vectors for $u$ and $v$ instead of the PCC, and min is used as the t-norm in (11). The authors compare their approach with the results of a nearest profile based recommendation approach (recommend the urls visited in the prototypical user session that is most similar to the ongoing user session) and with a $k$-nearest neighbor approach followed by top-$n$ recommendations (recommend the $n$ most frequently visited urls from the $k$ most similar prototypical user sessions). They report a small drop in precision which is more than compensated for by an increase in recall. The question whether perhaps even better results could be obtained with Formula (9) remains open.

## *5.3 Social Recommenders*

When a web application with a built-in recommender offers a social networking component which enables its users to form a trust network, it can generate more personalized recommendations by combining data from the user profiles (ratings) with information from the social network. These are the so-called trust-enhanced or social recommendation systems. Ratings are predicted in a style similar to collaborative filtering, with the similarity score $Sim(u,v)$ replaced by a trust score $Trust(u,v)$ corresponding to the degree to which user $u$ trusts user $v$. The trust-based versions of (9) and (10) are at the heart of the trust-enhanced recommendation algorithms of Golbeck et al. [35] and Massa et al. [61] respectively. If no direct trust score is available (because $u$ does not know $v$), then it can often still be derived through *trust propagation* and *aggregation* in the online network, inspired by the way in which humans often seek recommendations in real life. For instance, the trust score of $u$ in $v$ can be estimated as a weighted mean of the trust scores of other users in $v$, weighted by the trust of $v$ in those other users [35, 61]

$$\widehat{Trust}(u,v) = \frac{1}{\sum_{w \in \mathscr{U}} Trust(u,w)} \sum_{w \in \mathscr{U}} (Trust(u,w) \cdot Trust(w,v)) \tag{12}$$

Formula (12) only considers one step propagation, i.e., where $u$ and $v$ are directly connected through a third party $w$; extensions that take into account longer propagation paths are possible as well. Propagation is modeled in (12) by the product. A proposal has been made to generalize this to an arbitrary t-norm and to use ordered weighting averaging operators that can deal with gradual trust as well as distrust [106, 105]. Even though some of the initial experimental results are promising, a proper evaluation of the effect of different t-norms on the performance of a trust-enhanced recommender system is currently hampered by the lack of a publicly available benchmark dataset that contains both item ratings as well as a social network with gradual trust relations.

Moreover, one may wonder what the precise meaning of a trust degree is, why a formula such as (12) is compatible with this meaning, and how such degrees can be acquired in practice. The basic intuition seems to be that friends are more likely to have similar interests than random users, which would suggest to use trust mainly to adapt the Pearson correlation in the collaborative filtering model, such that the degree of similarity between friends is boosted. The notion of trust then takes a role which is similar in spirit to that of a prior probability in Bayesian decision theory.

## 6 Conclusions

In this chapter, we have looked at the use of fuzzy set theory in three research areas that are related to the world wide web: information retrieval, the semantic web, and recommender systems. While the motivation for using fuzzy techniques is very natural in each of these domains, the most commonly used techniques are nonetheless still based on other approaches. This can partly be explained by the fact that more experimental evidence is needed to demonstrate whether fuzzy methods are really able to outperform state-of-the-art approaches. In addition, the assessment of the impact of fuzzy methods on the web is obscured by the fact that sometimes methods are used which are based on its ideas, without making use of its vocabulary.

To stimulate the future impact of fuzzy approaches to web intelligence, we believe that more efforts are needed to lay bare what fuzzy set theory really has to offer in this domain, beyond the (important) fact that it allows to develop elegant and intuitively appealing methods.

Information retrieval research is dominated by algebraic (vector space model) and probabilistic (language models) approaches. Fuzzy set theory has mainly been applied to implement more flexible ways of formulating queries, and to develop semantically informed retrieval models for particular narrow domains. In addition, fuzzy rule based methods have sometimes proven useful for translating human intuitions on how search results should be manipulated, in domains where sufficient training data is missing.

In the last decade, Tim Berners-Lee's vision of a semantic web has drawn many researchers to work on fuzzy versions of its main components. In particular, research on fuzzy description logics has substantially progressed, both at the theoretical (more expressive formalisms) and at the practical level (more efficient reasoners). More recently, there has also been a renewed interest in fuzzy logic programming, in relation to the semantic web. There exists some confusion, however, between the need for flexible querying, the presence of vague concepts, the presence of uncertainty, and the need for fuzziness at the representation level. While convincing applications of fuzzy description logics have already been developed, we are not aware of any applications that are in the spirit of the semantic web, e.g. dealing with problems that result from linking different fuzzy description logics that have been developed independently. More work is needed to clarify the advantages of fuzzy description logics over extensions of traditional web information systems that are endowed with flexible querying capabilities.

Recommender systems aim to solve a problem that is familiar to the fuzzy set community, namely predicting the degree to which a target user might like a target item. Most solutions proposed in the fuzzy set community are very similar in structure to those proposed outside. One potential advantage that fuzzy set theory has to offer is its wider variety of operators, compared with traditional approaches that tend to limit themselves to the use of the product for conjunction and the use of the mean for aggregation. The proof of the pudding is in the eating though, in this case, whether some of these other operators can lead to more and better recommendations in practice. Since the first proposals for fuzzy logic recommendation techniques were made, a variety of benchmark datasets have become available. Even though the nature of these datasets does not allow yet to empirically evaluate the use of fuzzy methods for trust-enhanced recommender systems, an evaluation of fuzzy methods for content-based and collaborative filtering seems a feasible and logical next step.

**Acknowledgements.** Steven Schockaert was funded as a postdoctoral fellow of the Research Foundation – Flanders.

## References


1. Adomavicius, G., Tuzhilin, A.: Toward the next generation of recommender systems: a survey of the state-of-the-art and possible extensions. IEEE Transactions on Knowledge and Data Engineering 17(6), 734–749 (2005)
2. Arampatzis, A., van Kreveld, M., Reinbacher, I., Jones, C., Vaid, S., Clough, P., Joho, H., Sanderson, M.: Web-based delineation of imprecise regions. Computers, Environment and Urban Systems 30(4), 436–459 (2006)
3. Arotaritei, D., Mitra, S.: Web mining: a survey in the fuzzy framework. Fuzzy Sets and Systems 148(1), 5–19 (2004)
4. Baader, F., Calvanese, D., McGuinness, D.L., Nardi, D., Patel-Schneider, P.F.: The Description Logic Handbook. Cambridge University Press, Cambridge (2007)
5. Bauters, K., Schockaert, S., De Cock, M., Vermeir, D.: Possibilistic answer set programming revisited. In: Proceedings of the 26th Conference on Uncertainty in Artificial Intelligence (to appear)
6. Baziz, M., Boughanem, M., Loiseau, Y., Prade, H.: Fuzzy logic and ontology-based information retrieval. In: Wang, P., Ruan, D., Kerre, E. (eds.) Fuzzy Logic: A Spectrum of Theoretical & Practical Issues, pp. 193–218. Springer, Heidelberg (2007)
7. Baziz, M., Boughanem, M., Prade, H., Pasi, G.: A fuzzy logic approach to information retrieval using a ontology-based representation of documents. In: Fuzzy logic and the Semantic Web, pp. 363–377. Elsevier, Amsterdam (2006)
8. Benferhat, S., Dubois, D., Kaci, S., Prade, H.: Possibilistic merging and distance-based fusion of propositional information. Annals of Mathematics and Artificial Intelligence 34(1-3), 217–252 (2002)
9. Benferhat, S., Lagrue, S., Papini, O.: Reasoning with partially ordered information in a possibilistic logic framework. Fuzzy Sets and Systems 144(1), 25–41 (2004)
10. Benferhat, S., Sossai, C.: Reasoning with multiple-source information in a possibilistic logic framework. Information Fusion 7(1), 80–96 (2006)
11. Berners-Lee, T., Hendler, J., Lassila, O.: The semantic web. Scientific American 284(5), 28–37 (2001)
12. Binaghi, E., Gagliardi, I., Schettini, R.: Image retrieval using fuzzy evaluation of color similarity. International Journal of Pattern Recognition and Artificial Intelligence 8, 945 (1994)
13. Bobillo, F., Delgado, M., Gómez-Romero, J.: Optimizing the crisp representation of the fuzzy description logic SROIQ. In: da Costa, P.C.G., d'Amato, C., Fanizzi, N., Laskey, K.B., Laskey, K.J., Lukasiewicz, T., Nickles, M., Pool, M. (eds.) URSW 2005 - 2007. LNCS (LNAI), vol. 5327, pp. 189–206. Springer, Heidelberg (2008)
14. Bordogna, G., Pasi, G.: A fuzzy linguistic approach generalizing boolean information retrieval: A model and its evaluation. Journal of the American Society for Information Science 44(2), 70–82 (1993)
15. Bosteels, K., Kerre, E.E.: Fuzzy audio similarity measures based on spectrum histograms and fluctuation patterns. In: Proceedings of the 2007 International Conference on Multimedia and Ubiquitous Engineering, pp. 361–365 (2007)
16. Bosteels, K., Pampalk, E., Kerre, E.E.: On the benefits of representing music objects as fuzzy sets. In: Proceedings of the IFSA/EUSFLAT Conference, pp. 25–29 (2009)
17. Boughanem, M., Brini, A., Dubois, D.: Possibilistic networks for information retrieval. International Journal of Approximate Reasoning 50(7), 957–968 (2009)

18. Cao, Y., Li, Y.: An intelligent fuzzy-based recommendation system for consumer electronic products. Expert Systems with Applications 33, 230–240 (2007)
19. Chen, Y., Wang, J.: A region-based fuzzy feature matching approach to content-based image retrieval. IEEE Transactions on Pattern Analysis and Machine Intelligence 24(9), 1252–1267 (2002)
20. Confalonieri, R., Nieves, J.C., Vázquez-Salceda, J.: Pstable semantics for logic programs with possibilistic ordered disjunction. In: Proceedings of the Conference of the Italian Association for Artificial Intelligence, pp. 52–61 (2009)
21. Cordón, O., Moya, F., Zarco, C.: A new evolutionary algorithm combining simulated annealing and genetic programming for relevance feedback in fuzzy information retrieval systems. Soft Computing 6(5), 308–319 (2002)
22. Cornelis, C., Lu, J., Guo, X., Zhang, G.: One-and-only item recommendations with fuzzy logic techniques. Information Sciences 177, 4906–4921 (2007)
23. Crestani, F., Pasi, G.: Handling vagueness, subjectivity, and imprecision in information access: an introduction to the special issue. Information Processing & Management 39(2), 161–165 (2003)
24. Damásio, C.V., Pereira, L.M.: Hybrid probabilistic logic programs as residuated logic programs. In: Brewka, G., Moniz Pereira, L., Ojeda-Aciego, M., de Guzmán, I.P. (eds.) JELIA 2000. LNCS (LNAI), vol. 1919, pp. 57–72. Springer, Heidelberg (2000)
25. Dubois, D., Fargier, H., Prade, H.: Decision-making under ordinal preferences and comparative uncertainty. In: Proceedings of the 13th Conference on Uncertainty in Artificial Intelligence, pp. 157–164 (1997)
26. Dubois, D., Lang, J., Prade, H.: Towards possibilistic logic programming. In: Proceedings of the International Conference on Logic Programming, pp. 581–595 (1991)
27. Dubois, D., Mengin, J., Prade, H.: Possibilistic uncertainty and fuzzy features in description logic: A preliminary discussion. Fuzzy logic and the Semantic Web, 101–113 (2006)
28. Dubois, D., Ostasiewicz, W., Prade, H.: Fuzzy sets: History and basic notions. In: Dubois, D., Prade, H. (eds.) Fundamentals of Fuzzy Sets. The Handbooks of Fuzzy Sets Series, pp. 21–124. Kluwer Academic Publishers, Dordrecht (2000)
29. Dubois, D., Prade, H.: Possibility Theory: An Approach to Computerized Processing of Uncertainty. Plenum Press, New York (1988)
30. Dubois, D., Prade, H.: Fuzzy sets, probability and measurement. European Journal of Operational Research 40(2), 135–154 (1989)
31. Dubois, D., Prade, H.: The three semantics of fuzzy sets. Fuzzy Sets and Systems 90(2), 141–150 (1997)
32. Eiter, T., Ianni, G., Lukasiewicz, T., Schindlauer, R., Tompits, H.: Combining answer set programming with description logics for the semantic web. Artificial Intelligence 172(12-13), 1495–1539 (2008)
33. Gabrilovich, E., Markovitch, S.: Computing semantic relatedness using wikipedia-based explicit semantic analysis. In: Proceedings of the 20th International Joint Conference on Artifical Intelligence, pp. 1606–1611 (2007)
34. Gelfond, M., Lifschitz, V.: The stable model semantics for logic programming. In: Proceedings of the Fifth International Conference and Symposium on Logic Programming, pp. 1081–1086 (1988)
35. Golbeck, J., Hendler, J.: Filmtrust: movie recommendations using trust in web-based social networks. In: Proceedings of the 3rd IEEE Consumer Communications and Networking Conference, pp. 282–286 (2006)

36. Hájek, P.: On the Metamathematics of Fuzzy Logic. Kluwer Academic Press, Dordrecht (1998)
37. Hájek, P.: Making fuzzy description logic more general. Fuzzy Sets and Systems 154(1), 1–15 (2005)
38. Hedlund, T., Pirkola, A., Järvelin, K.: Aspects of swedish morphology and semantics from the perspective of mono- and cross-language information retrieval. Information Processing & Management 37(1), 147–161 (2001)
39. Herlocker, J., Konstan, J., Terveen, L., Riedl, J.: Evaluating collaborative filtering recommender systems. ACM Transactions on Information Systems 22, 5–53 (2004)
40. Heymans, S., Toma, I.: Ranking services using fuzzy hex programs. In: Proceedings of the 2nd International Conference on Web Reasoning and Rule Systems, pp. 181–196 (2008)
41. Hisdal, E.: Are grades of membership probabilities? Fuzzy Sets and Systems 25(3), 325–348 (1988)
42. Hollunder, B.: An alternative proof method for possibilistic logic and its application to terminological logics. International Journal of Approximate Reasoning 12(2), 85–109 (1995)
43. Höppner, F., Klawonn, F., Kruse, R., Runkler, T.: Fuzzy Cluster Analysis: Methods for Classification, Data Analysis and Image Recognition. Wiley, Chichester (1999)
44. Janssen, J., Schockaert, S., Vermeir, D., De Cock, M.: General fuzzy answer set programs. In: Proceedings of the 8th International Workshop on Fuzzy Logic and Applications (WILF), pp. 352–359 (2009)
45. Jing, Y., Croft, W.: An association thesaurus for information retrieval. In: Proceedings of RIAO 1994, pp. 146–160 (1994)
46. Kerre, E., Zenner, R., De Caluwe, R.: The use of fuzzy set theory in information retrieval and databases: A survey. Journal of the American Society for Information Science 37(5), 341–341 (1986)
47. Kraft, D., Chen, J., Martin-Bautista, M., Vila, M.: Textual information retrieval with user profiles using fuzzy clustering and inferencing. Studies In Fuzziness And Soft Computing, 152–165 (2003)
48. Krishnapuram, R., Joshi, A., Yi, L.: A fuzzy relative of the k-medoids algorithm with application to web document and snippet clustering. In: IEEE International Fuzzy Systems Conference Proceedings, vol. 3, pp. 1281–1286 (1999)
49. Krishnapuram, R., Medasani, S., Jung, S., Choi, Y., Balasubramaniam, R.: Content-based image retrieval based on a fuzzy approach. IEEE Transactions on Knowledge and Data Engineering 16(10), 1185–1199 (2004)
50. La Cascia, M., Sethi, S., Sclaroff, S.: Combining textual and visual cues for content-based image retrieval on the world wide web. In: IEEE Workshop on Content-Based Access of Image and Video Libraries, pp. 24–28 (1998) Citeseer
51. Lamere, P., Celma, O.: Music recommendation tutorial (2007)
52. Lukasiewicz, T.: Fuzzy description logic programs under the answer set semantics for the semantic web. Fundamenta Informaticae 82(3), 289–310 (2008)
53. Lukasiewicz, T., Straccia, U.: Tightly integrated fuzzy description logic programs under the answer set semantics for the semantic web. In: Proceedings of the 1st International Conference on Web Reasoning and Rule Systems, pp. 289–298 (2007)
54. Lukasiewicz, T., Straccia, U.: Top-k retrieval in description logic programs under vagueness for the semantic web. In: Prade, H., Subrahmanian, V.S. (eds.) SUM 2007. LNCS (LNAI), vol. 4772, pp. 16–30. Springer, Heidelberg (2007)

55. Lukasiewicz, T., Straccia, U.: Description logic programs under probabilistic uncertainty and fuzzy vagueness. International Journal of Approximate Reasoning 50(6), 837–853 (2009)
56. Mamdani, E., Assilian, S.: An experiment in linguistic synthesis with a fuzzy logic controller. International Journal of Man-Machine Studies 7(1), 1–13 (1975)
57. Martin-Bautista, M., Kraft, D., Vila, M., Chen, J., Cruz, J.: User profiles and fuzzy logic for web retrieval issues. Soft Computing 6(5), 365–372 (2002)
58. Martin-Bautista, M., Sanchez, D., Chamorro-Martinez, J., Serrano, J., Vila, M.: Mining web documents to find additional query terms using fuzzy association rules. Fuzzy Sets and Systems 148(1), 85–104 (2004)
59. Martin-Bautista, M., Vila, M., Larsen, H.: A fuzzy genetic algorithm approach to an adaptive information retrieval agent. Journal of the American Society for Information Science 50(9), 760–771 (1999)
60. Martin-Bautista, M.J., Sanchez, D., Vila, M.A., Delgado, M.: A new fusion strategy for web metasearch. Soft Computing 14(8), 847–855 (2009)
61. Massa, P., Avesani, P.: Trust metrics on controversial users: balancing between tyranny of the majority and echo chambers. International Journal on Semantic Web and Information Systems 3, 39–64 (2007)
62. McCloskey, M., Glucksberg, S.: Decision processes in verifying category membership statements: Implications for models of semantic memory. Cognitive Psychology 11(1), 1–37 (1979)
63. Medina, J., Ojeda-Aciego, M., Vojtá, P.: Similarity-based unification: a multi-adjoint approach. Fuzzy Sets and Systems 146(1), 43–62 (2004)
64. Medina, J., Ojeda-Aciego, M., Vojtáš, P.: Multi-adjoint logic programming with continuous semantics. In: Eiter, T., Faber, W., Truszczyński, M. (eds.) LPNMR 2001. LNCS (LNAI), vol. 2173, pp. 351–364. Springer, Heidelberg (2001)
65. Meghini, C., Sebastiani, F., Straccia, U.: A model of multimedia information retrieval. Journal of the ACM 48(5), 909–970 (2001)
66. Miyamoto, S.: Information retrieval based on fuzzy associations. Fuzzy Sets and Systems 38(2), 191–205 (1990)
67. Montello, D., Goodchild, M., Gottsegen, J., Fohl, P.: Where's downtown?: behavioral methods for determining referents of vague spatial queries. Spatial Cognition and Computation 3(2-3), 185–204 (2003)
68. Nachtegael, M., van der Weken, D., de Witte, V., Schulte, S., Melange, T., Kerre, E.: Color image retrieval using fuzzy similarity measures and fuzzy partitions. In: IEEE International Conference on Image Processing, pp. 13–16 (2007)
69. Petenes, C., Nasraoui, O.: Combining web usage mining and fuzzy inference for website personalization. In: Proceedings of WebKDD, pp. 37–46 (2003)
70. Nauck, D., Kruse, R.: A neuro-fuzzy method to learn fuzzy classification rules from data. Fuzzy Sets and Systems 89(3), 277–288 (1997)
71. Nicolas, P., Garcia, L., Stéphan, I., Lefèvre, C.: Possibilistic uncertainty handling for answer set programming. Annals of Mathematics and Artificial Intelligence 47(1-2), 139–181 (2006)
72. Nie, J.-Y., Ren, F.: Chinese information retrieval: using characters or words? Information Processing & Management 35(4), 443–462 (1999)
73. Nie, Z., Ma, Y., Shi, S., Wen, J.-R., Ma, W.-Y.: Web object retrieval. In: Proceedings of the 16th International Conference on World Wide Web, pp. 81–90 (2007)
74. Nieves, J.C., Osorio, M., Cortés, U.: Semantics for possibilistic disjunctive programs. In: Proceedings of the International Conference on Logic Programming and Nonmonotonic Reasoning, pp. 315–320 (2007)

75. Pan, J.Z., Stamou, G., Stoilos, G., Taylor, S., Thomas, E.: Scalable querying services over fuzzy ontologies. In: Proceedings of the 17th International Conference on World Wide Web, pp. 575–584 (2008)
76. Pavelka, J.: On fuzzy logic. Zeitschr. f. Math. logik und Grundl. der Math. 25, 45–52, 119–134, 447–464 (1979)
77. Pazzani, M., Billsus, D.: Content-based recommendation systems. In: Brusilovsky, P., Kobsa, A., Nejdl, W. (eds.) Adaptive Web 2007. LNCS, vol. 4321, pp. 325–341. Springer, Heidelberg (2007)
78. Pérez-Iglesias, J., Fresno, V., Pérez-Agüera, J.: FuzzyFresh: A Fuzzy Logic Approach to the Ranking of Structured Documents. In: Proceedings of the 2008 IEEE/WIC/ACM International Conference on Web Intelligence and Intelligent Agent Technology, pp. 755–758 (2008)
79. Zucker, J.D., Perny, P.: Collaborative filtering methods based on fuzzy preference relations. In: Proceedings of EUROFUSE-SIC, pp. 279–285 (1999)
80. Ponte, J.M., Croft, W.B.: A language modeling approach to information retrieval. In: Proceedings of the 21st Annual International ACM SIGIR Conference on Research and Development in Information Retrieval, pp. 275–281 (1998)
81. Porcel, C., López-Herrera, A., Herrera-Viedma, E.: A recommender system for research resources based on fuzzy linguistic modeling. Expert Systems with Applications 36, 5173–5183 (2009)
82. Qi, G., Pan, J.Z., Ji, Q.: Extending description logics with uncertainty reasoning in possibilistic logic. In: Mellouli, K. (ed.) ECSQARU 2007. LNCS (LNAI), vol. 4724, pp. 828–839. Springer, Heidelberg (2007)
83. Radecki, T.: Mathematical model of information retrieval system based on the concept of fuzzy thesaurus. Information Processing & Management 12(5), 313–318 (1976)
84. Ruspini, E.: Numerical methods for fuzzy clustering. Information Sciences 2(3), 319–350 (1970)
85. Salton, G., Wong, A., Yang, C.S.: A vector space model for automatic indexing. Communications of the ACM 18(11), 613–620 (1975)
86. Sanchez, E. (ed.): Fuzzy Logic and the Semantic Web. Capturing Intelligence. Elsevier, Amsterdam (2006)
87. Sarwar, B., Karypis, G., Konstan, J., Riedl, J.: Item-based collaborative filtering recommendation algorithms. In: Proceedings of the 10th International Conference on World Wide Web, pp. 285–295 (2001)
88. Schockaert, S., De Cock, M.: Neighborhood restrictions in geographic IR. In: Proceedings of the 30th Annual International ACM SIGIR Conference on Research and Development in Information Retrieval, pp. 167–174 (2007)
89. Schockaert, S., De Cock, M.: Temporal reasoning about fuzzy intervals. Artificial Intelligence 172, 1158–1193 (2008)
90. Schockaert, S., De Cock, M., Cornelis, C., Kerre, E.: Clustering web search results using fuzzy ants. International Journal of Intelligent Systems 22(5), 455–474 (2007)
91. Schockaert, S., De Cock, M., Kerre, E.: Automatic acquisition of fuzzy footprints. In: Meersman, R., Tari, Z., Herrero, P. (eds.) OTM-WS 2005. LNCS, vol. 3762, pp. 1077–1086. Springer, Heidelberg (2005)
92. Schockaert, S., De Cock, M., Kerre, E.E.: Reasoning about fuzzy temporal information from the web: towards retrieval of historical events. Soft. Computing 14(8), 869–886 (2010)
93. Shackle, G.: Decision, Order and Time in Human Affairs. Cambridge University Press, Cambridge (1961)

94. Sinclair, J., Cardew-Hall, M.: The folksonomy tag cloud: when is it useful? Journal of Information Science 34(1), 15–29 (2008)
95. Smeulders, A., Worring, M., Santini, S., Gupta, A., Jain, R.: Content-based image retrieval at the end of the early years. IEEE Transactions on Pattern Analysis and Machine Intelligence, 1349–1380 (2000)
96. Soto, A., Olivas, J., Prieto, M.: Fuzzy Approach of Synonymy and Polysemy for Information Retrieval. Granular Computing: At the Junction of Rough Sets and Fuzzy Sets, 179–198 (2008)
97. Stoilos, G., Stamou, G., Pan, J.Z., Tzouvaras, V., Horrocks, I.: Reasoning with very expressive fuzzy description logics. Journal of Artificial Intelligence Research 30(1), 273–320 (2007)
98. Stoilos, G., Stamou, G., Tzouvaras, V., Pan, J., Horrocks, I.: Reasoning with the fuzzy description logic f-SHIN: theory, practice and applications. In: da Costa, P.C.G., d'Amato, C., Fanizzi, N., Laskey, K.B., Laskey, K.J., Lukasiewicz, T., Nickles, M., Pool, M. (eds.) URSW 2005 - 2007. LNCS (LNAI), vol. 5327, pp. 262–281. Springer, Heidelberg (2008)
99. Stoilos, G., Straccia, U., Stamou, G., Pan, J.Z.: General concept inclusions in fuzzy description logics. In: Proceeding of the European Conference on Artificial Intelligence, pp. 457–461 (2006)
100. Straccia, U.: Reasoning within fuzzy description logics. Journal of Artificial Intelligence Research 4, 137–166 (2001)
101. Straccia, U.: Description logics with fuzzy concrete domains. In: 21st Conference on Uncertainty in Artificial Intelligence, pp. 559–567 (2005)
102. Straccia, U.: Annotated answer set programming. In: Proceedings of the 11th International Conference on Information Processing and Management of Uncertainty in Knowledge-Based Systems, IPMU 2006 (2006)
103. Straccia, U.: Fuzzy description logic programs. In: Proceedings of the 11th International Conference on Information Processing and Management of Uncertainty in Knowledge-Based Systems, pp. 1818–1825 (2006)
104. Straccia, U., Ojeda-Aciego, M., Damásio, C.V.: On fixed-points of multivalued functions on complete lattices and their application to generalized logic programs. SIAM Journal on Computing 38(5), 1881–1911 (2009)
105. Victor, P.: Trust Networks for Recommender Systems. PhD thesis, Ghent University (2010)
106. Victor, P., Cornelis, C., De Cock, M., Pinheiro da Silva, P.: Gradual trust and distrust in recommender systems. Fuzzy Sets and Systems 160(10), 1367–1382 (2009)
107. Vojtás, P.: Fuzzy logic programming. Fuzzy Sets and Systems 124(3), 361–370 (2001)
108. Vrettos, S., Stafylopatis, A.: A Fuzzy Rule-Based Agent for Web Retrieval-Filtering. In: Simons, B., Spector, A.Z. (eds.) Fault-Tolerant Distributed Computing. LNCS, vol. 448, p. 453. Springer, Heidelberg (1990)
109. Waters, T., Evans, A.: Tools for web–based GIS mapping of a "fuzzy" vernacular geography. In: Proceedings of the 7th International Conference on GeoComputation (2003)
110. Yager, R.R.: Fuzzy logic methods in recommender systems. Fuzzy Sets and Systems 136(2), 133–149 (2003)
111. Zadeh, L.A.: Fuzzy sets. Information and Control 8, 338–353 (1965)
112. Zadeh, L.A.: Fuzzy sets as a basis for a theory of possibility. Fuzzy Sets and Systems 1(1), 3–28 (1978)

113. Zadeh, L.A.: A note on web intelligence, world knowledge and fuzzy logic. Data & Knowledge Engineering 50(3), 291–304 (2004)
114. Zhao, Y., Karypis, G.: Soft clustering criterion functions for partitional document clustering: a summary of results. In: Proceedings of the Thirteenth ACM International Conference on Information and Knowledge Management, pp. 246–247 (2004)
115. Zhou, X., Hu, X., Zhang, X., Lin, X., Song, I.-Y.: Context-sensitive semantic smoothing for the language modeling approach to genomic ir. In: Proceedings of the 29th Annual International ACM SIGIR Conference on Research and Development in Information Retrieval, pp. 170–177 (2006)
116. Zobel, J., Moffat, A.: Exploring the similarity space. SIGIR Forum 32(1), 18–34 (1998)

# Using Fuzzy and Interval-Valued Fuzzy Sets in Automatic Text Categorization Based on a Fuzzy Information Retrieval Model

Sławomir Zadrożny, Janusz Kacprzyk, and Katarzyna Nowacka

**Abstract.** We consider the problem of categorization of the textual documents which is relevant and challenging both from the point of view of theory and applications. We assume a perspective (cf. Zadrożny and Nowacka [28]) that the problem is seen as a sort of the basic information retrieval task, that is, of finding documents relevant to a given query. Specifically, we employ here some extension of a fuzzy logic based information retrieval model due to Nowacka, Kacprzyk and Zadrożny [21] in which the representation of documents and queries is based on Zadeh's linguistic statements of the type $X$ IS $A$ and their matching is computed by pairs of the necessity and possibility measures. We show the use of interval valued fuzzy sets to implement the new method proposed. Moreover, these new concepts are proposed as tools to adapt an inductive learning method of Koriche and Quinqueton [15] for the purposes of text categorization in the case of imprecise (fuzzy) information.

**Keywords:** text categorization, information retrieval, information retrieval fuzzy model, fuzzy set, interval valued fuzzy set, version space, inductive learning.

## 1 Introduction

An overwhelming majority of information that exists in the present world is contained in textual documents. More and more of them are available in electronic form, notably via the Internet. An effective and efficient use of such vast information resources boils down to being able to quickly retrieve

Sławomir Zadrożny · Janusz Kacprzyk · Katarzyna Nowacka
Systems Research Institute
Polish Academy of Sciences
Newelska 6, 01-447 Warsaw, Poland
e-mail: {Slawomir.Zadrozny,Janusz.Kacprzyk}@ibspan.waw.pl

C. Cornelis et al. (Eds.): 35 Years of Fuzzy Set Theory, STUDFUZZ 261, pp. 267–291.
springerlink.com 

documents meeting the users' information needs, preferences and intentions. This is a problem traditionally studied in the area of *information retrieval.* The basic issues concern first a proper representation of documents and of (human) information needs (*queries*), and then proper ways of matching them against each other. Various approaches have been developed to solve these problems and they are usually referred to as *information retrieval models.* It is easy to see that, first, textual documents are usually strongly related to natural language, and – second – natural language is the only fully natural means of articulation and communication of the human beings. Since natural language is inherently imprecise, fuzzy fuzzy logic has been find useful in this area quite early. However there still does not seem to exist a comprehensive model dealing with the main tasks of information retrieval using the core of fuzzy logic meant in the wide sense, i.e., linguistic statements introduced by Zadeh [26, 25]. Recently we have made an attempt to close this gap and proposed such a model [28] the advantage of which has been shown as the possibility to simultaneously grasp the imprecision and uncertainty within the same theoretical framework.

Another important problem concerning the collections of documents is their proper organization. One of the key tasks is here a proper grouping of documents so that they form coherent subcollections. Due to the growing size of document collections, the often employed human grouping may be impossible due to financial and/or time constraints, and automatic methods may only be feasible. Usually, some classification techniques are employed to solve this problem. A set of documents correctly assigned to particular groups (categories) is presented to the system and it learns the rules of a proper classification of a new document without known assignment. This problem is usually, also in this paper, referred to as *text categorization.* It is easy to notice that some tasks are common for the text categorization and the basic information retrieval task of finding documents relevant with respect to a query. For example, categories may be treated as documents and the new documents can be classified as queries. Then, provided that some representation is assumed for the categories and documents, the assignment of a document to a given category may be equated with whether a document is relevant to a query.

In this paper we follow this line of reasoning and discuss similarities between the retrieval of relevant documents and the text categorization using our fuzzy information retrieval model. The construction of the categorization algorithm and its connection with the fuzzy information retrieval model are presented, together with results of some computational experiments.

In Section 2 we recall some basic concepts and properties concerning fuzzy logic, including the concept of the interval-valued fuzzy set. Section 3 briefly presents the essence of our fuzzy information retrieval model. Section 4 provides a brief introduction to text categorization. In Section 5 the idea of the inductive learning algorithm proposed in [15] is presented with its fuzzy extension. Section 6 presents our algorithm for text categorization, based on the inductive learning algorithm proposed in [15], as well as results of the computational experiments. The appendix contains a list of stopwords used in the computational experiments.

## 2 Basic Concepts of Fuzzy Logic

In this section we remind the basic concepts of the broadly meant fuzzy logic which will be later needed to describe our fuzzy information retrieval model and text categorization algorithm.

A *fuzzy set* $F$ in $X = \{x\}$, will be characterized and equated with its membership function $\mu_F : X \longrightarrow [0,1]$ such that $\mu_F(x) \in [0,1]$ is the grade of membership of $x \in X$ in $F$, from full membership to full non-membership, through all intermediate values. For a finite $X = \{x_1, \ldots, x_n\}$ we write $A = \mu_A(x_1)/x_1 + \cdots + \mu_A(x_n)/x_n$. Moreover, we denote $a \wedge b = \min(a,b)$ and $a \vee b = \max(a,b)$. Other, more specific notation will be given when needed.

The counting of elements is crucial for our discussion, and we will use the basic, classic concept of a *scalar cardinality* of a fuzzy set, so-called $\Sigma$-Count defined as follows:

$$\text{card}(F) = \Sigma\text{-Count}(F) = \sum_i \mu_F(x_i) \tag{1}$$

The term *fuzzy logic* is not uniquely meant and there are basic meanings adopted: fuzzy logic in the narrow sense, which is some sort of multiple valued logic with some operations and definitions that have some specific interpretations, and fuzzy logic in the broad sense meant as a foundation of imprecise reasoninh/ In this work will be follow the second view, which is the original Zadeh's one, and – to be more specific – we will consider fuzzy logic in terms of Zadeh's *calculus of linguistic statements.*

The atomic formula in that calculus of linguistic statements takes the form:

$$X \text{ IS } F \tag{2}$$

where $X$ denotes a (linguistic) variable [25] and $F \in \mathcal{F}(U)$ a fuzzy set in the universe of discourse $U$.

The truth of (2) is considered under a valuation (knowledge) of the variable $X$ expressed by another expression of type (2), say $X$ IS $E$, $E \in \mathcal{F}(U)$.

It is assumed that $X$ IS $E$ generates a *possibility distribution* $\pi_E : U \rightarrow [0,1]$ in the space of possible values of $X$, and (cf. Zadeh [26])

$$\pi_E(x) = \mu_E(x) \tag{3}$$

The truth of $X$ IS $F$ under the valuation (knowledge) $X$ IS $E$ may be evaluated by a pair of values of *possibility* and *necessity* measures, $\Pi_E$ and $\mathrm{N}_E$, related to the possibility distribution $\pi_E$ (3):

$$\Pi_E(F) = \sup_{x \in U} \min(\pi_E(x), \mu_F(x)) \tag{4}$$

$$\mathrm{N}_E(F) = \inf_{x \in U} \max(1 - \pi_E(x), \mu_F(x)) \tag{5}$$

More generally, we can assume a fuzzy truth value but this case will not be considered here.

The linguistic statements (2) may be combined using *logical connectives.* For example the *conjunction* $X_1$ IS $E_1 \wedge \ldots X_n$ IS $E_n$, $E_i \in \mathcal{F}(U_i)$, with each $X_i$ IS $E_i$ generating $\pi_{E_i}$ by (3), generates a *joint possibility distribution* $\pi$ on $U = U_1 \times \ldots \times U_n$ such that:

$$\pi(x) = \min(\pi_{E_1}(x_1), \ldots, \pi_{E_n}(x_n)), \quad x = (x_1, \ldots, x_n) \in U \tag{6}$$

assuming the *non-interactiveness* [26] of the particular variables $X_i$. Then the truth of the conjunction of $n$ linguistic statements $X_i$ IS $F_i$, $i = 1, \ldots, n$, $F_i \in \mathcal{F}(U_i)$, may be assessed by the pair of measures:

$$\Pi_{E_1 \times \ldots \times E_n}(F_1 \times \ldots \times F_n) = \min(\Pi_{E_1}(F_1), \ldots, \Pi_{E_n}(F_n)) \tag{7}$$

$$\mathrm{N}_{E_1 \times \ldots \times E_n}(F_1 \times \ldots \times F_n) = \min(\mathrm{N}_{E_1}(F_1), \ldots, \mathrm{N}_{E_n}(F_n)) \tag{8}$$

Zadeh introduced also the extended forms of statements (2) [26, 7], i.e., *qualified statements*, which involve some qualifiers exemplified by importances or hedges. These qualified statements are relevant for our fuzzy information retrieval model but will not be dealt with here due to lack of space.

Among many extensions of the basic concept of a fuzzy set, exemplified by an $l$-fuzzy set, type 2 ($n$, in general) fuzzy set, flou set, intuitionistic fuzzy set, etc, of a particular importance and usefulness for our purposes is an *interval-valued fuzzy set* $H$ in the universe $X$ is defined by two membership functions, lower $\underline{\mu}_H$ and upper $\overline{\mu}_H$:

$$\underline{\mu}_H, \overline{\mu}_H : X \longrightarrow [0, 1] \tag{9}$$

$$\forall x \in X \qquad 0 \leq \underline{\mu}_H(x) \leq \overline{\mu}_H(x) \leq 1 \tag{10}$$

Equivalently an interval-valued fuzzy set may be interpreted as having membership function, values of which are intervals $[\underline{\mu}(x), \overline{\mu}(x)]$ instead of single numbers. The basic operations of the complement, intersection and union on the interval-valued fuzzy sets are given as the operations on *interval-valued truth values* as:

$$\overline{[\underline{\mu}(x), \overline{\mu}(x)]} = [1 - \overline{\mu}(x), 1 - \underline{\mu}(x)] \tag{11}$$

$$[\underline{\mu}(x), \overline{\mu}(x)] \wedge [\underline{\mu}(y), \overline{\mu}(y)] = [\min(\underline{\mu}(x), \underline{\mu}(y)), \min(\overline{\mu}(x), \overline{\mu}(y)] \tag{12}$$

$$[\underline{\mu}(x), \overline{\mu}(x)] \vee [\underline{\mu}(y), \overline{\mu}(y)] = [\max(\underline{\mu}(x), \underline{\mu}(y)), \max(\overline{\mu}(x), \overline{\mu}(y)] \tag{13}$$

The cardinality of an interval-valued fuzzy set $H$ may be defined using the $\Sigma$-Count concept (1):

$$\text{card}(H) = \left[ \sum_i \underline{\mu}_H(x_i), \sum_i \overline{\mu}_H(x_i) \right] \tag{14}$$

We will not deal with more definitions and properties of interval-valued fuzzy sets, and for more information the reader is referred to, for instance, Cornelis, Deschrijver and Kerre [4], Deschrijver [5] or Deschrijver and Kerre [6].

## 3 A Fuzzy Information Retrieval Model

In our previous work [21] we proposed a fuzzy logic based information retrieval model which seemed to be a very general and comprehensive approach to information retrieval using fuzzy logic. We also refer the reader to our previous papers [29, 2, 13, 30] showing the steps leading to the model presented in [21]. It has been further developed in our next paper [28]. Our starting point is the classic Boolean model (cf., e.g., [1]) and its fuzzy logic based extension, notably due to Bordogna and Pasi [3].

To represent documents and queries, linguistic statements $X$ IS $A$ (2) are used, where $X$ is a linguistic variable [25] and $A$ its linguistic value modelled by a fuzzy set which will also be denoted as $A$. For each keyword $t_i$ a linguistic variable $X_i$ is defined which represents importance of this keyword in the representation of a query/document.

The statement $X_i$ IS $A$ is a generic form of the expressions exemplified by:

> "Keyword $t_i$ is *fairly important* for the representation of the content of the document (query)"

so that we can model imprecision concerning the actual importance of the keywords.

It is quite obvious that besides imprecision also uncertainty is usually related to the representation of both queries and documents. Thus the *certainty qualified statements* [26, 7] are employed that make it possible to take uncertainty into account. During the actual retrieval the certainty qualified statements are transformed to qualifier-free forms of (2) using rules discussed, e.g., by Dubois and Prade [7]. More details on the representation of uncertainty within our model may be found in our other works [21, 14, 28].

In the model considered a document is represented as a compound linguistic statement built of expressions (2). The most typical form of such a statement is the conjunction:

$$(X_1 \text{ IS } A_1) \wedge (X_2 \text{ IS } A_2) \wedge \ldots \wedge (X_n \text{ IS } A_n) \tag{15}$$

where $X_i$ is a linguistic variable expressing the importance of keyword $t_i$ and $A_i$ is its linguistic value such as "very important", "important to a degree around 0.8", etc.; $n$ is the number of all keywords used to index the documents.

The linguistic values $A_i$ are modelled by fuzzy sets defined in the interval $[0,1]$ which is assumed as the range of importance degrees. Thus each such a statement generates a possibility distribution $\pi_{A_j}$ on the space of importance degrees of a given keyword, according to (3). The whole document represented by (15) generates a multidimensional possibility distribution as in (6):

$$\pi_D(x_1,\ldots,x_n) = \min(\mu_{A_1}(x_1),\ldots,\mu_{A_n}(x_n)) \tag{16}$$

assuming the non-interactiveness of variables corresponding to the importances of particular keywords.

A query is also represented as a compound linguistic statement built of the statements $X$ IS $A$ (2). A query is treated as a fuzzy set $Q$ in a multidimensional space $[0,1]^n$. In case a query $Q$ is a conjunction such as (15), we have:

$$\mu_Q(x_1,\ldots,x_n) = \min(\mu_{A_1}(x_1),\ldots,\mu_{A_n}(x_n)) \tag{17}$$

The relevance of a document against a query is identified with the truth degree of the statement $Q$ representing the query under the assumption that the statement $D$ representing the document is true.

We use the pair of necessity and possibility values (4)–(5), and obtain:

$$(\mathrm{N}_D(Q), \Pi_D(Q)) \tag{18}$$

The documents obtained as a response to a query are lexicographically ordered on these pairs of values. Thus, in our model the matching degree expresses the possibility and necessity of matching between a document and a query.

This presentation of the theoretical foundations of our model is very brief, just containing the concept needed. For more information on other elements of the model as well as many pragmatic aspects of its use we refer the reader to our works [21, 28]. The most important is the question how the shapes of the fuzzy sets $A_i$ in (15) are to be determined. Here we are particularly interested in an automatic indexing of the documents with linguistic values $A_i$. In [21, 28] we proposed to use the term weighting schemes of the vector space model as the starting point. In what follows we present one of the variants of this approach which will be used in our categorization algorithm in Section 4.

The weights of keywords in the documents may be determined in many different ways. Numerous schemes were proposed in the framework of the vector space model in particular. A concise summarization of most of them is given in the paper by Salton and Buckley [23]. They distinguish 3 components of a compound keyword weight accounting respectively for:

- frequency of the keyword in the document,
- proportion of the documents in the collection in which the keyword appears at least once, and
- normalization.

Each of these components may be expressed in a different way and, in particular, may be omitted in an overall weighting scheme.

Tables 1-3 show some typical forms of the particular components. In order to obtain the overall weight the representations (values) of selected components should be multiplied. Different weighting schemes obtained in this way are denoted with a 3-letter code where each letter comes from the "Code" column of subsequent table. For example tfx denotes the popular tf × IDF scheme (without normalization).

All the weights obtained by a combination of the three components listed in Tables 1–3 are normalized by max, i.e., the weight of a keyword is divided by the maximum weight of all other keywords in the same document.

**Table 1** Variants of the first component of the compound term weighting scheme

| **Code** | **Value** | **Description** |
|---|---|---|
| $b$ | $0/1$ | binary weight: =1 if a keyword appears in a document or query and = 0 otherwise |
| $t$ | $tf$ | frequency of a keyword in a document/query, i.e. the number of times it appears |
| $n$ | $0.5 + 0.5\left(\frac{tf}{\max_i tf_i}\right)$ | normalized frequency of a keyword |

**Table 2** Variants of the second component of the compound term weighting scheme

| **Code** | **Value** | **Description** |
|---|---|---|
| $x$ | 1 | this component is omitted |
| $f$ | $\log \frac{N}{n}$ | inverse document frequency (IDF), $N$ - number of all documents in the collection, $n$ - number of documents containing the keyword |
| $p$ | $\log \frac{N-n}{n}$ | modified version of IDF having a sound probabilistic justification |

**Table 3** Variants of the third component of the compound term weighting scheme

| **Code** | **Value** | **Description** |
|---|---|---|
| $x$ | 1 | no normalization |
| $c$ | $\frac{1}{\sqrt{\sum_i w_i^2}}$ | normalization by the Euclidean norm of the vector; $w_i$ is the weight of an $i$-th keyword |

In [21] we proposed the following automatic indexing procedure leading to the representation of documents in the form of (15). For each keyword $t_i$ its weight $D_i$ is computed using the selected scheme, i.e., a standard representation of the vector space model is derived. Then the particular weights (vectors coordinates) are "fuzzified" using one of the proposed approaches. Here we present just one such an approach – more approaches, including also those concerning the query representation, are discussed in [21, 28].

In the indexing approach adopted also in our categorization algorithm the triangular membership functions are assumed for the linguistic values $A_i$ in (15). More specifically, we use the following membership functions:

$$\mu_{A_i}(x) = 1 - \lambda \mid x - D_i \mid \tag{19}$$

where $D_i$ denotes the weight of the keyword computed using a selected weighting scheme.

The linguistic term modelled by such a fuzzy set (its equivalent membership function) may be interpreted as "important to a degree around $D_i$". There is a parameter $\lambda$ which determines how "narrow" or "wide" the membership function is around $D_i$.

There are many practical aspects of the proposed model which may be addressed depending on the retrieval context. Besides the indexing procedure, as mentioned above, another important question is the use of a proper aggregation scheme. In the basic form of the model the theoretical considerations justify the use of the minimum operator to aggregate the matching degrees corresponding to particular keywords (cf. (7)–(8)). This works well also in case of the categorization algorithm proposed in this paper since the representation of documents is obtained using a very small set of carefully selected keywords. However, in a typical retrieval scenario there are no such strong hints as to the choice of the keywords for representation. Thus much larger sets of keywords are used and the simultaneous use of the minimum operator in (7)-(8) often leads to poor results. These and other pragmatic aspects of the model will be discussed in our forthcoming papers (cf. e.g. [21, 14]).

## 4 Basic Aspects of Text Categorization

Basically the task of text categorization task may be briefly stated as follows. There is a fixed set of categories and the set of documents. The problem consists in assigning the documents to the categories or, equivalently, assigning the categories to the documents. Some versions of this general task may be distinguished depending on the total number of categories and the number of categories that may be assigned to a given document. The former distinction leads to the single-class versus multi-class categorization and the latter to the single-label versus multi-label categorization.

The practical significance of this task is best exemplified with the Internet services such as Yahoo! Directory (http://dir.yahoo.com/) or Open Directory

Project (ODP, http://www.dmoz.org) which try to categorize all documents available on the Web. However many apparently unrelated tasks may be cast in the form fitting the general text categorization scenario (see, e.g., [24]). Categorization may be carried out manually by some hired experts. However it is often impractical a solution (cf. the ODP project as a kind of an exception) and some automatic means are desirable and looked for. In the knowledge engineering framework experts are still involved but instead of directly assigning categories to documents they are constructing a set of categorization rules (an expert system) to be automatically executed against the documents in question. This is still an expensive solution and recently the broadly meant machine learning based approaches are gaining a wider acceptance as a preferable approach.

Basically, the essence of this class of approaches boils down to the treating of the categorization task as an example of the general classification problem (supervised learning). Thus a training set of documents with known categories is assumed. It is used to determine for each category a criterion making it possible later to decide if a new document of an unknown category should be classified under this category.

In order to apply one of the possible classifier construction methods a representation of documents has to be established. Usually, directly or indirectly, each document is represented by a vector of keywords weights. Then, any of numerous classifier construction algorithms may be applied, including rule-based systems, decision trees, artificial neural networks, support vector machines etc.

In our computational experiments, cf. Section 6, we use the Reuters-21578 corpus [18] that is widely accepted as a testbed for text categorization algorithms. This is a collection of news that are usually classified to a number of categories. Thus, this calls for methods dealing with a multi-class and multi-label case. The multi-label categorization requires the solution of an additional problem while building a classifier. Namely, a classifier such as the one considered here produces for a document a list of categories to which it possibly belongs. These categories are ordered non-increasingly according to their matching with the document. Then, a decision has to be made which of them, or more precisely, how many of those from the top of the list are to be assigned to a document under consideration. This is referred to as a *thresholding strategy*. We will not deal here with the details related to thresholding strategies; see [27] for an approach proposed by us to apply here elements of fuzzy logic.

## 5 Description of Learning Categories

A distinguishing feature of our approach to text categorization is its use of concepts that have been introduced in our fuzzy information retrieval model, presented in Section 3. Moreover, our approach belong to to the class of typical machine learning approaches (cf. a comprehensive state of the art exposition in Koronacki *et al.* [16, 17]).

More specifically, our approach is inspired by the work of Koriche and Quinqueton [15] whose approach fits the general *version space* paradigm of inductive learning – cf. Mitchell [19]. Namely, a set of positive and negative examples $E$ as well as a space of the the concepts under consideration $C$ are assumed. The *version space* of a positive (negative) example is a set of concepts that are consistent (inconsistent) with this example. The concepts inferred from the set of examples $E$ are those belonging to the intersection of the version spaces of all examples. These concepts may be then treated as a description of the class of positive examples. The problem with the basic form of this paradigm is that the intersection of the atomic spaces of all examples may be empty. This happens when the data (set of examples) is noisy, not fully consistent which is rather a rule than exception in real life. Thus in [15] it is proposed to replace the intersection operation with a more general merging operator based on the notion of a distance between the concepts. Some theoretical results and an inductive learning algorithm for a specific class of concepts are provided.

Let us briefly recall the essence of Koriche and Quinqueton's [15] approach which is originally meant for the classical binary logic setting. We will use a modified terminology that will better suit our purpose and contaxt.

A finite set $V = \{v_1, \ldots, v_n\}$ of Boolean variables is considered. An *atom* $l_i$ is either a variable $v$ or its negation $\neg v$. A disjunctive normal form (DNF) formula $\phi$ is a disjunction $\phi = c_1 \vee\ c_2 \vee \ldots \vee c_m$ of several conjunctions of atoms $c_i = l_1 \wedge l_2 \wedge \ldots \wedge l_k$ (these conjunctions will be hereafter referred to as *conjuncts*). Thus a space of concepts under consideration is the space of the DNF formulas defined for a given alphabet $V$. In fact, due to the computational complexity of the resulting algorithms the space has to be constrained. A positive integer $k$ is assumed and only $k$*-conjuncts*, i.e., conjuncts that contain at most $k$ literals, are considered as components of the DNF formulas. Thus, eventually the concept space is here the space of $k$-DNF formulas, i.e., the DNF formulas composed of $k$-terms. We assume here and later on that $k$ is fixed.

In this approach to inductive learning it is assumed that there are two classes (categories): positive and negative. The examples are assumed to be binary vectors in the space $\{0, 1\}^{n+1}$, where the last coordinate corresponds to the class of the example: "1" and "0" denote a positive and negative example, respectively.

The examples are represented within the above mentioned logical framework of the $k$-DNF formulas in the following way. An *instance* is a map $I$ from $V$ to $\{0, 1\}$. Then, an example $e \in E$ is represented by a pair $(I_e, v_e)$ where $I_e$ is an instance and $v_e$ is the value of a variable corresponding to the class assigned to the example. An example $e$ is called positive if $v_e = 1$ and negative if $v_e = 0$. An example $e$ and a formula $\phi$ are said to be *consistent* if $I_e$ is a logical model of $\phi$. Otherwise, $e$ is said to be *inconsistent* with $\phi$.

For a positive (negative) example $e$, the atomic version space $VS(e)$, is set of all $k$-DNF formulas (concepts) that are consistent (inconsistent) with $e$. The version space $VS(E)$ of a set of examples $E$ is the set of all $k$-DNF formulas that are consistent with every positive example in $E$ and that are inconsistent with every negative example in $E$. It may be easily shown that:

$$VS(E) = \bigcap_{e \in E} VS(e) \tag{20}$$

Often for a real life set of examples its version space is empty. The following approach is proposed in [15] as a remedy to this problem. A distance $d(\phi, \psi)$ between two $k$-DNF formulas $\phi$ and $\psi$ is defined in terms of the number of conjuncts they differ at:

$$d(\phi, \psi) = \mid (\phi \cup \psi) - (\phi \cap \psi) \mid$$

where the formulas are treated as sets of their conjuncts and $| \cdot |$ denotes the cardinality of the set.

Next, the distance $d(\phi, e)$ between a $k$-DNF formula $\phi$ and an example $e$ is defined as

$$d(\phi, e) = \min_{\psi \in VS(e)} d(\phi, \psi)$$

and finally the distance $d(\phi, E)$ between a $k$-DNF formula and a set of examples $E$ is defined as

$$d(\phi, E) = \sum_{e \in E} d(\phi, e)$$

We use here the same symbol $d$ to denote all these distances even if they are defined in different spaces. However a proper interpretation is obvious due to the arguments consistently denoted with different symbols standing for the formulas $(\phi, \psi)$, examples $(e)$, and sets of examples $(E)$.

Now, instead of using the version space $VS(E)$ (see (20)) as a result of the inductive learning, the following set of the $k$-DNF formulas $\phi$ is used:

$$\Delta(E) = \{\phi : \phi = \arg\min_{\psi} d(\psi, E)\}$$

The set $\Delta(E)$ is guaranteed to be non-empty and reduces to $VS(E)$ if that is non-empty.

In order to devise an algorithm for determining the elements of the set $\Delta(E)$ some additional definitions are needed. Basically, the algorithm aims at characterizing the quality of conjuncts so as then the best of them could be selected to construct the formulas $\phi$ belonging to the set $\Delta(E)$. For a given $k$-conjunct $c$ its *extension* $E(c)$ in the set of examples $E$ is a subset of $E$ comprising examples that are consistent with $c$:

$$E(c) = \{(I_e, v_e) \in E : I_e \text{ is a model of } c\} \tag{21}$$

A *weight* $w(c, E)$ of $c$ in $E$ is the cardinality of the set $E(c)$:

$$w(c, E) = \mid E(c) \mid \tag{22}$$

Finally, a *cover* of $E$ is a list of $k$-conjuncts $\rho = (c_1, \ldots, c_r)$ such that every positive example in $E$ is consistent with at least one conjunct $c_i$ from $\rho$.

In Koriche and Quinqueton [15] it is shown how these notions are related to the distance $d(\phi, E)$ between the $k$-DNF formulas and sets of examples and further to construct an approximate algorithm to construct the formulas belonging to the set $\Delta(E)$. We will omit here the details, and will show how the idea of the above approach may be adapted to the fuzzy case, notably for the purposes of text categorization in the framework of our fuzzy information retrieval model. We will then present a version of the earlier mentioned algorithm adapted to our modified setting.

Our reinterpretation of this approach in terms of fuzzy logic, which is the base for our fuzzy information retrieval model briefly presented in section 3, is as follows.

The *atoms* take now the form of expressions:

$$X_i \text{ IS } A_j \tag{23}$$

where $X_i$ is a linguistic variable concerning the importance of a keyword $t_i$, $i = 1, \ldots, n$, and $A_j$ is its label represented by a fuzzy set in the interval [0,1]. The set of the linguistic variables $X_i$ will be denoted as $V$.

A $k$-conjunct $c$ is the conjunction of $k$ atoms:

$$c = (X_1 \text{ IS } A_1) \wedge (X_2 \text{ IS } A_2) \wedge \ldots \wedge (X_k \text{ IS } A_k) \tag{24}$$

where each $X_i$ refers to a different keyword.

A vector $(x_1, x_2, \ldots, x_n) \in [0,1]^n$ is compatible with the conjunct $c = X_1 \text{ IS } A_1 \wedge X_2 \text{ IS } A_2 \wedge \ldots \wedge X_k \text{ IS } A_k$ to the degree:

$$\mu_c(x_1, \ldots, x_n) = \min(\mu_{A_1}(x_1), \mu_{A_2}(x_2), \ldots, \mu_{A_k}(x_k)) \tag{25}$$

A $k$-DNF formula $\phi$ is the disjunction of $k$-conjuncts. These formulas comprise the concept space.

A vector $(x_1, x_2, \ldots, x_n) \in [0,1]^n$ is compatible with a DNF formula:

$$\phi = c_1 \vee c_2 \vee \ldots \vee c_m \tag{26}$$

to the degree:

$$\mu_\phi(x_1, \ldots, x_n) = \max(\mu_{c_1}(x_1, \ldots, x_m), \ldots, \mu_{c_m}(x_1, \ldots, x_m)) \tag{27}$$

A DNF formula $\phi$ will later be identified with a fuzzy set $\Omega_\phi$ in the universe $U = [0,1]^n$ with the membership function determined by using equation (27), i.e.:

$$\Omega_\phi \in \mathcal{F}([0,1]^n) \tag{28}$$

$$\mu_{\Omega_\phi}(x_1, \ldots, x_n) = \mu_\phi(x_1, \ldots, x_n) \tag{29}$$

where $\mathcal{F}(U)$ denotes the family of fuzzy sets defined in an universe $U$.

An *instance* $I$ is the conjunct:

$$c = X_1 \text{ IS } A_1 \wedge X_2 \text{ IS } A_2 \wedge \ldots \wedge X_n \text{ IS } A_n \tag{30}$$

where the number of atoms $n$ is equal to the number of keywords used to index the documents in given collection.

Such a conjunct generates a possibility distribution $\pi_I$ on $U = [0,1]^n$ defined as follows. Each atom (23) generates a possibility distribution:

$$\pi_i(x) = \pi_{X_i \text{ IS } A_j}(x) = \mu_{A_j}(x) \tag{31}$$

Then overall possibility distribution $\pi_I$ is given by:

$$\pi_I(x_1, x_2, \ldots, x_n) = \min(\pi_1(x_1), \pi_2(x_2), \ldots, \pi_n(x_n)) \tag{32}$$

assuming non-interactivity of particular variables. An instance $I$ may be identified with the possibility distribution $\pi_I$ defined by (32).

An instance $I$ is said to be *consistent* with a formula $\phi$ *to the degree* expressed by a pair

$$(\mathcal{N}_I(\phi), \Pi_I(\phi)) \tag{33}$$

where

$$\mathcal{N}_I(\phi) = \mathcal{N}_I(\Omega_\phi) \tag{34}$$

$$\Pi_I(\phi) = \Pi_I(\Omega_\phi) \tag{35}$$

and $\Omega_\phi$ is defined by (28)–(29) while $\Pi_I$ and $\mathcal{N}_I$ are, respectively, the possibility and necessity measures related to the possibility distribution $\pi_I$ defined by (32).

An instance $I$ is said to be *inconsistent* with a formula $\phi$ *to the degree* expressed by a pair

$$(\mathcal{N}_I(\overline{\phi}), \Pi_I(\overline{\phi})) \tag{36}$$

where

$$\mathcal{N}_I(\overline{\phi}) = \mathcal{N}_I(\overline{\Omega_\phi}) \tag{37}$$

$$\Pi_I(\overline{\phi}) = \Pi_I(\overline{\Omega_\phi}) \tag{38}$$

and $\overline{A}$ denotes a complement of a fuzzy set $A$.

It may be noticed that if (33) is treated as the membership degree to an interval-valued fuzzy set then (36) is its counterpart obtained using the complement operation (11) of interval-valued fuzzy sets.

For given category $g \in G$, an *example* $e$ is a pair $(I_e, v_e)$, where $I_e$ is an instance and $v_e$ is a valuation of a binary variable such that $v_e = 1$ if $I_e$ is assigned the category $g$ and $v_e = 0$ otherwise. For $v_e = 1$ and $v_e = 0$ the example is called *positive* and *negative*, respectively. An example $e = (I_e, v_e)$ is consistent with a formula $\phi$ to the degree to which the instance $I_e$ is consistent with $\phi$.

Notice that it may be useful to define the *extension* of a conjunct $c$, similarly to (21), as an interval-valued fuzzy set $E(c)$:

$$E(c) = \{(e, \underline{\mu}_{E(c)}, \overline{\mu}_{E(c)})\} \tag{39}$$

$$e = (I_e, v_e) \tag{40}$$

$$\underline{\mu}_{E(c)}(e) = \mathcal{N}_{I_e}(c) \tag{41}$$

$$\overline{\mu}_{E(c)}(e) = \Pi_{I_e}(c) \tag{42}$$

where the right hand sides of formulas (41)–(42) are special cases of (34)–(35) for $\phi$ composed of just one conjunct $c$.

Similarly to (22) we define the weight $w(c, E)$ of a conjunct $c$ in a set of examples $E$ as the cardinality of the set $E(c)$ using the cardinality of interval-valued fuzzy sets defined by (14):

$$w(c, E) = \left[\sum_i \underline{\mu}_{E(c)}(x_i), \sum_i \overline{\mu}_{E(c)}(x_i)\right] \tag{43}$$

Given a positive integer $k$ and a positive example $e^+$, the *fuzzy atomic version space* of $e^+$ with respect to $k$, denoted $VS(e^+)$, may be defined as an interval valued fuzzy set of $k$-DNF formulas $\phi$ with which $e^+$ is consistent to some degree, i.e., such that:

$$\mu_{VS(e^+)}(\phi) = (\mathcal{N}_{I_{e^+}}(\phi), \Pi_{I_{e^+}}(\phi)) \tag{44}$$

Similarly, for a negative example $e^-$ its fuzzy atomic version space with respect to $k$ is an interval valued fuzzy set of $k$-DNF formulas $\phi$ with which $e^-$ is inconsistent to some degree, i.e., such that:

$$\mu_{VS(e^-)}(\phi) = (\mathcal{N}_{I_{e^-}}(\overline{\phi}), \Pi_{I_{e^-}}(\overline{\phi})) \tag{45}$$

Given a set of examples $E$, the fuzzy version space of $E$ with respect to $k$, denoted $VS(E)$, is the interval-valued fuzzy set of the $k$-DNF formulas $\phi$ with the membership function determined by the degree to which every positive example in $E$ is consistent with $\phi$ and every negative example in $E$ is inconsistent with $\phi$. The set $VS(E)$ may be expressed as

$$VS(E) = \bigcap_{e \in E} VS(e) \tag{46}$$

where $\bigcap$ is an obvious generalization of the interval-valued fuzzy sets intersection (conjunction) (12).

A set of examples (a training set) $E$ is called consistent with respect to the set of the $k$-DNF formulas if there exists $\phi$ with a *high enough* membership degree to $\in VS(E)$, and inconsistent otherwise. The threshold value related to the notion of "high enough" is to be a parameter of the proposed methods. When $E$ is consistent then a concept $\phi$ with a maximum membership degree should be reasonably assumed as describing $E$.

## 6 The Algorithm

In our proposed approach to the text categorization we proceed as follows. In the learning phase, for each category $g \in G$ a $k$-DNF formula describing this category is constructed. The set of documents to which a given category is assigned in the training set forms the set of positive examples. The remaining documents form the set of negative examples and the algorithm described below is applied to induce a $k$-DNF formula. The same is repeated for each category $g \in G$. During the classification phase a new document with unknown category is treated as an instance and its consistency with all $k$-DNF formulas representing particular categories is checked. This consistency is expressed by an interval defined by (33). Then the categories are ordered in such a way that most consistent of them are at the top of the list. The details of all steps are given below.

### *6.1 Learning Phase*

In order to construct a $k$-DNF formula we adopt the basic idea of the greedy algorithm proposed in Koriche and Quinqueton [15]. Namely, the best conjuncts are iteratively selected to form a $k$-DNF formula where the quality of a conjunct is defined by how they are consistent with the positive examples and inconsistent with the negative.

The particular steps of the algorithm, repeated for each category $g \in G$ in turn, may be described as follows:

Input: the training set of documents of known membership to particular categories $g \in G$; parameters $k$, $L$ and $S$
Output: $k$-DNF formulas describing each category

1. create representations of all documents using selected weighting scheme [23]
2. create a set of positive $E_p$ and negative $E_n$ examples
3. select a subset of keywords to describe the documents
4. generate all $k$-conjuncts to be used to construct a $k$-DNF formula
5. while $E_p$ is non-empty do the following

a. find a best conjunct and add it to the resulting $k$-DNF formula $\phi$
b. remove from $E_p$ all positive examples consistent enough with the selected conjunct
c. remove the selected conjunct from the list of all considered conjuncts

6. return $\phi$ – the disjunction of all selected conjuncts

This is a very general description of the algorithm. Now we will present details of the particular steps, taking into account specificity of the data set that was used for the computational experiments, namely the Reuters corpus [18].

Step 1

In order to construct a classifier, first the representation of documents in a training data set have to be established. Here we use the representation proposed in our fuzzy information retrieval model, cf. section 3. Thus documents are represented using statements (15) that are interpreted for the purposes of the inductive learning as instances (30). As suggested in Section 3, practically the linguistic values in (15) (and thus also in (30)) in the representation of a document may be determined using one of the weighting schemes presented in Tables 1–3 (in our computational experiment we have chosen one of them; see Subsection 6.3).

Step 2

Basically, all documents in the training data set belonging to a given category form the set of positive examples and all remaining documents form the set of negative examples. For some algorithms, including the one proposed in [15], and some data sets it may be impractical due to large differences in the resulting sets of positive and negative examples. For example in case of the Reuters-21578 collection [18] the cardinality of the categories ranges from 1 to a couple of thousands. Thus for small categories it may be reasonable to reduce the size of the sets of negative examples.

It is also worth noticing that in the case of multi-label categorization the same document may be a positive example for many categories.

Step 3

Usually the number of different words appearing in document collections are huge even after stopword elimination and stemming. This requires an aggressive dimensionality reduction to be applied. In our approach a full search of the space of $k$-conjuncts (24) is done so that their number be relatively small. It may be obtained by assuming a small $k$ and a very small number of keywords to be used to form the conjuncts, for each category. Thus, in our algorithm besides $k$ there are two other parameters: $L$ determining the number of keywords used to describe a category (i.e., to build a $k$-DNF formula for it) and $S$ which will be discussed in the description of Step 4.

Thus, for each category $L$ keywords are selected in the following way. For each keyword $t \in T_g$ that appeared at least once in the documents of given category $g$ the following weight tf $\times$ ICF is computed ($T_g$ denotes the set of all documents assigned in the training set to the category $g$):

$$\text{tf } s \text{ ICF}(t, g) = \frac{cfreq(t, g)}{\max_{u \in T_g} cfreq(u, g)} * \log(\frac{\mid G \mid}{\mid G(t) \mid}) \tag{47}$$

where $cfreq(t, g)$ denotes the number of times the keyword $t$ appeared in all documents $T_g$ belonging to the category $g$, $G$ is the set of all categories, $G(t)$ is the set of categories in documents of which the keyword $t$ appeared at least once (i.e., at least once in at least one document from $T_g$) and $\mid \cdot \mid$ denotes the cardinality.

The rationale for the tf $\times$ ICF weight is similar as in case of the popular tf $\times$ IDF weighting scheme: the weight for a keyword $t$ and category $g$ is the higher the more frequently $t$ appears in documents of category $g$ and the lower the more there are categories in which documents it appears at least once. The keywords $t$ having higher tf $\times$ ICF$(t, g)$ weight are more specific for the category $g$. Thus for the purposes of our algorithm the $L$ keywords with highest value of tf x ICF are selected for each category.

Step 4

Obviously there is an infinite number of conjuncts in the form given by (24). Thus in the algorithm a limited number of the linguistic values $A_i$ has to be assumed and the particular conjuncts are built out of them. Basically, various shapes of the membership functions of fuzzy sets representing the $A_i$'s may be assumed as mentioned in Ssection 3. In our experiments (cf. Section 6.3) we assumed the triangular membership functions with one parameter $x_0 \in [0, 1]$ and then considered their instances for a number of values of this parameter determined by an algorithm – cf Section 6.3.

Step 5

Concerning this step the main point to be clarified is how the best conjunct is defined and selected. In the algorithm proposed in [15] the conjunct $c$ minimizing the following quality index $QI$ is selected:

$$QI(c, E_p, E_n) = \frac{\min(w(c, E_n), w(c, E_p))}{w(c, E_p)} \tag{48}$$

where $w(\cdot, \cdot)$ is defined in (22). Thus the best are the conjuncts:

"Consistent with lowest possible number of negative examples and highest possible number of positive examples" (49)

A semantically similar, though formally different, inductive learning problem formulation was proposed in our works (cf. Kacprzyk and Iwański [10], and Kacprzyk and Szkatuła [11, 12].

In our setting a direct adoption of (48) is rather cumbersome requiring the use of interval arithmetic as the weights $w(c,\cdot)$ are intervals instead of single numbers. Thus, we propose to use another conjunct quality index preserving the spirit of (48). Namely, we employ a direct logical interpretation of the linguistic formulation of the conjunct quality index defining truth value of (49) as follows:

$$\begin{aligned} QI(c,E_p,E_n) &= w'(c,E_p) \wedge \overline{w'(c,E_n)} = \\ &= [\min(\sum_{e_i \in E_p} \frac{\underline{\mu}_{E_p(c)}(e_i)}{\mid E_p \mid}, 1 - \sum_{e_j \in E_n} \frac{\overline{\mu}_{E_n(c)}(e_j)}{\mid E_n \mid}), \\ &\quad \min(\sum_{e_i \in E_p} \frac{\overline{\mu}_{E_p(c)}(e_i)}{\mid E_p \mid}, 1 - \sum_{e_j \in E_n} \frac{\underline{\mu}_{E_n(c)}(e_j))}{\mid E_n \mid})] = \\ &= [\min(\sum_{e_i \in E_p} \frac{\mathcal{N}_{I_{e_i}}(c)}{\mid E_p \mid}, 1 - \sum_{e_j \in E_n} \frac{\Pi_{I_{e_j}}(c)}{\mid E_n \mid}), \\ &\quad \min(\sum_{e_i \in E_p} \frac{\Pi_{I_{e_i}}(c)}{\mid E_p \mid}, 1 - \sum_{e_j \in E_n} \frac{\mathcal{N}_{I_{e_j}}(c)}{\mid E_n \mid})] \end{aligned} \tag{50}$$

where $w'(c,E)$ is a normalized version of the formula (43) for the conjunct weight, i.e., $w'(c,E) = w(c,E)/\mid E \mid$. Thanks to this normalization the sets of positive and negative examples may be of significantly different size what may be convenient in case of certain data sets.

Thus we can assess the quality of each conjunct generated in Step 4 using the formula (50). It should be noticed that in a naive basic approach $QI$ has to be computed for all remaining conjuncts in each iteration of the loop appearing in Step 5. This is due to the changes in the size of the set of positive examples $E_p$ in each iteration.

The quality index $QI$ takes intervals as its value so that we need to impose a certain ordering on the intervals so as to be in a position to choose (define) the best conjunct. The standard order on intervals is defined as follows:

$$[x_1,x_2] \prec_1 [y_1,y_2] \iff (x_1 < y_1) \wedge (x_2 < y_2) \tag{51}$$

Unfortunately, (51) defines only a partial order. Thus we supplement it with another one [8]:

$$[x_1,x_2] \prec_2 [y_1,y_2] \iff (x_1 + x_2 < y_1 + y_2) \wedge (x_2 - x_1 > y_2 - y_1) \tag{52}$$

However, even the combination of both these order definitions does not ensure the comparability of any two intervals. Therefore, to finally compare two conjuncts $c_1$ and $c_2$ we use the following sequence of comparisons:

1. use order $\prec_1$ to compare $QI(c_1, E_p, E_n)$ and $QI(c_2, E_p, E_n)$; if it does not work
2. use order $\prec_2$ to compare $QI(c_1, E_p, E_n)$ and $QI(c_2, E_p, E_n)$; if it does not work
3. choose a conjunct $c$ for which $\sum_{e_i \in E_p} \frac{\mu_{E_p(c)}(e_i)}{|E_p|}$ is higher; if it is equal for $c_1$ and $c_2$
4. choose a conjunct $c$ for which $\sum_{e_i \in E_p} \frac{\Pi_{I_{e_i}}(c)}{|E_p|}$ is higher; if it is equal for $c_1$ and $c_2$
5. choose $c_1$

Thus first we try to compare the values of the quality index $QI$ for the conjuncts using orders $\prec_1$ and $\prec_2$. If the conjuncts are incomparable then the lower and upper bounds on the number of positive examples consistent (cf. eq. (33)) with the conjuncts are taken into account – the conjunct with a higher bound is selected. If both bounds are equal, then the first conjunct is selected.

Finally in Substep b) of Step 5 we check the consistency of each positive example remaining in the set $E_p$ with the best conjunct selected in the above mentioned procedure. The formula (33) is used to compute the consistency degree in the form of an interval. Then in our experiments we use a simple criterion to decide if the conjunct and the example are consistent. Namely only the possibility measure related component, i.e., the upper bound of the interval is used, and it is checked if it is greater than a prespecified threshold value. If it is the case the example is taken as consistent, i.e., well enough described by the conjunct, and is removed from the set $E_p$. It may happen then there is no example consistent with the selected conjunct to a sufficiently high degree determined by the threshold value. Then the example(s) consistent to a highest degree are removed.

## 6.2 Categorization Phase

First of all the representation of a tested document to be categorized is determined. This is done in exactly the same way as for the training documents in Step 1 of the learning phase. Then the consistency between each category $g$, represented by a $k$-DNF formula $\phi_g$ and the tested document is checked using the formula (33). A straightforward interpretation is to treat the tested document as an example and use the formula (33) exactly in the same way as in the learning phase. However, in some experiments we obtained better results when the roles were reversed, i.e., the $k$-DNF formula is treated as inducing a possibility distribution (with an obvious interpretation of disjunction via the maximum operator) and the tested document is treated as a formula, namely a conjunction such as given by (15) or equivalently by (30).

A final result of our algorithm for a tested document is the ordered list of categories starting with the categories with the highest consistency degree with the document. The ordering is obtained using the same procedure which is used in Step 5 of the learning phase, described in Subsection 6.1. In the single-label variant of categorization it is reasonable to assign the first category to the document. In the multi-label categorization case some additional techniques have to be used, cf. Section 4.

### *6.3 A Computational Experiment*

In the computational experiments we used the Reuters-21578 corpus [18]. In particular we used the Modified Apte ("ModApte") Split of the data, i.e., for the training phase a subset of news characterized by the attributes LEWISSPLIT="TRAIN" and TOPICS="YES" and for testing phase a subset characterized by LEWISSPLIT="TEST"; TOPICS="YES". We only used news containing topics and body of the text or at least the title. This gives rise to 7728 training, 3005 test documents and 114 categories. The title of the news and its body are concatenated to produce the document. The documents are preprocessed by removing stop words and numbers. Stemming is done using the standard Porters algorithm [22]. No additional dimensionality reduction is done during the preprocessing concerning the keywords as an aggressive local (i.e., for each category separately) reduction is a part of the algorithm (cf. Step 3, Section 6).

We have taken into account all categories – even those which are represented in the training data set by just one document. Around half of them do not even appear in the test dataset thus they just make the categorization more difficult. We obtained just a few misclassifications to these "one-document" categories while in case of one, COTTON-OIL, we obtained two correct classification for two documents with this category appearing in the test data set (i.e., 100% recall) and two misclassification (i.e., in total the 50% precision for this extremely small category).

The sets of positive and negative examples in the learning phase have been constructed as follows. We basically took all documents assigned to given category as the positive examples and the same number of documents from outside of this category as the negative examples. However we set a lower bound on the size of the set of negative examples to 10, thus for categories underrepresented in the training data set (with less than 10 documents) exactly 10 documents belonging to other categories have been selected. We have tested a few strategies for the selection of the negative examples. The results reported later in this section have been obtained for the simplest strategy: first, $m$ documents not belonging to a given category have been selected

($m$ is the required number of negative examples and the order of training documents is such as in the original collection).

Our fuzzy information retrieval model (cf. Section 3) assumes a generic form of the document representation given by (15). In the experiments we have used its various specific forms. The results obtained are reported below for the document representation given by (19) with the parameter $\lambda = 1$ and the *bfx* weighting scheme (cf. Tables 1–3) to produce the $D_j$'s. The same representation is assumed for the training and testing documents.

Moreover, in the results reported below, we have assumed the following values for other parameters mentioned in previous subsections:

- $k = 3$, i.e., only the conjuncts built of maximum 3 atoms (cf., (23)) are used to construct $k$-DNF formulas (cf. (26)) for categories
- $L = 5$, i.e., only the 5 best (in the sense of (47)) keywords are used to characterize a category by a $k$-DNF formula
- $S = 0.1$, i.e., for each keyword the atoms (23) with $A_i$ defined using (19) for $D_j = 0, 0.1, 0.2, \ldots 1.0$ are considered.

Summarizing the above values of parameters $L$ and $S$: in Step 4 of the learning phase all conjuncts which are then generated are built out of 11 atoms (as 11 distinct values for $D_j$ are assumed) for each of 5 ($L$) selected keywords.

In Substep b) of the learning phase in Step 5 we use the threshold value of 0.75 to decide if a conjunct and the positive example are consistent. In the experiment the results of which are reported below the $k$-DNF formulas representing the particular categories are treated as inducing a possibility distribution and the tested document is treated as a formula (a conjunction such as given by (15)).

The final assignment of categories to a given test document in the multi-label case, such as the case of the Reuters dataset considered, requires the use of a specific thresholding strategy (cf. Section 4). Then the evaluation of the results in terms of (exact) precision and recall is possible. Precision is the proportion of correctly assigned categories among all assigned categories, while recall is the proportion of correctly assigned categories among all known correct categories for a given document.

Here we use the effectiveness measures that evaluate the whole ordering of the categories with respect to a given test document. Basically, the closer the correct (known in advance) categories to the beginning of the ordering the better.

We use the following measures of this type in our experiments:

- RPrecision: proportion of correctly assigned categories among all assigned categories assuming that $\theta$ categories first in the ordering produced by the algorithm are assigned, where $\theta$ is the number of known correct categories for the document,
- Average Precision At Seen Relevant: the precision measure is computed at the positions where in the ordering produced by the algorithm known

correct categories appear and then so computed precision values are averaged

- AVP-11: the places in the ordering where the recall measure is interpolated to reach the values 0, 0.1, 0.2,..., 0.9, 1.0 are identified and the precision measure is computed at each of them; the average of the values of the precision measure obtained in this way is taken
- precision-recall breakeven point: a place in ordering is found where (interpolated) values of precision and recall measure are equal and one of them is taken as the effectiveness measure

The last of these measures is most widely used in the categorization experiments for the Reuters dataset. In one of our experiments we have obtained the following results: RPrecision = 0.5047, Average Precision At Seen Relevant = 0.6238, AVP-11 = 0.6266, breakeven point = 0.6453. As reported by Joachims [9] the values of the breakeven point for various methods are in the range of 0.72 (Bayes) to 0.86 (SVM). Thus they are definitely higher than those we obtained. However we have not yet tested the whole spectrum of possible configurations of our algorithm. In particular we are carrying out tests concerning the underlying fuzzy information retrieval model. Many different keyword weighting schemes may be applied as a starting point for deriving the representation of documents given by (15). Moreover, various shapes of the membership functions may be adopted – we have quite thoroughly tested the triangular membership functions and are in the process of doing the same for the trapezoidal ones. Different aggregation operators may be applied (cf. 3) and in the standard information retrieval scenario the best results have been obtained for the averaging operators. Finally there are still many parameters of the categorization algorithm itself to be tested, including $k$, $L$ and $S$. Thus we hope to obtain better results and establish a procedure for the proper selection of all these parameters.

On the other hand, one of the advantages of the proposed categorization approach is related to the flexibility of the underlying fuzzy information retrieval model. Namely, it makes it possible to grasp both imprecision and uncertainty while determining the representation of documents and queries. In the scenario of the regular information retrieval it is especially attractive with respect to the "manual" procedure. A user supported by a properly constructed interface may express his or her information needs in a more natural way. In case of the documents the manual indexing is much less attractive as usually there are huge collections of documents. However, manual descriptions of the categories using $k$-DNF formulas become an option and may be of interest in some scenarios.

The presented categorization algorithm has been implemented in Perl without much focus on efficiency. The computational complexity is determined by exhaustive search of the space of possible conjuncts. Small values of parameters $k$ and $L$ make it manageable. The experiments have been carried out on a PC computer with Intel Dual Core 2.13 GHz processor and 2 GB RAM.

The training and categorization phase took 250 and 1 minute, respectively (with document preprocessed in advance).

## 7 Concluding Remarks

We have presented a concept of a novel text categorization algorithm inspired by the inductive learning algorithm by Koriche and Quinqueton [15] and employing our fuzzy information retrieval model as given in Nowacka, Zadrożny and Kacprzyk [21], and Zadrożny and Nowacka [28]. Its essence consists in constructing for each category a fuzzy linguistic statement in the disjunctive normal form describing, in terms of the keyword weights, the documents belonging to this category. Both during the training and classification the matching between documents and linguistic statements is computed as in our fuzzy information retrieval model.

The advantage of the proposed algorithms is related to its close links with our model which offers a flexible environment for representing and comparing documents and queries. The results of the computational experiments we have carried out so far show no advantage over the known state of the art approaches. On the other hand, within the framework of the underlying retrieval model it is possible to grasp both imprecision and uncertainty, and this may be a subject of the future research in the text categorization algorithm discussed. Fine tuning of the parameters of the model proposed, as well as increasing its numerical efficiency, is another relevant future subject of investigation.

## References

1. Baeza-Yates, R., Ribeiro-Neto, B.: Modern information retrieval. ACM Press and Addison Wesley (1999)
2. Bieniek, K., Gola, M., Kacprzyk, J., Zadrożny, S.: An Approach to Use Possibility Theory in Information Retrieval. In: Proceedings of the 12th Zittau East-West Fuzzy Colloquium, Zittau, Germany, pp. 1–5 (2005)
3. Bordogna, G., Carrara, P., Pasi, G.: Fuzzy approaches to extend Boolean information retrieval. In: Bosc, P., Kacprzyk, J. (eds.) Fuzziness in Database Management Systems, pp. 231–274. Physica Verlag, Heidelberg (1995)
4. Cornelis, C., Deschrijver, G., Kerre, E.E.: Implication in intuitionistic fuzzy and interval-valued fuzzy set theory: construction, classification, application. International Journal of Approximate Reasoning 35, 55–95 (2004)
5. Deschrijver, G.: Arithmetic operators in interval-valued fuzzy set theory. Information Sciences 177, 2906–2924 (2007)
6. Deschrijver, G., Kerre, E.: On the relationship between some extensions of fuzzy set theory. Fuzzy Sets and Systems 133, 227–235 (2004)
7. Dubois, D., Prade, H.: Fuzzy Sets in Approximate Reasoning, part 1: Inference with possibility distributions. Fuzzy Sets and Systems 40, 143–202 (1991)

8. Ishibuchi, H., Tanaka, H.: Multiobjective Programming in Optimisation of the Interval Objective Function. European Journal of Operational Research 48, 219–225 (1990)
9. Joachims, T.: Text Categorization with Support Vector Machines: Learning with Many Relevant Features. In: Nédellec, C., Rouveirol, C. (eds.) ECML 1998. LNCS, vol. 1398, pp. 137–142. Springer, Heidelberg (1998)
10. Kacprzyk, J., Iwański, C.: Fuzzy logic with linguistic quantifiers in inductive learning. In: Zadeh, L.A., Kacprzyk, J. (eds.) Fuzzy Logic for the Management of Uncertainty, pp. 465–478. Wiley, New York (1992)
11. Kacprzyk, J., Szkatuła, G.: An algorithm for learning from erroneous and incorrigible examples. International Journal of Intelligent Systems 11, 565–582 (1996)
12. Kacprzyk, J., Szkatuła, G.: An inductive learning algorithm with a preanalysis of data. International Journal of Knowledge-Based Intelligent Engineering Systems 3, 135–146 (1999)
13. Kacprzyk, J., Nowacka, K., Zadrożny, S.: A Possibilistic-logic-based Information Retrieval Model with Various Term-weighting Approaches. In: Rutkowski, L., Tadeusiewicz, R., Zadeh, L.A., Żurada, J.M. (eds.) ICAISC 2006. LNCS (LNAI), vol. 4029, pp. 1120–1129. Springer, Heidelberg (2006)
14. Kacprzyk, J., Zadrożny, S., Nowacka, K.: An Experimental Comparison of Various Aggregation Operators in a Fuzzy Information Retrieval Model. In: North American Fuzzy Information Processing Society Annual Conference (NAFIPS 2008), New York, USA (2008) (submitted)
15. Koriche, F., Quinqueton, J.: Robust -DNF Learning via Inductive Belief Merging. In: Lavrač, N., Gamberger, D., Todorovski, L., Blockeel, H. (eds.) ECML 2003. LNCS (LNAI), vol. 2837, pp. 229–240. Springer, Heidelberg (2003)
16. Koronacki, J., Raś, Z., Wierzchoń, S.T., Kacprzyk, J. (eds.): Advances in Machine Learning I. Springer, Heidelberg (2010)
17. Koronacki, J., Raś, Z., Wierzchoń, S.T., Kacprzyk, J. (eds.): Advances in Machine Learning II. Springer, Heidelberg (2010)
18. Lewis, D.D.: Reuters-21578, Distribution 1.0., `http://www.daviddlewis.com/resources/testcollections/reuters21578`
19. Mitchell, T.M.: Generalization as Search. Artificial Intelligence 18, 203–226 (1982)
20. Neumaier, A.: Clouds, fuzzy sets and probability intervals. Reliable Computing 10, 249–272 (2004)
21. Nowacka, K., Zadrożny, S., Kacprzyk, J.: A New Fuzzy Logic Based Information Retrieval Model. In: 12th International Conference on Information Processing and Management of Uncertainty in Knowledge-Based Systems (IPMU 2008), Malaga, Spain (2008) (submitted)
22. Porter, M.F.: An Algorithm for Suffix Stripping. Program 14(3), 130–137 (1980)
23. Salton, G., Buckley, C.: Term Weighting Approaches in Automatic Text Retrieval. Information Processing and Management 24, 513–523 (1988)
24. Sebastiani, F.: A Tutorial on Automated Text Categorisation. In: Amandi, A., Zunino, A. (eds.) Proceedings of ASAI 1999, 1st Argentinian Symposium on Artificial Intelligence, Buenos Aires, Argentina, pp. 7–35 (1999)
25. Zadeh, L.A.: The Concept of a Linguistic Variable and its Application to Approximate Reasoning (Part I-III). Information Sciences 8(8,9), 199–249, 301–357, 43–80 (1975)

26. Zadeh, L.A.: Fuzzy Sets as a Basis for a Theory of Possibility. Fuzzy Sets and Systems 1, 3–28 (1978)
27. Zadrożny, S., Kacprzyk, J.: Computing with Words for Text Processing: An Approach to the Text Categorization. Information Sciences 176(4), 415–437 (2006)
28. Zadrożny, S., Nowacka, K.: Fuzzy Information Retrieval Model Revisited. Fuzzy Sets and Systems 160(15), 2173–2191 (2008)
29. Zadrożny, S., Kacprzyk, J.: An Extended Fuzzy Boolean Model of Information Retrieval Revisited. In: The 14th Annual IEEE International Conference on FUZZY Systems, Reno, USA, pp. 1020–1025 (2005)
30. Zadrożny, S., Nowacka, K., Kacprzyk, J.: A Concept of a Possibilistic Logic Based Information Retrieval Model. In: 11th International Conference on Information Processing and Management of Uncertainty in Knowledge-Based Systems (IPMU 2006), Paris, France, pp. 992–999 (2006)

# Author Index